NANOMATERIALS FOR WASTEWATER REMEDIATION

NANOMATERIALS FOR WASTEWATER REMEDIATION

RAVINDRA KUMAR GAUTAM

Department of Chemistry
Environmental Chemistry Research Laboratory
University of Allahabad, Allahabad
Uttar Pradesh, India

MAHESH CHANDRA CHATTOPADHYAYA

Professor and Former Head
Department of Chemistry
Environmental Chemistry Research Laboratory
University of Allahabad, Allahabad
Uttar Pradesh, India

ELSEVIER

Amsterdam • Boston • Heidelberg • London
New York • Oxford • Paris • San Diego
San Francisco • Singapore • Sydney • Tokyo
Butterworth-Heinemann is an imprint of Elsevier

BH

Butterworth-Heinemann is an imprint of Elsevier
The Boulevard, Langford Lane, Kidlington, Oxford OX5 1GB, UK
50 Hampshire Street, 5th Floor, Cambridge, MA 02139, USA

British Library Cataloguing-in-Publication Data
A catalogue record for this book is available from the British Library

Library of Congress Cataloging-in-Publication Data
A catalog record for this book is available from the Library of Congress

ISBN: 978-0-12-804609-8

For information on all Butterworth-Heinemann publications visit our website at http://www.elsevier.com/

Publisher: Joe Hayton
Senior Acquisition Editor: Ken McCombs
Editorial Project Manager: Peter Jardim
Production Project Manager: Kiruthika Govindaraju
Designer: Victoria Pearson Esser

Dedicated to

My loving sister late Pratiksha Gautam

(R.K. Gautam)

And our

Maa & Baba

CONTENTS

ABOUT THE AUTHORS

R.K. Gautam did his postgraduate in environmental science at the University of Allahabad, India, in 2009. He earned a postgraduate diploma in disaster management from Indira Gandhi National Open University, New Delhi, in 2010. Thereafter, Gautam worked for 1 year at the National Environmental Engineering Research Institute, Council of Scientific & Industrial Research (NEERI-CSIR), Nagpur, India. He qualified for the CSIR-UGC National Eligibility Test for Junior Research Fellowship. He has published 66 research papers, including original research articles, reviews, books, book chapters, and conference proceedings. He has written a book entitled, *Environmental Magnetism: Fundamentals and Applications* (ISBN-10: 3659209090 | ISBN-13: 978-3659209093) which was published by LAP Lambert Academic Publishing, Saarbrucken, Germany, and still five research articles and two books are in the pipeline. Gautam has been selected as a fellow of the Indian Chemical Society and life member of the Indian Science Congress Association in 2013. He is a member of the editorial boards of the *International Journal of Nanoscience and Nanoengineering*, *American Journal of Environmental Engineering and Science*, *International Journal of Environmental Monitoring and Protection*, and *International Journal of Industrial Chemistry and Biotechnology*. He also serves as a reviewer for more than 20 journals of international repute. Currently, he is engaged in doctoral work in the Department of Chemistry, University of Allahabad in Allahabad, India. His areas of interest are adsorption and nanomaterials and their analogs for water/wastewater remediation.

M.C. Chattopadhyaya is a professor in the Department of Chemistry, University of Allahabad, Allahabad (UP), India. He did his postgraduate in chemistry at Gorakhpur University, with specialization in inorganic chemistry in 1967. He was selected for postgraduate course in radiological physics by Bhabha Atomic Research Centre. Chattopadhyaya obtained a PhD degree from the Indian Institute of Technology Bombay, Mumbai. He did a short course of the American Chemical Society on the "Interpretation of Infra-red Spectra." Thereafter, he joined the University of Allahabad as lecturer in 1974 and subsequently became reader and professor of chemistry. He served the University as head of the Department of Chemistry during the period 2008–2010. Besides teaching inorganic and analytical chemistry, Chattopadhyaya also taught environmental chemistry at the University. He was elected as fellow of Cambridge Philosophical Society and fellow of the Chemical Society, London, Indian Chemical Society, Institution of Chemists, India. He was also president of the Indian Chemical Society during the years 2011–2013. He has published more than 150 research papers in different national and international journals. Under his supervision, 28 research scholars have been awarded doctoral degrees. Currently, Chattopadhyaya is working on nanomaterials and nanocomposites and their applications for environmental remediation, solid oxide fuel cells, synthesis of inorganic materials, and development of sensors for estimation of surfactants.

FOREWORD

It gives me immense pleasure to write a note of Foreword to "*Nanomaterials for Wastewater Remediation*." The field of nanocomposites and nanomaterials has progressed greatly over the last few years, as shown by the widespread use of nanomaterials for environmental applications as superadsorbents. Water pollution has become an important issue and posses an enormous challenge to the environmental scientists and engineers. Recently, nanoscale materials and nanotechnology received much attention in the sensing, monitoring, assessment, and separation and purification of hazardous contaminants in water. During the last five years, various types of nanomaterials such as magnetic nanosorbents, titanium dioxide, alumina, layered double hydroxides, and graphene-based nanosorbents have been widely applied in the remediation of environmental contaminants from wastewater.

Due to the large amount of up-to-date and down-to-earth information, the book will be appropriate for use as a reference book for students and professionals who are interested in the development of new generation superadsorbent nanomaterials and nanocomposites for solving the emerging wastewater problems. Due to the tutorial style, basic concept coverage over kinetics and equilibrium modeling, review analysis of heavy metals, metalloids, and dyes sequestration potential of advanced nanomaterials, the book will be suitable for use as a reference textbook for graduate and research students, and various academia and industry that have had a keen interests in inorganic to organic pollutants removal and recovery from aqueous media.

Nanotechnology is a broad, highly interdisciplinary, and still evolving field. Covering even the most important aspects of nanotechnology in a single book that reaches readers ranging from students to active researchers in academia and industry is an enormous challenge. To prepare such a book on emerging nanomaterials for solving the wastewater problems, *Gautam* and *Chattopadhyaya* have harnessed their own knowledge and experience, gained in several institutions and universities, and have assembled and written the present thirteen chapters monograph.

This comprehensive book focused on advanced nanomaterials and nanocomposites is intended to serve both as a textbook for university courses as well as a reference book for researchers. It is a timely addition to the literature on emerging field of nanotechnology, which I anticipate will stimulate further interest in this important new field and serve as an invaluable resource to members of the international scientific and industrial community.

C.L. Khetrapal
Former Professor of Chemistry
Indian Institute of Sciences
Bangalore, India

Distinguished Professor
Centre of Biomedical Research (CBMR)
Sanjay Gandhi Post-Graduate Institute of Medical Sciences Campus
Raibareli Road, Lucknow, Uttar Pradesh, India

PREFACE

"Water is the driving force in nature."

—Leonardo da Vinci

Nanotechnology is an important emerging field in several applications such as optoelectronics, chemistry, device storage, physics, medicine, biology, and environmental restoration. Among the wide group of nanotechnology, the use of nanoparticles for environmental remediation can be considered a novel application of nanotechnology. Nanoparticles-based remediation has a specific potential to minimize the need for treatment and disposal of contaminated water, by removing or transforming organic or inorganic contaminants into harmless forms. Although a large number of publications and studies are available in literature, they usually present certain gaps that prevent the use of these novel materials in full-scale real-case applications, while some initiatives have already taken place.

A wide range of nanomaterials such as advanced oxidation photocatalysts, magnetic nanoparticles, layered double hydroxides, and graphene and graphene-based nanocomposites have been extensively used in the separation and purification of hazardous contaminants from water and wastewater.

Graphene is a one-atom layer of graphite with a unique two-dimensional structure possesses excellent physicochemical and mechanical properties has attracted an increasing attention in recent years since its discovery in 2004 by Geim and coworkers. Their adsorptive properties are due to their high surface area to volume ratio, high degree of surface functional groups, ease in tunability and a high degree of surface reactivity with inorganic and organic moieties, and wonderful stability in extreme environmental conditions. Therefore, the graphene and graphene-based nanocomposite materials have been widely used to separate, concentrate, and purify the hazardous environmental pollutants from various aqueous matrices. Consequently, graphene and graphene-based nanocomposites are of immense interest to many industrial and civil sectors such as treatment of drinking water contaminated and polluted by dyes, pigments, heavy metals, metalloids, pharmaceuticals, and metal rich acid mine drainage operations in industrial and urban wastewater.

Interest in graphene as adsorbents received a big boost due to the increased environmental pollution of aqueous media and because of the continuous stringent rules and regulations by the governmental and environmental authorities to discharge the environmental pollutants in the aquatic system, after proper treatment under prescribed permissible limits. Hence, during the last five years several research papers dealing with adsorption, separation and preconcentration of pollutants on graphene and graphene-based adsorbents have been reported. However, these results are still new and scattered in various journals, magazines, and proceedings of the conferences. It was, therefore, thought worthwhile and opportune to prepare a reference book that describes the synthesis, fabrication, and application of nanoadsorbents, their adsorption phenomenon, and the adsorption of hazardous dyes, organic pollutants, and heavy metals and metalloids from aqueous solutions. The book also contains consistent explanation for adsorption kinetics, equilibrium isotherm modelings and thermodynamic analysis of various pollutants on nanomaterials of various categories.

The present book will fill the gap in understanding the mechanism of remediation on to the magnetic nanoparticles, catalysts, layered double hydroxides, graphene and graphene-based nanocomposites. It starts with an introductory chapter on nanoadsorbents frequently used in the remediation of toxic materials. A short overview on remediation technologies have been presented in Chapter 2. Chapter 3 deals with the advanced oxidation processes used in the environmental remediation of the aquatic ecosystem. Chapter 4 describes the role of graphene-based nanocomposites as nanosorbents, and Chapter 5 explain the removal of heavy metals and metalloids from water and wastewater and their adsorption mechanisms such as kinetics, diffusion in pores, and equilibrium isotherm modeling. Chapter 6 gives a detail account of the use of graphene-based nanocomposites for the separation and purification of dyes from liquid phases and their removal mechanism. Chapter 7 deals with the functionalized magnetic nanomaterials as new generation nanosorbents. A detail account of layered double hydroxide has been compiled in the Chapter 8. Magnetic nanophotocatalysts for remediation of environmental contaminants have been discussed in Chapter 9. Chapter 10 focuses on the use of alumina nanomaterials in treatment of water and wastewater. A new type of adsorption and catalytic material "bimetallic nanomaterials" has been described in Chapter 11. Desorption, regeneration, recovery, and reuse of nanomaterials is a very crucial factor in real wastewater treatment. Therefore, a concise study on regeneration and reuse of nanomaterials has been presented in Chapter 12. Any materials being manufactured by the engineered processes may produce some toxic behavior after their release in the environment. Thus, an effort has been carried out to compile the data on the release of nanomaterials in the environment and their toxic phenomenon on living organisms. Chapter 13 deals with the hazardous impact of nanomaterials in the environment. It is our hope that this book will be used as a reference book as well as a teaching book by academicians and engineers who wish to carry out research and teachings in the field of separation, adsorption, and purification. During the preparation of this monograph, we repeatedly felt that the much work remains to be done in this field and that many directions of research and development are waiting for newcomers to seek out. Because of our knowledge and experience, we know that many important problems which require discussion are not included in this monograph, if possible these should be addressed in the future edition.

First of all, we are thankful to Dr Kenneth P. McCombs, Senior Acquisitions Editor, Elsevier for inviting us to write this book and accepting our proposal. Without the kind email, guidance, and cooperation of Ken, it was not possible for us to complete this task. Thanks Ken! We are also grateful to all the editors and technical staffs, at Elsevier for their kind emails and messages. A special thanks to Dr Sushmita Banerjee and Dr P.K. Gautam for writing an expert chapter on layered double hydroxides for wastewater remediation (Chapter 8). We are again thankful to Dr Sushmita Banerjee for writing Chapter 10 on alumina nanoparticles and alumina-based adsorbents for wastewater treatment. Thank you so much friends for your kind support and cooperation. The support and encouragement of Emeritus Professor J.D. Pandey and Professor V.S. Tripathi from the Department of Chemistry, University of Allahabad is also appreciated. We are also thankful to Mrs Kokila Banerjee, Mr Ajit K. Banerjee, and Ms Tuhina Banerjee for their kind support and cooperation. Chattopadhyaya is thankful to his wife Mrs Alpana Chattopadhyaya and sons Mr Shushant and Mr Shashank Chattopadhayaya for making this journey beautiful and memorable. Finally, no words can express the feelings toward our family members who have contributed and sacrificed a lot for us to accomplish this task and will always remain a sole source of inspiration in our life to achieve higher goals. Last, but certainly not the least, we thank every reader of this book, and solicit your comments to our email ravindragautam1987@gmail.com and mcchattopadhyaya@gmail.com. Please let us know what you think of this edition; we will earnestly try to incorporate your suggestions to strengthen future editions. Enjoy this book!

R.K. Gautam
M.C. Chattopadhyaya

January 31, 2016, Allahabad

CHAPTER 1

Nanotechnology for Water Cleanup

1.1 INTRODUCTION

Recently, nanoscience and nanotechnology have produced important roles for the environmental restoration of aquatic ecosystems with an important number of published papers reporting successful experiments related to the removal of several typical pollutants, such as heavy metals, dyes, pharmaceuticals, phenols, organic matters, and pesticides. The amount of production and discharge of wastewater due to industrialization and urbanization have increased all over the world. Industrial wastewater, when discharged into nearby water bodies, show harmful effects which may lead to mutagenesis and carcinogenesis in aquatic creatures. Discharge of dye-bearing effluents into water bodies has raised much concern because of potential health hazards associated with toxicity and their by-products into the food chains of humans and aquatic animals. The synthetic dyes when entering the water alter the aqueous chemistry by changing the solution pH, color, and chemical oxygen demand and hinder the growth of microbial organisms. Dye also impedes the penetration of solar light, thus changing the photosynthetic activity. Therefore, there is an urgent need to remove environmental contaminants before effluent is discharged into the receiving aquatic system.

Nanotechnology is an important emerging field in several applications, such as optoelectronics, chemistry, device storage, physics, medicine, biology, and environmental restoration. Among the wide group of nanotechnology, the use of nanoparticles for environmental remediation can be considered a novel application of nanotechnology. Nanoparticle-based remediation has specific potential to minimize the need for treatment and disposal of contaminated water by removing or transforming organic or inorganic contaminants into harmless forms. Although a large number of publications and studies are available in literature, they usually present certain gaps that prevent the use of these novel materials in full-scale, real-case applications, while some initiatives have already taken place.

Nanomaterials for Wastewater Remediation
http://dx.doi.org/10.1016/B978-0-12-804609-8.00001-7

Nanomaterials have many attractive properties that can be harnessed for environmental remediation. Their size provides them with appropriate dispersibility, high surface area, porosity, and tunable reactivity. They can be dispersed appropriately. Under the right formulation and in the presence of stabilizing agents, nanoparticles can travel just like water in soil, and when in the liquid phase, they quickly scrutinize large volumes. Likely, their reactivity can be tuned to be enough to attack organic and inorganic molecules, but not enough to attack life forms. Additionally, their reactivity is linked to a short life, where the nanoparticles will blend into micrometric composites or dissolve once and their work is done. Their nonpersistency should ease their introduction into waters and land. Thus, nanoparticles can be designed to take advantage of their special physicochemical properties, such as the presence of a higher surface area per unit of volume, or the ability to be functional with a number of surfactants to enhance their affinity toward target molecules. In addition, their small size and capacity to be suspended in groundwater allows them to travel further than larger macrosized particles, achieving a wider distribution.

Recently, a variety of nanomaterials such as iron oxide–based magnetic nanoparticles, photocatalysts based on titanium dioxide, carbon nanotubes, graphene based nanoadsorbents, bimetallic nanoparticles, metal organic frameworks, and layered double hydroxides have been widely explored for their efficient application in the treatment and restoration of water. In this chapter, we are going to briefly discuss the application of various types of nanomaterials for the treatment and remediation of organic and inorganic pollutants from water and wastewater.

1.2 MAGNETIC NANOPARTICLES

During the last 5 years, magnetic nanoparticles, due to their magnetic properties, high chemical stability, low toxicity, ease in synthesis, and excellent recycling capability, have aroused great attention and have been extensively studied to remove toxic metal ions and organic pollutants from water. These materials show higher removal capacities than the bulk material. Their nanoscale-sized overall structure provides the necessary mechanical robustness against wear and tear, and provides the high surface area as well as the high removal capacity for heavy metal ions and dyes. Magnetic nanoparticles have a wide range of application including magnetic fluids, catalysis, biomedicine, drug delivery, magnetic resonance imaging, data storage, and environmental remediation (Jung et al., 2011; Singh et al., 2011; Ambashta and Sillanpaa, 2010; Shen et al., 2009; Lu et al., 2007).

Several appropriate methods have been developed for the synthesis of magnetic nanoparticles of a variety of different compositions. In the majority of the envisaged applications, the particles perform best when the size of the nanoparticles is below a critical value, which is dependent on the source material but is normally around 10–20 nm (Lu et al., 2007). The design and synthesis of nanoparticle-based adsorbents has generated great interest in a variety of scientific communities, ranging from chemical, biological, and environmental science to engineering. Magnetic nanoparticle-based adsorbents can be used in the separation and purification of heavy metals and toxic dyes from aqueous solutions with high precision and accuracy (Wang et al., 2012; Singh et al., 2011; Sun et al., 2011; Girginova et al., 2010; Koehler et al., 2009; Lee et al., 2009).

However, the bare magnetic nanoparticles are very prone to atmospheric conditions as they are easily oxidized in open air and water. Hence, scientists have coated or functionalized the magnetic nanoparticles to enhance the functional groups and to stabilize the synthesized magnetic nanoparticles in the extreme environmental conditions. Functionalized magnetic nanoparticles are very promising for applications in catalysis, reduction, and oxidation of pollutants and for the separation of heavy metals and hazardous dyes. Extraction of heavy metals and dyes from their liquid phase with such small and magnetically separable particles may be useful as they combine the advantages of high dispersion, high reactivity, high stability under acidic conditions, and easy recycling of used adsorbents in separation.

Several accepted methods such as coprecipitation, microemulsion, thermal decomposition, sonochemical, microwave synthesis, chemical vapor deposition, combustion synthesis, carbon arc, laser pyrolysis synthesis, hydrothermal synthesis, sol–gel synthesis, and solvothermal methods have been reported for synthesis of magnetic nanoparticles (Laurent et al., 2008; Lu et al., 2007; Zhou et al., 2009; Tai et al., 2011; Willis et al., 2005; Hong et al., 2008). However, the synthesis of magnetic nanoparticles is a complex process because of their colloidal nature and high reactivity. In the last decades, much attention has been devoted to the synthesis of magnetic nanoparticles. Particularly during the last few years, many publications have described efficient synthetic routes to shape-controlled, highly stable, functionalized, and monodispersed magnetic nanoparticles with homogeneous composition and narrow size distribution. However, the most common method for the synthesis of magnetite nanoparticles is the coprecipitation technique of iron salts.

Recently, Gautam et al. (2014) synthesized the superparamagnetic Fe_3O_4 nanoparticles via a coprecipitation route for the adsorption of Ni(II) ions from aqueous solutions. The transmission electron microscopy (TEM) images showed that the synthesized nanomaterials are spherical in shape and the material is highly crystalline in nature (Fig. 1.1a–f). Most of the prepared nanoparticles were in the size range of ~7–15 nm. pH_{pzc} was determined by measuring the zeta potentials for magnetite nanoparticles adsorbent at varied pH. pH_{pzc} indicates the electrical neutrality of the adsorbent and the

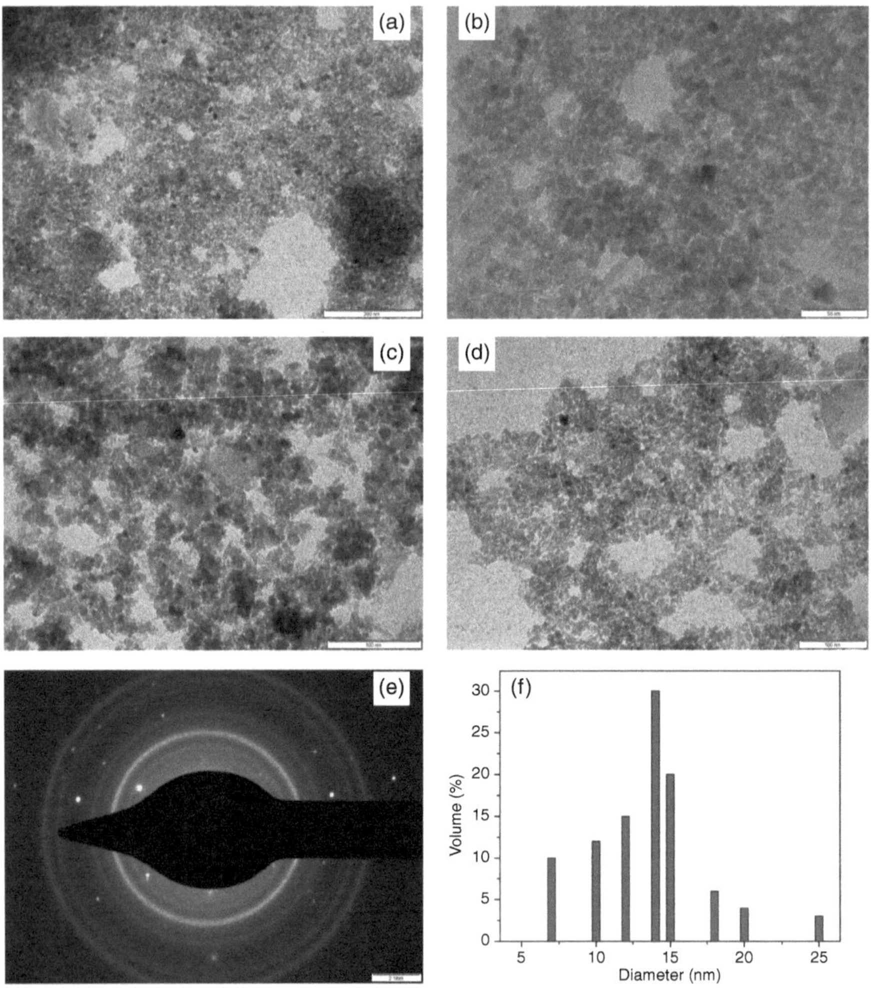

Figure 1.1 TEM image of magnetite magnetic nanoparticles (a, b); Ni(II) loaded magnetite magnetic nanoparticles (c); recycled magnetic nanoparticles after three adsorption–desorption cycles (d); SAED pattern image of magnetite magnetic nanoparticles (e); and size histogram of suspension (magnetite was dispersed in hexane) of magnetic nanoparticles obtained from DLS measurements (f). *(Source: Gautam et al., 2014, © Elsevier.)*

surface at a particular value of pH. The pH of zero point charge (pH_{pzc}) of Fe_3O_4 was 6.2. Magnetic measurements of magnetite nanoparticles were investigated with a vibrating sample magnetometer (VSM) at room temperature in the applied magnetic field sweeping from −50 to +50 kOe. The magnetic saturation value of the magnetite was 84 emu/g. Such an excellent magnetic property means that as-prepared magnetic nanomaterials have strong magnetic responsivity and can be separated easily from the aqueous solution with the help of an external magnetic force.

Recently, Rai et al. (2015) synthesized the $SnFe_2O_4$@activated carbon magnetic nanocomposite via coprecipitation and applied for the adsorptive removal of the cationic dye crystal violet from aqueous solutions. The effect of various parameters such as initial dye concentration, contact time, pH, and temperature were investigated, and the adsorption equilibrium data showed that the adsorption process was fitted to the Langmuir model. The maximum sorption capacity for the removal of crystal violet was 158.73 mg/g at 323 K. The adsorption equilibrium was established within 80 min of contact time (Fig. 1.2), and the pH 8 was suitable for the maximum removal of dye (Fig. 1.3). The adsorption kinetics was better described

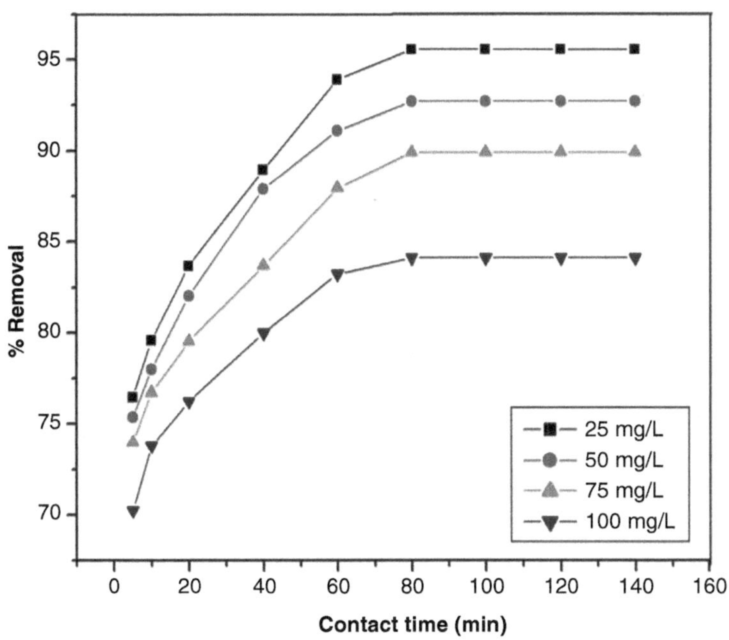

Figure 1.2 Effect of contact time and initial concentration on adsorption of crystal violet on magnetic nanocomposite (pH 8, adsorbent dose = 2.0 g/L, shaking speed = 180 rpm, and temperature = 303 K). *(Source: Rai et al., 2015, © Elsevier.)*

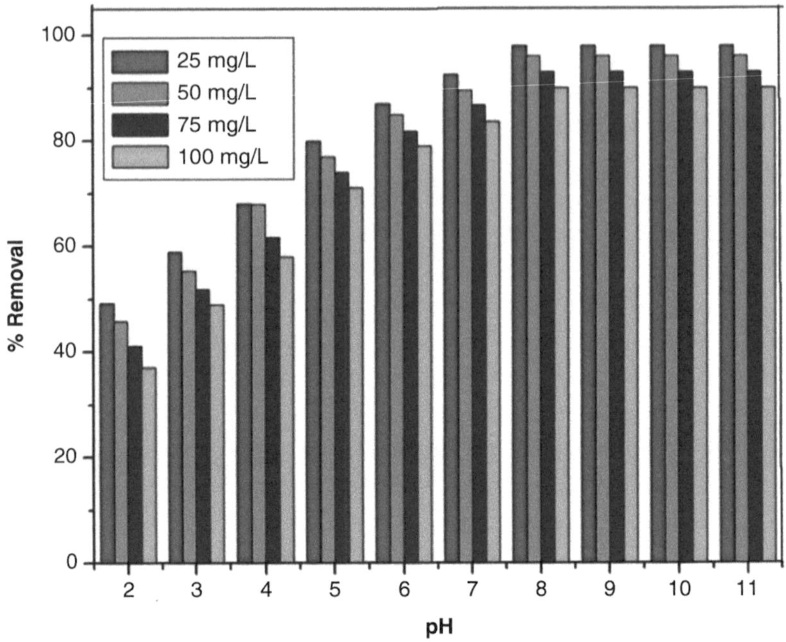

Figure 1.3 Effect of pH on the removal of crystal violet by magnetic nanocomposite (adsorbent dose = 2.0 g/L, shaking speed = 180 rpm, and temperature = 303 K). *(Source: Rai et al., 2015, © Elsevier.)*

with pseudo-second order model. The thermodynamic parameters show that the adsorption process was endothermic and spontaneous in nature.

Singh et al. (2014) synthesized citric acid–coated magnetite nanoparticles (Fe_3O_4-Cit) for the removal of cadmium from aqueous solutions. Several factors affecting the adsorption of Cd(II) ions, such as pH, temperature, and contact time, were investigated. The adsorption equilibrium was established within 35 min at the optimized pH 5. Kinetic study shows the Cd(II) adsorption onto Fe_3O_4-Cit followed the pseudo-second order kinetic model. The isotherm modeling showed that the adsorption process was fitted with the Langmuir isotherm models. The thermodynamic analysis showed that the adsorption system was spontaneous and feasible in nature.

1.3 LAYERED DOUBLE HYDROXIDES (LDHs) FOR ENVIRONMENTAL APPLICATIONS

Layered double hydroxides (LDHs) have been widely used as adsorbents for decontaminating the environment and preventing the dispersion of pollutants in nature. A wide range of pollutants can be easily removed by

LDHs from wastewater by anion exchange, adsorption process, and catalytic remediation. Recently, many molecules such as pesticides, toxic organic chemicals, heavy metals, and some other undesirable inorganic substances are more targeted for the remediation by LDHs. Application of LDHs in water treatment can be explained by the following points.

1.4 REMOVAL OF INORGANIC CONTAMINANTS BY LDHs

Due to their anion exchange properties LDHs have been widely used for environmental remediation. They have been studied extensively to characterize the adsorption properties of the materials under vigorous solid–liquid interface conditions. The effect of sorbent composition, surface and bulk adsorption, and concentration of adsorption site have been assessed. The adsorption capacity is deeply affected by the nature of the counter-anion of the LDH layer.

Miyata (1983) first reported the ion exchange isotherm of a series of LDHs leading to the ion selectivity of layered double hydroxides for monovalent and divalent anions, $OH^- > F^- > Cl^- > Br^- > NO_3^- > I^-$, and $CO_3^{2-} > SO_4^{2-}$. Parker et al. (1995) have pointed out that the high selectivity for carbonate anion prevents LDHs from being used as an anion-exchange material unless further treatment is made. They evaluated the anion-exchange capacities of hydrotalcite in the single and mixed anion solutions by observing the amount of anion adsorbed by freshly prepared hydrotalcite. They have also compared the relative preference for $SO_4^{2-} > F^- > HPO_4^{2-} > Cl^- > B(OH)_4^- > NO_3^-$ within 24 h.

1.4.1 Uptake of Heavy Metal Cations

LDHs can be used as precipitating agents of heavy metal cations for the decontamination of wastewater. Mn^{2+}, Fe^{2+}, and Cu^{2+} cations were removed by synthetic hydrotalcite-like compounds; zaccagnaite (Jaiswal and Chattopadhyaya, 2009) and hydrotalcite thin films are used for the remediation of aqueous wastes containing hazardous metal ions. A chromotropic acid (CTA) intercalated, layered double hydroxide (LDH) supramolecular assembly of Zn_4Al–CTA–LDH was prepared using separate nucleation and aging steps (Chen and Song, 2013). The as-prepared Zn_4Al–CTA–LDH material was used as adsorbents for a number of metal ions from aqueous solutions. The results showed that Zn_4Al–CTA–LDH was highly selective and efficient for the adsorption of both Cr(VI) and Cu(II), and the sorption kinetic curves were well fitted with the pseudo-second-order model. The maximum adsorption capacity of Zn_4Al–CTA–LDH toward Cr(VI) was ~782 mg/g and toward Cu(II) was ~450 mg/g, respectively.

1.5 REMOVAL OF NUCLEAR WASTES

Hydrocalumite or ettringite, the calcium form of hydrotalcite, have been evidenced as early alteration products of cements or basaltic glasses (Zhang and Reardon, 2003) and studied for their adsorption properties of actinides (Th, U, Np, Pu, Se, and Am) from aqueous solutions (Zhang et al., 2012; Chen and An, 2012). Kang et al. (1999) used the LDHs for the removal of Tc, Re, and Mo in their anionic form from radioactive wastewater. Synthetic hydrotalcite has been investigated for the sorption of iodine and iodine-containing anionic species (Theiss et al., 2012).

1.5.1 Removal of Organic Pollutants by LDHs

Recently, LDHs and their oxides have been investigated as scavengers to treat wastewater-containing organic contaminants. The materials are known to function as effective anion adsorbents in the uptake of phenols, terephthalate, anionic surfactants, ionizable pesticides and herbicides, humic and fulvic acid, anionic dyes, and colored organics from aqueous solutions (Ulibarri et al., 1995; Hermosin et al., 1996; Pavlovic et al., 1997; Valente et al., 2009; Setti et al., 2010; Sun et al., 2010).

1.5.2 Removal of Organic Compounds and Pesticides

Functionalization or intercalation of LDHs enhances the functional groups and porosity in the materials. You et al. (2002) examined the surfactant-enhanced adsorption of organic compounds by Mg–Al LDHs. Organo-Mg–Al LDHs were prepared by incorporating anionic surfactants such as octylsulfate, DS, 4-octylbenzenesulfonate, and dodecylbenzene sulfonate via ion exchange. The octylsulfate formed bimolecular films, and other surfactants resulted in monolayer structure. Intercalation of surfactants into Mg–Al LDH decreased the surface area, whereas surfactants dramatically enhanced the LDH affinity for 1,2,4-trichlorobenzene and 1,1,1-trichloroethane in aqueous solutions.

1.5.3 Removal of Dyes

The discharged effluents of industries contain a large number of dyes, increasing the total COD of wastewaters (Rajeshwarisivaraj et al., 2001). LDHs have high adsorption capacities for dye molecules and can be very competitive with other sorbents. Recently, adsorptive removal of organic dye methyl orange from aqueous solutions with hydrothermally synthesized Mg-Al LDH was carried out in batch experiments (Ai et al., 2011). The adsorption kinetics was well described by the pseudo-second-order model. The equilibrium adsorption data were fitted well to both the Langmuir and Freundlich models.

1.6 GRAPHENE-BASED ADSORBENTS

Mishra and Ramaprabhu (2011) synthesized the graphene sheets by hydrogen-induced exfoliation of graphitic oxide followed by functionalization. These functionalized graphene sheets were used for simultaneous removal of high concentrations of inorganic species of arsenic and sodium from aqueous solution using supercapacitor-based water filters. In addition, these functionalized graphene sheet–based water filters were used for desalination of sea water. Maximum adsorption capacities using Langmuir isotherms for arsenate, arsenite, and sodium were found to be nearly 142, 139, and 122 mg/g, respectively.

Lee and Yang (2012) synthesized self-assembled flower-like TiO_2 on exfoliated graphite oxide for heavy metal removal from aqueous solutions. The surface-attached, oxygenated functional groups of exfoliated graphite oxide showed a high removal capacity of heavy metals. The flower-like TiO_2 on the graphene oxide (GO) structure significantly improved the removal efficiency of heavy metals. The GO-TiO_2 hybrid adsorption capacities of heavy metal ions, after 6 and 12 h of hydrothermal treatment at 100°C, were respectively 44.8 ± 3.4 and 88.9 ± 3.3 mg/g for removing Zn^{2+}, 65.1 ± 4.4 and 72.8 ± 1.6 mg/g for removing Cd^{2+}, and 45.0 ± 3.8 and 65.6 ± 2.7 mg/g for removing Pb^{2+} at pH 5.6. Comparatively, colloidal GO under identical condition showed removal capacities of 30.1 ± 2.5, 14.9 ± 1.5, and 35.6 ± 1.3 mg/g for Zn^{2+}, Cd^{2+}, and Pb^{2+}, respectively. TiO_2 blossoms markedly formed upon GO as the hydrothermal treatment time at 100°C increased from 6 h to 12 h. Longer treatment times resulted in an increase in the surface area of GO–TiO_2 hybrid and thus increased removal capacity of heavy metal from aqueous solutions.

Zhang et al. (2011) synthesized ordered porous chitosan–gelatin/graphene oxide (CGGO) monoliths with over 97% porosity by a unidirectional freeze-drying method and used them as adsorbents for metal ions. The adsorption behavior of the CGGO monoliths and influencing factors such as pH, GO concentration, and metal ion concentration, as well as the effect of ethylenediaminetetraacetic acid (EDTA), were investigated. The incorporation of GO significantly increased the compressive strength of the CGGO monoliths in both their wet and dry states and changed their porous structure. They exhibited an extremely high adsorbing ability for metal ions, which decreased at low pH but increased from 20 to 88% upon the addition of EDTA at low pH. The CGGO monoliths have good stability and can be recycled several times with only a slight loss in adsorption ability. The adsorption ability of the porous CGGO monoliths for Cu^{2+} was greatly influenced by the pH of the Cu^{2+} solution. The adsorption of Cu^{2+} dramatically increased from 15.6 to 88.6% as the pH changed from 2.0 to 6.0. The CGGO monolith was doped with only 0.5 wt.% GO, and the maximum adsorption capacity of CGGO for Pb^{2+} increased from 68.5 to 75.5 mg/g. As the GO concentration was increased further, the maximum adsorption capacity for Pb^{2+} gradually increased to a maximum value of about 100 mg/g.

Graphene oxide aerogels were prepared by a unidirectional freeze-drying method, their structure was observed, and their adsorption performance for Cu^{2+} in aqueous solution was evaluated by Mi et al. (2012). The results showed that graphene oxide aerogels were unidirectional porous structures and had good adsorption ability for Cu^{2+}. The adsorption system depended on the initial Cu^{2+} ion concentration and followed a pseudo second-order kinetic model. An adsorbent dose of 0.06 g of graphene oxide aerogels was sufficient to remove the maximum Cu^{2+} from 100 mL of aqueous solution within 15 min of equilibrium contact time. The adsorption isotherms were simulated by the Langmuir, Freundlich, and Temkin models and fitted well with classical Langmuir adsorption isotherms with a maximum capacity of 29.59 mg/g at 40°C. The adsorption of Cu^{2+} on graphene oxide aerogels was strongly dependent on pH, indicating an ion exchange mechanism. The results obtained by Mi et al. (2012) have demonstrated that the graphene oxide aerogels can be used as an effective adsorbent for Cu^{2+} removal from water. However, graphene oxide can be aggregated by Cu^{2+} interaction in aqueous solution with a huge Cu^{2+} absorption capacity (Yang et al., 2010). The Cu^{2+} causes graphene oxide sheets to be folded and to form large aggregates. The folding and aggregation were most likely triggered by the coordination between graphene oxide and Cu^{2+} in aqueous solutions. The average distance between the folded layers was around 2 nm, which was higher than the typical single layer thickness of graphene oxide (around 0.9 nm). The equilibrium Cu^{2+} concentrations and equilibrium absorption capacity of graphene oxide were measured to estimate the maximum absorption capacity of graphene oxide for Cu^{2+} using Langmuir and Freundlich isotherm models. The adsorption of Cu^{2+} by graphene oxide well fitted to the Langmuir isotherm model with a maximum adsorption capacity of 46.6 mg/g. Graphene oxide has a huge adsorption capacity for Cu^{2+}, which is around 10 times that of active carbon.

1.7 METAL ORGANIC FRAMEWORKS (MOFs)

Metal organic framework (MOF) material has found great application in energy storage and environmental science and technology (Tella et al., 2014; Schneemann et al., 2014; Hasan et al., 2012; Zhao et al., 2011) due to its unique properties, such as high surface area to volume ratio, high physicochemical stability, increased functionality, biocompatibility, and highly tailorable microporous properties (Furukawa et al., 2014; Shultz et al., 2011). During the last 5 years, due to their excellent properties the MOFs have been successfully applied in the pollutants' separation and remediation from both gaseous and liquid phases (de Voorde et al., 2014; Zhu and Xu, 2014; Haque et al., 2014; Adeyemo et al., 2012; Paul et al., 2009).

Huang et al. (2012) applied hierarchically mesostructured MIL-101 MOFs for the adsorptive removal of methylene blue from aqueous media. Recently, Chen et al. (2012) synthesized MOF material based on chromium-benzenedicarboxylates (MIL-101) and applied it to the adsorption of xylenol orange from aqueous solution. MOF-5 had been synthesized on silk fiber through electrostatic layer-by-layer assembly, and adsorptive removal and recovery of hazardous anionic dye Congo red was performed (Khanjani and Morsali, 2014). There are very few reports on the copper-coordinated dithiooxamide (Cu-DTO) MOFs for wastewater treatment. Li et al. (2013) used Cu-DTO MOFs for the adsorptive removal of basic dye crystal violet from aquatic media.

1.8 BIMETALLIC NANOPARTICLES

Recently, zero valent iron metal (Fe^0) and magnetic nanoparticles have been successfully utilized for the removal of dyestuffs and metal ions from aqueous media due to their low cost, high efficiency, ease in operation, effectiveness, and biocompatibility in nature (Fan et al., 2009; Hu et al., 2007; Fang et al., 2011). It has been observed that iron-oxidized organic compound when used in the presence of oxygen can remediate a variety of contaminants (Joo et al., 2005). Fan et al. (2009) used the nanoscale zerovalent iron particles for the decolorization of methyl orange from aqueous solutions, with a decolorization efficiency of 100% within 60 min without any pH adjustment because of its high surface area, small particles size, and large density of intrinsic surface sites (Nurmi et al., 2005). The use of nanoscale iron particles have been well established for separation and purification of toxic metal ions and organic compounds (Gautam et al., 2014). However, it has been reported that the pollutants' adsorption rate was very slow by iron oxide nanoparticles (Joo et al., 2005). Furthermore, the addition of a second catalytic metal such as Ni, Cu, Pd, and Pt may enhance the adsorption rate as compared to nanoscale iron alone (Wang et al., 2004; Schrick et al., 2002) and also protects from oxidation in the air. Although various combinations of nanoscale iron-based bimetallic compounds have been utilized for the degradation of phenol, bisphenol A, and chlorinated organic compounds (Lee et al., 2014; He and Zhao, 2005; Lin et al., 2004; Wang et al., 2009), very few studies are reported on the removal of dyestuffs from aqueous media. An Fe–Cu bimetallic system was developed for decolorization of methylene blue in aqueous solution (Ma et al., 2004), and the degradation of orange G in aqueous solution was investigated using Fe–Ni bimetallic nanoparticles (Bokare et al., 2007). Furthermore, the addition of a second metal (Cu, Ni, and Pd) with iron may produce toxic effects, and the incorporation of Zn with iron enhances the biocompatibility, stability, and adsorption capacity (Hu et al., 2007).

Recently, a novel bimetallic Fe–Zn nanoparticle was synthesized via a coprecipitation method and applied for the adsorptive removal of carcinogenic dyes malachite green and Congo red in water (Gautam et al., 2015). Batch adsorption experiments were conducted to evaluate the adsorption process by varying the optimization parameters such as the solution pH, time, and initial dye concentration. Fig. 1.4 shows the effect of pH on the removal of malachite green and Congo red. The kinetics of removal was showed the best agreement with the pseudo-second order model. Adsorption isotherms are a good fit with the Langmuir isotherm model. The Langmuir maximum adsorption capacity for malachite green and Congo red was found to be 21.74 and 28.56 mg/g, respectively.

Desorption of malachite green from malachite green-sorbed nanoparticles was performed in methanol containing 5% acetic acid solution as the eluent. About a 95.0% adsorption efficiency was achieved when 5% (v/v)

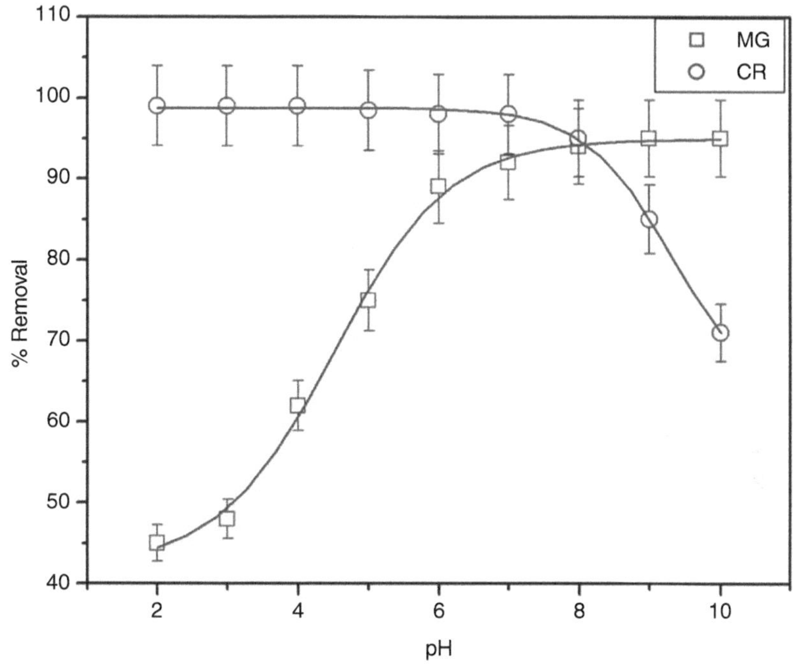

Figure 1.4 Effect of pH on the removal of light gray (malachite green in the web version) and dark gray (Congo red in the web version) (initial dye concentration = 25 mg/L, agitation speed = 120 rpm, adsorbent dose = 0.15 g, contact time = 60 min, temperature = 303 K). *(Source: Gautam et al., 2015, © Elsevier.)*

acetic acid was used in methanol solution (Fig. 1.5). Desorption of Congo red from Fe–Zn bimetallic nanoparticles was demonstrated using three different eluents: 0.01 M NaOH, 0.1 M NaOH, and 0.5 M NaOH. The quantitative desorption efficiencies were determined to be 82.6, 96.3, and 88.0%, respectively. The reusability was checked by following the adsorption–desorption process for the three eluents. The 0.1 M NaOH solution was the optimum eluent.

1.9 CONCLUSIONS

Nanotechnology has made an important niche in the advanced water remediation technologies, and this area is increasing quickly toward real applications in industrial wastewater treatment. A variety of nanoscale materials such as magnetic nanoparticles, graphene-based composites, LDHs, MOFs, advanced photocatalysts, and bimetallic nanoparticles have proven their ability to remediate heavy metals and organic dyes from wastewater

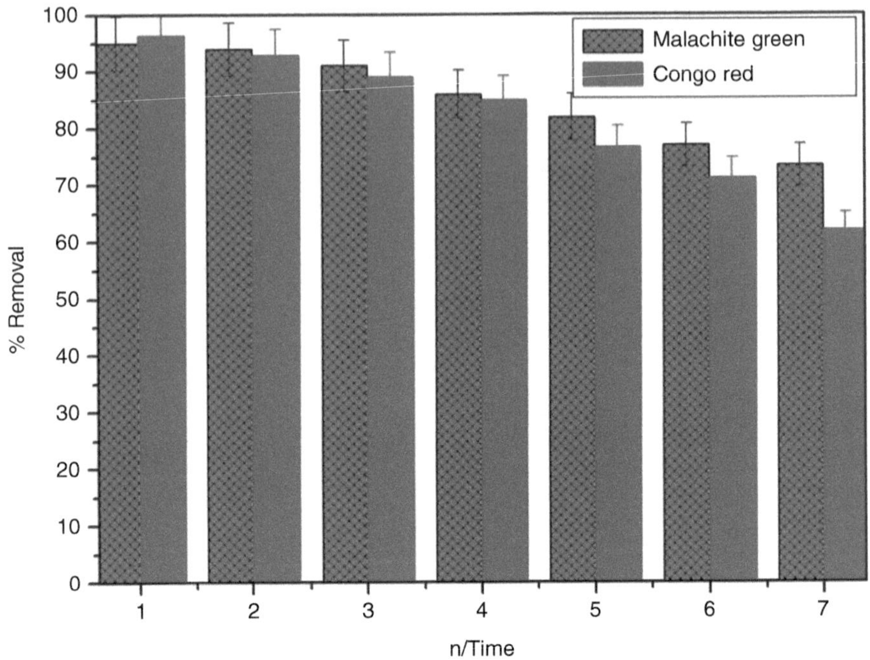

Figure 1.5 Effect of recycling Fe–Zn bimetallic nanoparticles on the light gray (malachite green in the web version) and dark gray (Congo red in the web version) adsorption (temperature = 303 K, agitation speed = 120 rpm, contact time = 60 min, solution pH malachite green = 9, solution pH Congo red = 4). *(Source: Gautam et al., 2015, © Elsevier.)*

of industrial effluents. However, the bare nanomaterials of iron oxide have high surface area to volume ratio but are easily oxidized in acidic environmental conditions. So there is a need to enhance the stability of iron oxide nanoparticles in harsh environmental conditions. Hence, many of the nanomaterials have been synthesized and functionalized with organic or inorganic moieties to enhance the functional groups on the surface and to stabilize the nanomaterials in extreme environmental conditions. Furthermore, it has been noted that the functionalization of nanomaterials by the organic or inorganic molecules may decrease the surface area in comparison with their bare nanoparticles. LDHs are also very promising nanomaterials, as they have high surface area and porosity, which is greatly helpful in the adsorption and diffusion of water pollutants.

Advanced oxidation processes and photocatalysts have gained much attention during the past few decades in the degradation and remediation of recalcitrant pollutants from water. However, there is a risk that degradation by-products may be more toxic than the parent contaminants. Although nanomaterials offer a lot of unique features to solve environmental problems, at the same time the regeneration and reuse of exhausted nanosorbents is also a major challenge. Most of the nanomaterials used till now are not cost effective and their synthesis and fabrication are still controlled by most of the developed countries. Further, the release of nanomaterials into the environment may pose unmeasured risk to the health of humans and animals.

REFERENCES

Adeyemo, A.A., Adeoye, I.O., Bello, O.S., 2012. Metal organic frameworks as adsorbents for dye adsorption: overview, prospects and future challenges. Toxicol. Environ. Chem. 94, 1846–1863.

Ai, L., Zhang, C., Meng, L., 2011. Adsorption of methyl orange from aqueous solution on hydrothermal synthesized Mg–Al layered double hydroxide. J. Chem. Eng. Data 56, 4217–4225.

Ambashta, R.D., Sillanpaa, M., 2010. Water purification using magnetic assistance: a review. J. Hazard. Mater. 180, 38–49.

Bokare, A.D., Chikate, R.C., Rode, C.V., Paknikar, K.M., 2007. Effect of surface chemical of Fe–Ni nanoparticle on mechanistic pathway of azo dye degradation. Environ. Sci. Technol. 41, 7437–7443.

Chen, C., Zhang, M., Guan, Q., Li, W., 2012. Kinetic and thermodynamic studies on the adsorption of xylenol orange onto MIL-101(Cr). Chem. Eng. J. 183, 60–67.

Chen, M.L., An, M., 2012. Selenium adsorption and speciation with Mg–$FeCO_3$ layered double hydroxides loaded cellulose fibre. Talanta 95, 31–35.

Chen, Y., Song, Y.F., 2013. Highly selective and efficient removal of Cr(VI) and Cu(II) by the chromotropic acid-intercalated Zn−Al layered double hydroxides. Ind. Eng. Chem. Res. 52, 4436–4442.

de Voorde, B.V., Bueken, B., Denayer, J., Vos, D.D., 2014. Adsorptive separation on metal–organic frameworks in the liquid phase. Chem. Soc. Rev. 43, 5766–5788.

Fan, J., Guo, Y., Wang, J., Fan, M., 2009. Rapid decolorization of azo dye methyl orange in aqueous solution by nanoscale zerovalent iron particles. J. Hazard. Mater. 166, 904–910.

Fang, Z., Qiu, X., Chen, J., Qiu, X., 2011. Debromination of polybrominated diphenyl ethers by Ni/Fe bimetallic nanoparticles: influencing factors, kinetics, and mechanism. J. Hazard. Mater. 185, 958–969.

Furukawa, S., Reboul, J., Diring, S., Sumida, K., Kitagawa, S., 2014. Structuring of metal–organic frameworks at the mesoscopic/macroscopic scale. Chem. Soc. Rev. 43, 5700–5734.

Gautam, R.K., Jaiswal, A., Chattopadhyaya, M.C., 2014. Functionalized magnetic nanoparticles for heavy metal removal from aqueous solutions: kinetics and equilibrium modeling. In: Tiwari, A., Syvajarvi, M. (Eds.), Advanced Materials for Agriculture, Food, and Environmental Safety. WILEY-Scrivener Publishing LLC, Massachusetts, pp. 291–331.

Gautam, R.K., Gautam, P.K., Banerjee, S., Soni, S., Singh, S.K., Chattopadhyaya, M.C., 2015. Removal of Ni(II) by magnetic nanoparticles. J. Mol. Liq. 204, 60–69.

Gautam, R.K., Rawat, V., Banerjee, S., Sanroman, M.A., Soni, S., Singh, S.K., Chattopadhyaya, M.C., 2015. Synthesis of bimetallic Fe–Zn nanoparticles and its application towards adsorptive removal of carcinogenic dye malachite green and Congo red in water. J. Mol. Liq. 212, 227–236.

Girginova, P.I., Daniel-da-Silva, A.L., Lopes, C.B., Figueira, P., Otero, M., Amaral, V.S., Pereira, E., Trindade, T., 2010. Silica coated magnetite particles for magnetic removal of Hg^{2+} from water. J. Colloid Interface Sci. 345, 234–240.

Haque, E., Lo, V., Minett, A.I., Harris, A.T., Church, T.L., 2014. Dichotomous adsorption behaviour of dyes on an amino-functionalised metal–organic framework, amino-MIL-101(Al). J. Mater. Chem. A 2, 193–203.

Hasan, Z., Jeon, J., Jhung, S.H., 2012. Adsorptive removal of naproxen and clofibric acid from water using metal-organic frameworks. J. Hazard. Mater. 209-210, 151–157.

He, F., Zhao, D., 2005. Preparation and characterization of a new class of starch-stabilized bimetallic nanoparticles for degradation of chlorinated hydrocarbons in water. Environ. Sci. Technol. 39, 3314–3320.

Hermosin, M.C., Pavlovic, I., Ulibarri, M.A., Cornejo, J., 1996. Hydrotalcite as sorbent for trinitrophenol: sorption capacity and mechanism. Water Res. 30, 171–177.

Hong, R.Y., Li, J.H., Li, H.Z., Ding, J., Zheng, Y., Wei, D.G., 2008. Synthesis of Fe_3O_4 nanoparticles without inert gas protection used as precursors of magnetic fluids. J. Mag. Mag. Mater. 320, 1605–1614.

Hu, J., Lo, I.M.C., Chen, G., 2007. Comparative study of various magnetic nanoparticles for Cr(VI) removal. Sep. Purif. Technol. 56, 249–256.

Huang, X.X., Qiu, L.G., Zhang, W., Yuan, Y.P., Jiang, X., Xie, A.J., Shen, Y.H., Zhu, J.F., 2012. Hierarchically mesostructured MIL-101 metal–organic frameworks: supramolecular template-directed synthesis and accelerated adsorption kinetics for dye removal. CrystEngComm 14, 1613–1617.

Jaiswal, A., Chattopadhyaya, M.C., 2009. Removal of zinc ion from industrial effluents by hydrotalcite-like compound. Desalin. Water Treat. 12, 127–132.

Joo, S.H., Faitz, A.J., Sedleck, D.L., Waite, T.D., 2005. Quantification of the oxidizing capacity of nanoparticulate zero valent iron. Ind. Environ. Sci. Technol. 39, 1263–1268.

Jung, J.H., Lee, J.H., Shinkai, S., 2011. Functionalized magnetic nanoparticles as chemosensors and adsorbents for toxic metal ions in environmental and biological fields. Chem. Soc. Rev. 40, 4464–4474.

Kang, M.J., Chun, K.S., Rhee, S.W., Do, Y., 1999. Comparison of sorption behavior of I^- and TcO_4^- on Mg/Al layered double hydroxide. Radiochim. Acta 85, 57–63.

Khanjani, S., Morsali, A., 2014. Ultrasound-promoted coating of MOF-5 on silk fiber and study of adsorptive removal and recovery of hazardous anionic dye "congo red". Ultrason. Sonochem. 21, 1424–1429.

Koehler, F.M., Rossier, M., Waelle, M., Athanassiou, E.K., Limbach, L.K., Grass, R.N., Gunther, D., Stark, W.J., 2009. Magnetic EDTA: coupling heavy metal chelators to metal nanomagnets for rapid removal of cadmium, lead and copper from contaminated water. Chem. Comm. 5, 4862–4864.

Laurent, S., Forge, D., Port, M., Roch, A., Robic, C., Elst, L.V., Muller, R.N., 2008. Magnetic iron oxide nanoparticles: synthesis, stabilization, vectorization, physicochemical characterizations, and biological applications. Chem. Rev. 108, 2064–2110.

Lee, Y.C., Yang, J.W., 2012. Self-assembled flower-like TiO_2 on exfoliated graphite oxide for heavy metal removal. J. Ind. Eng. Chem. 18, 1178–1185.

Lee, P.L., Sun, Y.C., Ling, Y.C., 2009. Magnetic nano-adsorbent integrated with lab-on-valve system for trace analysis of multiple heavy metals. J. Anal. Atom. Spect. 24, 320–327.

Lee, H., Yoo, H.Y., Choi, J., Nam, I.H., Lee, S., Lee, S., Kim, J.H., Lee, C., Lee, J., 2014. Oxidizing capacity of periodate activated with iron-based bimetallic nanoparticles. Environ. Sci. Technol. 48, 8086–8093.

Li, X., Zheng, L., Huang, L., Zheng, O., Lin, Z., Guo, L., Qiu, B., Chen, G., 2013. Adsorption removal of crystal violet from aqueous solution using a metal-organic frameworks material, copper coordination polymer with dithiooxamide. J. Appl. Polym. Sci. 129, 2857–2864.

Lin, C.J., Lo, S.L., Liou, Y.H., 2004. Dechlorination of trichloroethylene in aqueous solution by noble metal-modified iron. J. Hazard. Mater. B116, 219–228.
Lu, A.H., Salabas, E.L., Schuth, F., 2007. Magnetic nanoparticles: synthesis, protection, functionalization, and application. Ang. Chem. Inter. Ed. 46, 1222–1244.
Ma, L.M., Ding, Z.G., Gao, T.Y., Zhou, R.F., Xu, W.Y., Liu, J., 2004. Discoloration of methylene blue and wastewater from a plant by a Fe/Cu bimetallic system. Chemosphere 55, 1207–1212.
Mi, X., Huang, G., Xie, W., Wang, W., Liu, Y., Gao, J., 2012. Preparation of graphene oxide aerogel and its adsorption for Cu^{2+} ions. Carbon 50, 4856–4864.
Mishra, A.K., Ramaprabhu, S., 2011. Functionalized graphene sheets for arsenic removal and desalination of sea water. Desalination 282, 39–45.
Miyata, S., 1983. Anion-exchange properties of hydrotalcite-like compounds. Clays Clay Miner. 31, 305–311.
Nurmi, J.T., Tratnyek, P.G., Sarathy, V., Baer, D.R., Amonette, J.E., Pecher, K., Wang, C., Linehan, J.C., Matson, D.W., Penn, R.L., Driessen, M.D., 2005. Characterization and properties of metallic iron nanoparticles: spectroscopy, electrochemistry, and kinetics. Environ. Sci. Technol. 39, 1221–1230.
Parker, L.M., Milestone, N.B., Newman, R.H., 1995. The use of hydrotalcite as an anion absorbent. Ind. Eng. Chem. Res. 34, 1196–1202.
Paul, A.K., Madras, G., Natarajan, S., 2009. Adsorption–desorption and photocatalytic properties of inorganic–organic hybrid cadmium thiosulfate compounds. Phys. Chem. Chem. Phys. 11, 11285–11296.
Pavlovic, I., Ulibarri, M.A., Hermosin, M.C., Cornejo, J., 1997. Sorption of an anionic surfactant from water by a calcined hydrotalcite-like sorbent. Fres. Environ. Bull. 6, 266–271.
Rai, P., Gautam, R.K., Banerjee, S., Rawat, V., Chattopadhyaya, M.C., 2015. Synthesis and characterization of a novel $SnFe_2O_4$@activated carbon magnetic nanocomposite and its effectiveness in the removal of crystal violet from aqueous solution. J. Environ. Chem. Eng. 3, 2281–2291.
Rajeshwarisivaraj, S., Sivakumar, P., Senthilkumar, V., Subburam, 2001. Carbon from Cassava peel, an agricultural waste, as an adsorbent in the removal of dyes and metal ions from aqueous solution. Bioresour. Technol. 80, 233–235.
Schneemann, A., Bon, V., Schwedler, I., Senkovska, I., Kaskel, S., Fischer, R.A., 2014. Flexible metal–organic frameworks. Chem. Soc. Rev. 43, 6062–6096.
Schrick, B., Blough, J.L., Jones, A.D., Mallouk, T.E., 2002. Hydrodechlorination of trichloroethylene to hydrocarbons using bimetallic nickel-iron nanoparticles. Chem. Mater. 14, 5140–5147.
Setti, N.D., Jouini, N., Derriche, Z., 2010. Sorption study of an anionic dye – benzopurpurine 4B – on calcined and uncalcined Mg–Al layered double hydroxides. J. Phys. Chem. Solids 71, 556–559.
Shen, Y.F., Tang, J., Nie, Z.H., Wang, Y.D., Ren, Y., Zuo, L., 2009. Tailoring size and structural distortion of Fe_3O_4 nanoparticles for the purification of contaminated water. Bioresour. Technol. 100, 4139–4146.
Shultz, A.M., Sarjeant, A.A., Farha, O.K., Hupp, J.T., Nguyen, S.B.T., 2011. Post-synthesis modification of a metal-organic framework to form metallosalen-containing MOF materials. J. Am. Chem. Soc. 133, 13252–13255.
Singh, K.P., Gupta, S., Singh, A.K., Sinha, S., 2011. Optimizing adsorption of crystal violet dye from water by magnetic nanocomposite using response surface modeling approach. J. Hazard. Mater. 186, 1462–1473.
Singh, D., Gautam, R.K., Kumar, R., Shukla, B.K., Shankar, V., Krishna, V., 2014. Citric acid coated magnetic nanoparticles: Synthesis, characterization and application in removal of Cd (II) ions from aqueous solution. J. Water Process Eng. 4, 233–241.

Sun, Z., Jin, L., Shi, W., Wei, M., Duan, X., 2010. Preparation of an anion dye intercalated into layered double hydroxides and its controllable luminescence properties. Chem. Eng. J. 161, 293–300.
Sun, L., Tian, C., Wang, L., Zou, J., Mu, G., Fu, H., 2011. Magnetically separable porous graphitic carbon with large surface area as excellent adsorbents for metal ions and dye. J. Hazard. Mater. 21, 7232–7239.
Tai, Y., Wang, L., Gao, J., Amer, W.A., Ding, W., Yu, H., 2011. Synthesis of Fe_3O_4@ poly(methylmethacrylate-co-divinylbenzene) magnetic porous microspheres and their application in the separation of phenol from aqueous solutions. J. Colloid Interface Sci. 360, 731–738.
Tella, A.C., Owalude, S.O., Ojekanmi, C.A., Oluwafemi, O.S., 2014. Synthesis of copper–isonicotinate metal–organic frameworks simply by mixing solid reactants and investigation of their adsorptive properties for the removal of the fluorescein dye. New J. Chem. 38, 4494–4500.
Theiss, F.L., Sear-Hall, M.J., Palmer, S.J., Frost, R.L., 2012. Zinc aluminium layered double hydroxides for the removal of iodine and iodide from aqueous solutions 39, 166–175.
Ulibarri, M.A., Pavlovic, I., Hermosin, M.C., Cornejo, J., 1995. Hydrotalcite-like compounds as potential sorbents of phenols from water. Appl. Clay Sci. 10, 131–145.
Valente, J.S., Tzompantzi, F., Prince, J., Cortez, J.G.H., Gomez, R., 2009. Adsorption and photocatalytic degradation of phenol and 2,4 dichlorophenoxiacetic acid by Mg–Zn–Al layered double hydroxides. Appl. Catal. B 90, 330–338.
Wang, J., Blowers, P., Farell, J., 2004. Understanding reduction of carbon tetrachloride at nickel surfaces. Environ. Sci. Technol. 38, 1576–1581.
Wang, X., Chen, C., Chang, Y., Liu, H., 2009. Dechlorination of chlorinated methanes by Pd/Fe bimetallic nanoparticles. J. Hazard. Mater. 161, 815–823.
Wang, H., Yan, N., Li, Y., Zhou, X., Chen, J., Yu, B., Gong, M., Chen, Q., 2012. Fe nanoparticle-functionalized multi-walled carbon nanotubes: one-pot synthesis and their applications in magnetic removal of heavy metal ions. J. Hazard. Mater. 22, 9230–9236.
Willis, A.L., Turro, N.J., O'Brien, S., 2005. Spectroscopic characterization of the surface of iron oxide nanocrystals. Chem. Mater. 17, 5970–5975.
Yang, S.T., Chang, Y., Wang, H., Liu, G., Chen, S., Wang, Y., Liu, Y., Cao, A., 2010. Folding/aggregation of graphene oxide and its application in Cu^{2+} removal. J. Colloid Interface Sci. 351, 122–127.
You, Y., Zhao, H., Vance, G.F., 2002. Surfactant-enhanced adsorption of organic compounds by layered double hydroxides. Colloids Surf. A 205, 161–172.
Zhang, M., Reardon, E.J., 2003. Removal of B, Cr, Mo, and Se from wastewater by incorporation into hydrocalumite and ettringite. Environ. Sci. Technol. 37, 2947–2952.
Zhang, N., Qiu, H., Si, Y., Wang, W., Gao, J., 2011. Fabrication of highly porous biodegradable monoliths strengthened by graphene oxide and their adsorption of metal ions. Carbon 49, 827–837.
Zhang, X., Ji, L., Wang, J., Li, R., Liu, Q., Zhang, M., Liu, L., 2012. Removal of uranium(VI) from aqueous solutions by magnetic Mg–Al layered double hydroxide intercalated with citrate: Kinetic and thermodynamic investigation. Colloids Surf. A 414, 220–227.
Zhao, Z., Li, X., Huang, S., Xia, Q., Li, Z., 2011. Adsorption and diffusion of benzene on chromium-based metal organic framework MIL-101 synthesized by microwave irradiation. Ind. Eng. Chem. Res. 50, 2254–2261.
Zhou, Y.T., Nie, H.L., Branford-White, C., He, Z.Y., Zhu, L.M., 2009. Removal of Cu^{2+} from aqueous solution by chitosan-coated magnetic nanoparticles modified with α-ketoglutaric acid. J. Colloid Interface Sci. 330, 29–37.
Zhu, Q.L., Xu, Q., 2014. Metal–organic framework composites. Chem. Soc. Rev. 43, 5468–5512.

CHAPTER 2

Remediation Technologies for Water Cleanup: New Trends

2.1 INTRODUCTION

Rapid industrialization and urbanization coupled with tremendous growth in the human population have continuously deteriorated the global climate and natural ecosystem. Increased discharge of industrial effluents in the near water bodies has imposed a great risk for the creatures of rivers, lakes, estuaries, and oceans. There are various sources of environmental pollution such as leather and tannery industries, dye manufacturing, fertilizers, sewage and domestic wastes, agricultural discharges, and mining and metallurgical processes, which directly or indirectly discharge the generated industrial effluent and wastewater in the river, lakes or near water bodies (Fig. 2.1). Environmental pollution, such as dyes, persistent organic pollutants, heavy metals, metalloids, and volatile organic pollutants, is widely contaminating the aquatic ecosystems. Due to their increased mobilization in water and sediments of rivers and lakes, the bioaccumulation and magnification have tremendously increased during the last few decades. These toxic materials not only accumulate in the human body but also pose enough toxicity to cause cancer and mutation. Thus, the removal of hazardous substances from water is really a significant environmental challenge; hence, during the last few decades, scientists have developed many tools and techniques to identify, quantify, monitor, assess, and rapidly treat aqueous ecosystems. Several techniques, such as coagulation, flocculation, adsorption, bioremediation, membrane separation, ion exchange, photo-Fenton oxidation, ozonation, chlorination, and electrochemical and thermal degradation, have been developed and widely used to eliminate the toxic and hazardous materials. Our aim is to provide a brief overview on the developed techniques that have been widely utilized in the decontamination of environmental pollutants from aquatic media.

Nanomaterials for Wastewater Remediation
http://dx.doi.org/10.1016/B978-0-12-804609-8.00002-9

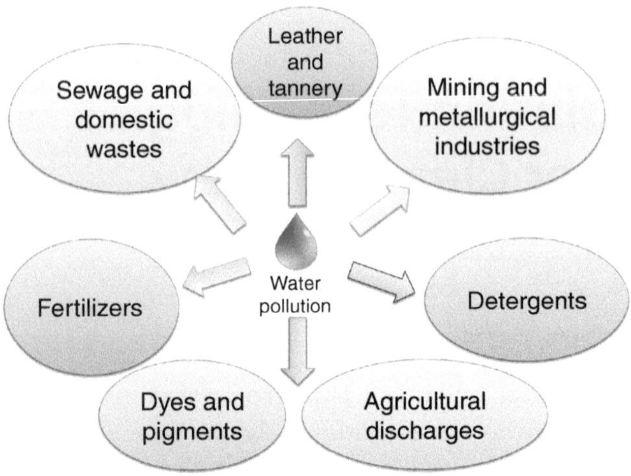

Figure 2.1 ***Various sources for the discharge of environmental pollutants into the aquatic ecosystem.***

2.2 REMEDIATION TECHNOLOGIES FOR EMERGING POLLUTANTS

Heavy metals are dangerous because they tend to bioaccumulate. Heavy metals accumulate in living things any time they are taken up and stored faster than they are broken down or excreted. Heavy metals can enter a water supply by industrial and consumer waste, or even from acidic rain breaking down soils and releasing heavy metals into streams, lakes, rivers, and groundwater. Many industries, such as dye, textile, paper, and plastic, consume substantial volumes of water and use dyes to color their products. As a result, they generate a considerable amount of colored wastewater. The presence of small amounts of dyes is highly visible and undesirable. Many of these dyes are also toxic and even carcinogenic and pose a serious threat to human beings and living organisms; hence, there is a need to treat wastewaters containing toxic dyes and metals before they are discharged into water bodies. Most heavy metals and dyes can be removed from drinking water very easily with reliable technology. Several technologies available in the market remove a huge range of metals and dyes commonly found in water and wastewater effluents. There are several remediation technologies which have been used for the removal of heavy metals and dyes from water/wastewater. These remediation technologies are presented in Fig. 2.2 and summarized in the following sections.

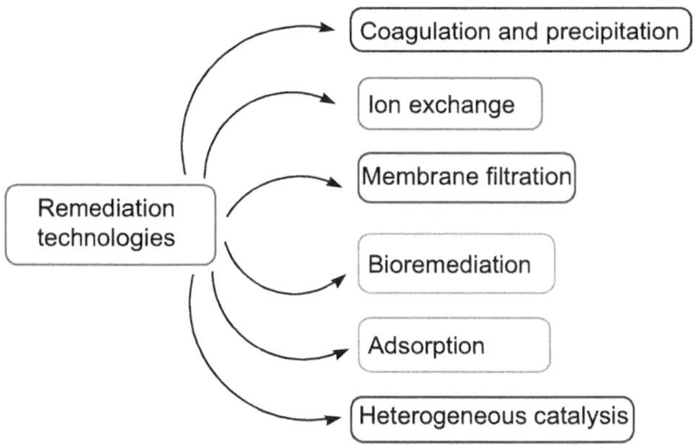

Figure 2.2 ***Schematic representation of various technologies used for the remediation of environmental pollutants from wastewater.***

2.2.1 Membrane Filtration

Membranes are complex structures that contain active elements on the nanometer scale. Modern day reverse osmosis membranes are typically homogeneous polymer thin films supported by porous support structures. Partitioning water and dissolved salts between the membrane and bulk solution, and transport of water and salts across the membrane, depends on the chemical properties of the membrane as well as physical structures at nano- to microscales. The nanometer length scale falls between the scale of macroscopic particles suspended in water and that of dissolved atomic and molecular species. From a filtration perspective, this intermediate range contains colloidal solids, large organic and biological molecules, polymers, and viruses. It also corresponds to the dimensions at which we recognize distinct modes of material transport across a membrane. For a larger dimension of porous membranes transport is described in terms of convective flow through pores. On the other hand, transport for dense reverse osmosis membranes is typically described in terms of diffusive flow through a homogeneous material.

2.2.2 Bioremediation and Phytoremediation

Bioremediation is the technological process whereby biological systems—plants and animals, including microorganisms—are harnessed to effectively clean up pollutants from environmental matrices (Head, 1998). Currently, microbial systems are widely employed in bioremediation programs, generally in the treatment of soils and waters contaminated with organic and inorganic pollutants, including heavy metals and metalloids.

Phytoremediation is a low-cost, low-tech, and emerging cleanup technology for contaminated soils, groundwater, and wastewater (Dushenkov et al., 1995). Plants are very sensitive to metals, but phytoremediation mainly uses wild and genetically modified plants, including grasses, herbs, forbs, and woody species. The plants receive heavy metals and metalloids through the process of phytostabilization, phytoextraction, phytofiltration, or rhizoremediation. But in contrast to organic compounds, the heavy metals and metalloids cannot be metabolized but instead accumulate in the plant biomass (Nunez-Lopez et al., 2008). The biomass generated by phytoremediation remains very limited in the form of persistence, whereas all biomass can be utilized in the forms of fertilizer, forage, mulch, and for the production of biogas (Polprasert, 1999). It is well known that metals are toxic for many plants, but plants have developed some internal mechanisms that allow for the uptake, tolerance, and accumulation of high metal concentration that would be toxic to other organisms. Many researchers have reported that aquatic macrophytes, namely *Typha*, *Phragmites*, *Eichhornia*, *Azolla*, and *Lemna*, are potential wetland plants for heavy metal and metalloid removal due to their morphological change (Dipu et al., 2012). Being cost effective and easy to apply technique, phytoremediation is implemented to enhance metals accumulation and translocation. In general, there are two strategies of phytoextraction that have been developed. These strategies are (1) continuous phytoextraction, which depends on the natural ability of some plants to accumulate, translocate, and resist high amounts of metals over the complete growth cycle (Garbisu and Alkorta, 2001; Salt et al., 1998; Huang et al., 1997) and (2) chemical induced phytoextraction techniques to cleanup contaminated water by using metal-tolerant plants to remove heavy metals and metalloids (Baker and Brooks, 1989). Efficiency of phytoextraction can be increased by using more biomass-producing plant species and applying suitable chelates. Hyperaccumulators or hyperaccumulating plants are capable of accumulating large amounts of heavy metals and metalloids, including Ni, As, Zn, Cd, and Pb in their aboveground tissues without any toxic symptoms (Baker, 1981).

Three aquatic plants, *Eichhornia crassipes, Lemna minor,* and *Spirodela polyrhiza* were, used for the phytoremediation of heavy metals from the coal mining effluent (Mishra et al., 2008). The obtained results revealed that the combination of *E. crassipes* and *L. minor* was the most efficient for the removal of heavy metals, while *E. crassipes* was the most efficient in monoculture. Translocation factor indicated that the highest metal accumulation was in the root and not in the shoot. The removal of arsenic by the water hyacinth (*E. crassipes*) and lesser duckweed (*L. minor*) was carried out in the laboratory (Alvarado et al., 2008). The removal rate for *L. minor* and *E. crassipes* was 140 and 600 mg As/ha d with a removal recovery of 5% and 18%, respectively, from water. The removal efficiency of *E. crassipes* was higher due to the biomass production and the more favorable climatic conditions during the removal periods. Due to catabolic efficiency, the Mycobacteria have been used as excellent candidates for remediating PAH contaminated sites (Bogan et al., 2003).

2.2.3 Heterogeneous Catalysts and Catalysis

Heterogeneous catalysis refers to the form of catalysis in which the phase of the catalyst differs from that of the reactants. Phase here refers not only to solid, liquid, and gas, but also immiscible liquids. The great majority of practical heterogeneous catalysts are solids, and the great majority of reactants are gases or liquids. Heterogeneous catalysis is of paramount importance in many areas of the chemical and energy industries.

The photocatalytic oxidation of organic compounds in water has received the most attention, but there is a rapidly increasing amount of work on the oxidation of volatile organic or inorganic compounds in the gas phase. Photocatalytic reduction of organic compounds and metal-containing ions is also well established. Studies of cell killing and disinfection by illuminated TiO_2 have also been increasing in number. Researchers have found that, in the presence of TiO_2, a wide range of surfactants are mineralized and a variety of bacteria, viruses, cancer cells, and other cell types are killed by irradiation with near-ultraviolet light. A wide range of high-molecular weight compounds representative of surface soiling agents have been studied as part of the work on self-cleaning surfaces. The breadth of work attests to the very wide range of applications that are being considered for the technology. The self-cleaning and wetting properties of irradiated titanium dioxide surfaces are now the subjects of a high level of interest because of the potential for commercial applications.

The heterogeneous photocatalytic reduction of Cr(VI) using semiconductors, an advanced oxidation technology, has been proposed as an economical and simple method of treatment (Chenthamarakshan et al., 2000; Litter, 1999). It is worthwhile to remark that surface-catalyzed Cr(VI) reduction, although very slow, is described as a feasible process in the presence of oxide surfaces such as TiO_2 (Deng and Stone, 1996). Furthermore, it has been largely demonstrated that the addition of organic donors able to chelate the TiO_2 surface accelerates the reduction of Cr(VI) in photocatalytic systems, the synergy being very dependent on the nature of the reducing agent (Colon et al., 2001a,b; Prairie et al., 1993; Ku and Jung, 2001).

2.2.4 Electrocoagulation

Electrocoagulation is an efficient method used for treating various process effluents containing textile dyes, nitrate, fluoride, restaurant wastewater, arsenic, phenolic compounds, latex particles, polyaromatic wastes, and tannery effluents (Can et al., 2003; Daneshvar et al., 2003; Koparal and Ogutveren, 2002; Shen et al., 2003; Hu et al., 2005; Chen et al., 2000; Kumar et al., 2004; Adhoum and Monser, 2004; Larue et al., 2003; Panizza et al., 2000; Murugananthan et al., 2004). Recently various workers have investigated electrocoagulation for removal of heavy metals (Cr, Zn, Cu, Pb, etc.) from wastewater (Kongsricharoern and Polprasert, 1995; Adhoum et al., 2004; Gao et al., 2005). The single bath chrome tanning process uses basic chrome sulfate solution for tanning of animal skins/hides. Almost all local tanneries use this process and their effluent contains Cr^{3+}. It is therefore important and necessary to find out technical and economic feasibility of electrocoagulation for removal of Cr^{3+} from wastewater.

Removal of Cr^{3+} from aqueous solution by electrocoagulation using iron electrodes is a feasible process. Golder et al. (2007) investigated the removal of Cr^{3+} from aqueous solutions in a batch stirred electrocoagulator with a mild steel electrode pair. Effects of operating time, stirrer rpm, current density, initial pH, initial concentration of Cr^{3+}, and supportive electrolyte NaCl concentration was investigated in their study. Removal of Cr^{3+} during electrocoagulation was due to the combined effect of chemical precipitation, coprecipitation, sweep coagulation, and adsorption. They reported remarkable removal of Cr^{3+} from aqueous solutions at higher current density and solution pH.

2.2.5 Clays/Layered Double Hydroxides

Among all the different adsorptive materials, clays have been used widely to capture heavy metal ions from aqueous solutions due to their unique properties such as high surface area associated with small particle size, low cost, and the ubiquitous occurrence in most soil and sediment environments (Liang et al., 2011; Bradl, 2004). Clays can adsorb heavy metal ions through ion exchange or surface complexation. However for natural clay minerals, there are still some limitations such as limited adsorption capacity, relatively small metal binding constants, and low selectivity to the type of metals. In recent years, surface modification of natural clay minerals with reagents containing metal chelating groups have been explored in an effort to enhance the heavy metal binding constants and the selectivity of the type of metals (Bradl, 2004; Cruz-Guzman et al., 2006; Mercier and Pinnavaia, 1998).

The approaches used for the surface modification of clays include intercalation of organic moieties into the interlayer space, grafting of organic moieties, and one-step synthesis. The intercalation refers to a process that a guest molecule or ion is inserted into the interlayer space by adsorption or ion exchange. The structure of the host and intercalated compound are only slightly changed by the guest species. For instance, four organic-modified clays based on montmorillonite were prepared by embedding ammonium organic derivatives with different chelating functionalities into the interlayer space of montmorillonite as heavy metal ion adsorbents for environmental remediation (Stathi et al., 2007). Montmorillonite intercalated with poly-hydroxyl Fe(III) complexes was used for the sorption of Cd(II) (Wu et al., 2009). Sodium dodecyl sulfate modified iron-pillared montmorillonite was applied for the removal of aqueous Cu(II) and Co(II) (Li and Wu, 2010). Smectite intercalated with nonionic surfactant performs well for removing heavy metal (Deng et al., 2003). Grafting is a process that links the inorganic and organic components via strong bonds such as covalent or ionic-covalent to obtain functionalized clays. The approach enables a durable immobilization of the reactive organic groups, preventing their leaching when they are used in liquid media.

2.2.6 Biomass-Based Biosorption

Over the past decades, many new processes have been tested successfully; a lot of them have gone into operation and a great number of papers have been published in this field. Biosorption has been defined as the "property of certain biomolecules to bind and concentrate selected ions or other molecules from aqueous solutions" (Volesky, 2007; Wang and Chen, 2006). As opposed to a much more complex phenomenon of bioaccumulation based on active metabolic transport, biosorption by dead biomass is passive and based mainly on the "affinity" between the sorbent and sorbate. Biosorption of heavy metal ions and dyes have become a popular environmentally driven research topic, and it represents only one particular type of a concentration-removal aspect of the sorption process (Gautam, 2010; Gautam et al., 2013a,b, 2014; Gautam and Chattopadhyaya, 2013). The methodology of studying biosorption is based on an interdisciplinary approach, whereby the phenomenon can be studied, examined, and analyzed from different angles and perspectives—by chemists, chemical engineers, and environmentalists.

In recent years, research on the mechanisms of biosorption has intensified since biomass can be employed to sequester heavy metal ions and dyes from industrial effluents or to recover precious metals from waste aqueous solutions. Of the many types of biosorbents (ie, fungi, bacteria, yeasts, weeds, and industrial and agricultural wastes) recently investigated for their ability to sequester heavy metal ions, these low-cost biomasses have proven to be highly effective as well as reliable and predictable in the removal of, for example, Pb^{2+}, Cu^{2+}, Cd^{2+}, Cr^{6+}, and Zn^{2+} ions from aqueous solutions. There has been an equally intense research effort aimed at characterizing the metal binding properties of various forms of biomass. These have included fresh and saltwater algae, bacteria, fungi and yeasts, and industrial and agricultural waste/by-products. Biomass is also used efficiently for the removal of various metals from wastewater effluents. The application of nonliving biomass in this technology has the following merits. It is not subjected to toxicity limitations due to metal concentration. Biomass from the fermentation industry can be available at a cheap rate, which is waste for the fermentation industry. The process is very rapid and the kinetic rate is faster. Due to nonliving nature of the biosorbent-operating conditions, such as pH and temperature, the metal concentration evenly fluctuates. Aseptic maintenance is not required. Metal desorption can be done with stronger reagents as compared to live biomass. Thus recovery is faster and more concentrated.

Activated charcoal was prepared from *Melocanna baccifera* raw charcoal by chemical treatment and used in various experiments to test its functions as adsorbents for removal of Ni(II) and Zn(II) from aqueous solution (Lalhruaitluanga et al., 2011). Freundlich isotherm constants for Ni and Zn were 28.53 and 27.01 mg/g, respectively, on *M. baccifera* activated carbon. While the Langmuir isotherm constants for Ni and Zn were 52.91 and 40.48 mg/g, respectively, on *M. baccifera* activated carbon. Hydroxyapatite/chitosan composite was synthesized by precipitation method for the adsorptive removal of Pb^{2+}, Co^{2+}, and Ni^{2+} by hydroxyapatite/chitosan composite from aqueous solution (Gupta et al., 2012). The Langmuir constant for Pb^{2+}, Co^{2+}, and Ni^{2+} removal onto hydroxyapatite/chitosan composite were 12.04, 10.63, and 8.54 mg/L, respectively. Equilibrium experimental data were fitted well to the Langmuir and Freundlich isotherm models.

2.2.7 Magnetic Nanoparticles as Nanosorbents

During the last few years, hierarchically structured magnetic nanoparticles, due to their magnetic properties, high chemical stability, low toxicity, ease in synthesis, and excellent recycling capability, have aroused widespread attention and have been extensively studied to remove toxic metal ions and organic pollutants from water, and these materials show higher removal capacities than the bulk material. Their micrometer-sized overall structure provides the necessary mechanical robustness against wear and tear, while their nanostructure provides the high surface area as well as the high removal capacity for heavy metal ions and dyes.

Magnetic nanoparticles are of great interest for researchers from a wide range of disciplines, including magnetic fluids, catalysis, biomedicine, drug delivery, magnetic resonance imaging, data storage, and environmental remediation (Ambashta and Sillanpaa, 2010; Jung et al., 2011). Although several suitable methods have been developed for the synthesis of magnetic nanoparticles of a variety of different compositions, successful application of such magnetic nanoparticles in the areas listed above is extremely dependent on the stability of the particles under a range of different conditions. In the majority of the envisaged applications, the particles perform best when the size of the nanoparticles is below a critical value, which is dependent on the source material but is typically around 10–20 nm (Lu et al., 2007). The design and fabrication of nanoparticle-based adsorbents has generated great interest in a variety of scientific communities, ranging from chemical, biological, and environmental science to engineering. Magnetic nanoparticle-based adsorbents can be used in the separation and purification of biologically as well as environmentally relevant target species with high precision and accuracy (Koehler et al., 2009; Girginova et al., 2010).

2.2.8 Advanced Oxidation Processes

Advanced oxidation processes are an attractive alternative for the treatment of contaminated ground water and industrial wastewater resources containing nonbiodegradable anthropogenic substances as well as purification and disinfection of drinking water. Advanced oxidation processes, which involve in situ generation of highly potent chemical oxidants such as hydroxyl radicals, have recently emerged as an important class of technologies for accelerating the oxidation and destruction of a wide range of organic contaminants from wastewater. Advanced oxidation process refers to a set of designed chemical processes that are used to detoxify or remove environmental pollutants from aquatic media by the generation of highly reactive hydroxyl radicals (•OH). The •OH radicals are powerful, nonselective, and highly effective oxidants and react effectively with most of the organic compounds. Recently, the carbon doped titanium dioxide photocatalysts were prepared and applied for the photocatalytic degradation of 4-chlorophenol under UV and solar irradiation (Kavitha and Devi, 2014). The pretreatment of carbon doped titanium dioxide with UV light reduced the visible light activity due to the removal of surface carbonaceous species, and it has been postulated that the electron transfer rate to oxygen molecules adsorbed on the TiO_2 surface increases the dioxygen reduction with subsequent formation of free radicals, which accelerates the rate of 4-chlorophenol degradation (Tian et al., 2008).

2.2.9 Graphene-Based Nanosorbents

Nowadays, the immense focus of the scientific community is on developing low-cost adsorbents which have high surface area to volume ratio, have surface tunable properties, are well dispersible in liquids and easily regenerated, and have recycling properties and abilities in harsh environments. Hence, carbon nanotubes have been applied for the purpose of adsorbents to remove and recover the toxic metal ions and dyes from liquids (Stafiej and Pyrzynska, 2007; Rao et al., 2007). Lee and Yang (2012) synthesized self-assembled flowerlike TiO_2 on exfoliated graphite oxide for heavy metal removal from aqueous solutions. The GO-TiO_2 hybrid adsorption capacities of heavy metal ions, after 6 and 12 h of hydrothermal treatment at 100°C, were respectively 44.8 ± 3.4 and 88.9 ± 3.3 mg/g for removing Zn^{2+}, 65.1 ± 4.4 and 72.8 ± 1.6 mg/g for removing Cd^{2+}, and 45.0 ± 3.8 and 65.6 ± 2.7 mg/g for removing Pb^{2+} at pH 5.6. Zhao et al. (2011) synthesized few layered graphene oxide nanosheets and used them as sorbents for the removal of Cd(II) and Co(II) ions from aqueous solutions. The maximum adsorption capacities of Cd(II) and Co(II) on graphene oxide nanosheets at pH 6.0 and T = 303 K were about 106.3 and 68.2 mg/g, respectively.

2.3 CONCLUSIONS

Contaminations of aquatic media by continuous discharge of environmental pollutants are imposing great challenges to scientific communities to develop new and effective treatment and removal processes. Various methods such as advanced oxidation processes, coagulation, flocculation, bioremediation, electrochemical separation, Fenton oxidation, and adsorption have been extensively applied for the removal of heavy metals, dyes, and volatile organic pollutants.

Several high efficient and chemically stable catalysts such as TiO_2, Fe_3O_4, alumina, ZnO, and CdS nanoparticles have been used either in bare form or functionalized with different surface functional groups. Catalytic degradation of environmental pollutants has been found to be a highly effective method; however, it may produce some toxic by-products in the degradation of parent environmental pollutants. Further, the degradation of pollutants on heterogeneous catalysts produces large amounts of sludge and requires additional energy.

Adsorption has been found a suitable method for the separation of hazardous pollutants from water. Various efforts have been made to produce activated carbons with high surface area to remove the heavy metals, dyes, and volatile organic pollutants. However, the high cost of activated carbons, energy requirements, and sludge generation hamper the application in developing countries. Recently, biosorption of pollutants on biomass derived adsorbents has been investigated to lower down the treatment costs, although the adsorption efficiency is not up to the mark for the separation of hazardous pollutants on biosorbents.

Very recently, efforts have been made to develop magnetic nanosorbents for the separation of pollutants. Magnetically responsive adsorbents can be easily recycled from the water by an external magnet, and the regeneration of adsorbents is possible.

Graphene and chitosan-coated magnetic nanosorbents have gained much attention in the new generation of adsorbents for the separation and purification of contaminants from water.

REFERENCES

Adhoum, N., Monser, L., 2004. Decolourization and removal of phenolic compounds from olive mill wastewater by electrocoagulation. Chem. Eng. Process 43, 1281–1287.

Adhoum, N., Monser, L., Bellakhal, N., Belgaied, J.E., 2004. Treatment of electroplating wastewater containing Cu^{2+}, Zn^{2+} and Cr(VI) by electrocoagulation. J. Hazard. Mater. B112, 207–213.

Alvarado, S., Guédez, M., Lué-Merú, M.P., Nelson, G., Alvaro, A., Jesús, A.C., Gyula, Z., 2008. Arsenic removal from waters by bioremediation with the aquatic plants water hyacinth (*Eichhornia crassipes*) and lesser duckweed (*Lemna minor*). Bioresour. Technol. 99, 8436–8440.

Ambashta, R.D., Sillanpaa, M., 2010. Water purification using magnetic assistance: A review. J. Hazard. Mater. 180, 38–49.

Baker, A.J.M., 1981. Accumulation and exchangers-strategy in the response of plants to heavy metals. J. Plant Nutr. 3, 643–645.

Baker, A.J.M., Brooks, R.R., 1989. Terrestrial higher plants which hyperaccumulate metallic elements—a review of their distribution, ecology and phytochemistry. Biorecovery 1, 81–126.
Bogan, B.W., Lahner, L.M., Sullivan, W.R., Paterek, J.R., 2003. Degradation of polycyclic aromatic and straight-chain aliphatic hydrocarbons by a strain of *Mycobacterium austroafricanum*. J. Appl. Microbiol. 94, 230–239.
Bradl, H.B., 2004. Adsorption of heavy metal ions on soils and soils constituents. J. Colloid Interface Sci. 277, 1–18.
Can, O.T., Bayramoglu, M., Kobya, M., 2003. Decolorization of reactive dye solutions by electrocoagulation using aluminum electrodes. Ind. Eng. Chem. Res. 42, 3391–3396.
Chen, X., Chen, G., Yue, P.L., 2000. Separation of pollutants from restaurant wastewater by electrocoagulation. Sep. Purif. Technol. 19, 65–76.
Chenthamarakshan, C.R., Rajeshwar, K., Wolfrum, E.J., 2000. Heterogeneous photocatalytic reduction of Cr(VI) in UV-irradiated titania suspensions: effect of protons, ammonium ions, and other interfacial aspects. Langmuir 16, 2715–2721.
Colon, G., Hidalgo, M.C., Navio, J.A., 2001a. Photocatalytic deactivation of commercial TiO_2 samples during simultaneous photoreduction of Cr(VI) and photooxidation of salicylic acid. J. Photochem. Photobiol. A 138, 79–85.
Colon, G., Hidalgo, M.C., Navio, J.A., 2001b. Influence of carboxylic acid on the photocatalytic reduction of Cr(VI) using commercial TiO_2. Langmuir 17, 7174–7177.
Cruz-Guzman, M., Celis, R., Hermosin, M.C., Koskinen, W.C., Nater, E.A., Cornejo, J., 2006. Heavy metal adsorption by montmorillonites modified with natural organic cations. Soil Sci. Soc. Am. J. 70, 215–221.
Daneshvar, N., Ashassi-Sorkhabi, H., Tizpar, A., 2003. Decolorization of orange II by electrocoagulation method. Sep. Purif. Technol. 31, 153–162.
Deng, B., Stone, A.T., 1996. Surface-catalyzed chromium(VI) reduction: reactivity comparisons of different organic reductants and different oxide surfaces. Environ. Sci. Technol. 30, 2484–2494.
Deng, Y., Dixon, J.B., White, G.N., 2003. Intercalation and surface modification of smectite by two non-ionic surfactants. Clays Clay Miner. 51, 150–161.
Dipu, S., Kumar, A.A., Thanga, S.G., 2012. Effect of chelating agents in phytoremediation of heavy metals. Remediation 22, 133–146.
Dushenkov, P.B., Kumar, A.N., Motto, H., Raskin, I., 1995. Rhizofiltration: the use of plants to remove heavy metals from aqueous streams. Environ. Sci. Technol. 29, 1239–1245.
Gao, P., Chen, X., Shen, F., Chen, G., 2005. Removal of chromium(VI) from wastewater by combined electrocoagulation–electroflotation without a filter. Sep. Purif. Technol. 43 (2), 117–123.
Garbisu, C., Alkorta, I., 2001. Phytoextraction: a cost-effective plant-based technology for the removal of metals from the environment. Bioresour. Technol. 77 (3), 229–236.
Gautam, R.K., 2010. Environmental Magnetism: Fundamentals and Applications, first ed. LAP Lambert Academic Publishing, Saarbrucken.
Gautam, R.K., Chattopadhyaya, M.C., 2013. Magnetic nanoparticles for environmental pollution management. In: Daniels, J.A. (Ed.), Advances in Environmental Research, 29, Nova Science Publishers, New York, pp. 1–38.
Gautam, R.K., Chattopadhyaya, M.C., Sharma, S.K., 2013a. Biosorption of heavy metals: recent trends and challenges. In: Sharma, S.K., Sanghi, R. (Eds.), Wastewater Reuse and Management. Springer, London, pp. 305–322.
Gautam, R.K., Mudhoo, A., Chattopadhyaya, M.C., 2013b. Kinetic, equilibrium, thermodynamic studies and spectroscopic analysis of Alizarin Red S removal by mustard husk. J. Environ. Chem. Eng. 1, 1283–1291.
Gautam, R.K., Mudhoo, A., Lofrano, G., Chattopadhyaya, M.C., 2014. Biomass-derived biosorbents for metal ions sequestration: Adsorbent modification and activation methods and adsorbent regeneration. J. Environ. Chem. Eng. 2, 239–259.

Girginova, P.I., Daniel-da-Silva, A.L., Lopes, C.B., Figueira, P., Otero, M., Amaral, V.S., Pereira, E., Trindade, T., 2010. Silica coated magnetite particles for magnetic removal of Hg^{2+} from water. J. Colloid Interface Sci. 345, 234–240.

Golder, A.K., Samanta, A.N., Ray, S., 2007. Removal of trivalent chromium by electrocoagulation. Sep. Purif. Technol. 53, 33–41.

Gupta, N., Kushwaha, A.K., Chattopadhyaya, M.C., 2012. Adsorptive removal of Pb^{2+}, Co^{2+} and Ni^{2+} by hydroxyapatite/chitosan composite from aqueous solution. J. Taiwan Inst. Chem. Eng. 43, 125–131.

Head, I.M., 1998. Bioremediation: towards a credible technology. Microbiology 144, 599–608.

Hu, C.Y., Lo, S.L., Kuan, W.H., Lee, Y.D., 2005. Removal of fluoride from semiconductor wastewater by electrocoagulation–flotation. Water Res. 39, 895–901.

Huang, J.W., Chen, J.J., William, R.B., Scott, D.C., 1997. Phytoremediation of lead-contaminated soils: role of synthetic chelates in lead phytoextraction. Environ. Sci. Technol. 31, 800–805.

Jung, J.H., Lee, J.H., Shinkai, S., 2011. Functionalized magnetic nanoparticles as chemosensors and adsorbents for toxic metal ions in environmental and biological fields. Chem. Soc. Rev. 40, 4464–4474.

Kavitha, R., Devi, L.G., 2014. Synergistic effect between carbon dopant in titania lattice and surface carbonaceous species for enhancing the visible light photocatalysis. J. Environ. Chem. Eng. 2, 857–867.

Koehler, F.M., Rossier, M., Waelle, M., Athanassiou, E.K., Limbach, L.K., Grass, R.N., Gunther, D., Stark, W.J., 2009. Magnetic EDTA: coupling heavy metal chelators to metal nanomagnets for rapid removal of cadmium, lead and copper from contaminated water. Chem. Commun. 32, 4862–4864.

Kongsricharoern, N., Polprasert, C., 1995. Electrochemical precipitation of chromium (Cr^{6+}) from an electroplating wastewater. Water Sci. Technol. 31 (9), 109–117.

Koparal, A.S., Ogutveren, U.B., 2002. Removal of nitrate from water by electroreduction and electrocoagulation. J. Hazard. Mater. B89, 83–94.

Ku, Y., Jung, I.L., 2001. Photocatalytic reduction of Cr(VI) in aqueous solutions by UV irradiation with the presence of titanium dioxide. Water Res. 35, 135–142.

Kumar, P.R., Chaudhari, S., Khilar, K.C., Mahajan, S.P., 2004. Removal of arsenic from water by electrocoagulation. Chemosphere 55, 1245–1252.

Lalhruaitluanga, H., Prasad, M.N.V., Radha, K., 2011. Potential of chemically activated and raw charcoals of *Melocanna baccifera* for removal of Ni(II) and Zn(II) from aqueous solutions. Desalination 271, 301–308.

Larue, O., Vorobiev, E., Vu, C., Durand, B., 2003. Electrocoagulation and coagulation by iron of latex particles in aqueous suspensions. Sep. Purif. Technol. 31, 177–192.

Lee, Y.C., Yang, J.W., 2012. Self-assembled flower-like TiO_2 on exfoliated graphite oxide for heavy metal removal. Ind. Eng. Chem. 18, 1178–1185.

Li, S.Z., Wu, P.X., 2010. Characterization of sodium dodecyl sulfate modified iron pillared montmorillonite and its application for the removal of aqueous Cu(II) and Co(II). J. Hazard. Mater. 173, 62–70.

Liang, X., Xu, Y., Sun, G., Wang, L., Sun, Y., Sun, Y., Qin, X., 2011. Preparation and characterization of mercapto functionalized sepiolite and their application for sorption of lead and cadmium. Chem. Eng. J. 174, 436–444.

Litter, M.I., 1999. Heterogeneous photocatalysis: transition metal ions in photocatalytic systems. Appl. Catal. B 23, 89–114.

Lu, A.H., Salabas, E.L., Schuth, F., 2007. Magnetic nanoparticles: synthesis, protection, functionalization, and application. Angew. Chem. Int. Ed. 46, 1222–1244.

Mercier, L., Pinnavaia, T.J., 1998. Heavy metal ion adsorbents formed by the grafting of a thiol functionality to mesoporous silica molecular sieves: factors affecting Hg(II) uptake. Environ. Sci. Technol. 32, 2749–2754.

Mishra, V.K., Upadhyaya, A.R., Pandey, S.K., Tripathi, B.D., 2008. Heavy metal pollution induced due to coal mining effluent on surrounding aquatic ecosystem and its management through naturally occurring aquatic macrophytes. Bioresour. Technol. 99, 930–936.
Muruganathan, M., Raju, G.B., Prabhakar, S., 2004. Separation of pollutants from tannery effluents by electro-flotation. Sep. Purif. Technol 40, 69–75.
Nunez-Lopez, R.A., Meas, Y., Gama, S.C., Borges, R.O., Olguin, E.J., 2008. Leaching of lead by ammonium salts and EDTA from *Salvinia minima* biomass produced during aquatic phytoremediation. J. Hazard. Mater. 154, 623–632.
Panizza, M., Bocca, C., Cerisola, G., 2000. Electrochemical treatment of wastewater containing polyaromatic organic pollutants. Water Res. 34 (9), 2601–2605.
Polprasert, C., 1999. Organic Waste Recycling, second ed. John Wiley & Sons, Chichester, UK.
Prairie, M.R., Evans, L.R., Stange, B.M., Martinez, S.L., 1993. An investigation of titanium dioxide photocatalysis for the treatment of water contaminated with metals and organic chemicals. Environ. Sci. Technol. 27, 1776–1782.
Rao, G.P., Lu, C., Su, F., 2007. Sorption of divalent metal ions from aqueous solution by carbon nanotubes: a review. Sep. Purif. Technol. 58, 224–231.
Salt, D.E., Smith, R.D., Taskin, I., 1998. Phytoremediation. Annu. Rev. Plant Physiol. Plant Mol. Biol. 49, 643–668.
Shen, F., Chen, X., Gao, P., Chen, G., 2003. Electrochemical removal of fluoride ions from industrial wastewater. Chem. Eng. Sci. 58, 987–993.
Stafiej, A., Pyrzynska, K., 2007. Adsorption of heavy metal ions with carbon nanotubes. Sep. Purif. Technol. 58, 49–52.
Stathi, P., Litina, K., Gournis, D., Giannopoulos, T.S., Deligiannakis, Y., 2007. Physicochemical study of novel organoclays as heavy metal ion adsorbents for environmental remediation. J. Colloid Interface Sci. 316, 298–309.
Tian, B., Zhang, J., Tong, T., Chen, F., 2008. Preparation of Au/TiO_2 catalysts from Au(I)-thiosulfate complex and study of their photocatalytic activity for the degradation of methyl orange. Appl. Catal. Part B 79, 394–401.
Volesky, B., 2007. Biosorption and me. Water Res. 41, 4017–4029.
Wang, J., Chen, C., 2006. Biosorption of heavy metals by *Saccharomyces cerevisiae*: a review. Biotechnol. Adv. 24, 427–451.
Wu, P., Wu, W., Li, S., Xing, N., Zhu, N., Li, P., Wu, J., Yang, C., Dang, Z., 2009. Removal of Cd^{2+} from aqueous solution by adsorption using Fe-montmorillonite. J. Hazard. Mater. 169, 824–830.
Zhao, G., Li, J., Ren, X., Chen, C., Wang, X., 2011. Few-layered graphene oxide nanosheets as superior sorbents for heavy metal ion pollution management. Environ. Sci. Technol. 45, 10454–10462.

CHAPTER 3

Advanced Oxidation Process–Based Nanomaterials for the Remediation of Recalcitrant Pollutants

3.1 ADVANCED OXIDATION PROCESSES

Advanced oxidation processes are an attractive alternative for the treatment of contaminated ground water and industrial wastewater resources containing nonbiodegradable anthropogenic substances as well as purification and disinfection of drinking water. Advanced oxidation processes, which involve the in situ generation of highly potent chemical oxidants such as hydroxyl radicals, have recently emerged as an important class of technologies for accelerating the oxidation and destruction of a wide range of organic contaminants in polluted water and soil.

3.2 MAIN ADVANCED OXIDATION PROCESSES

An advanced oxidation process refers to a set of designed chemical processes that is used to detoxify or remove environmental pollutants from aquatic media by the generation of highly reactive hydroxyl radicals (•OH). The •OH radicals are powerful, nonselective, and highly effective oxidants and react effectively with most of the organic compounds. The advanced oxidation process falls under the category of chemical treatment processes that usually use catalysts, ozone (O_3), hydrogen peroxide (H_2O_2), and UV light. Table 3.1 shows the widely used advanced oxidation processes for the degradation of wastewater pollutants.

3.2.1 Sonochemical Oxidation Methods

Advanced oxidation processes generally uses UV light in the presence of H_2O_2, O_3, Fenton reagents, or a catalyst to generate highly reactive oxidizing agents to degrade recalcitrant chemical compounds in aqueous media. Very few studies have been reported on γ-radiolysis or sonication to

Nanomaterials for Wastewater Remediation
http://dx.doi.org/10.1016/B978-0-12-804609-8.00003-0

Table 3.1 Widely used advanced oxidation processes for degradation of wastewater pollutants (Litter and Quici, 2010; Munter, 2001)

Photochemical processes		
Type of process	**Irradiation wavelength range (λ/nm)**	**Nonphotochemical processes**
Heterogeneous photocatalysis (HP) using TiO_2	UV: up to 380–400	Ferrate
Zerovalent iron plus UV light	UV range	Zerovalent iron (ZVI)
Photo-Fenton (PF) and related processes	UV–Vis up to 450	Wet air oxidation
UV/ozone (UV/O_3)	280–315	Electrohydraulic discharge-ultrasound (US, cavitation)
UV/hydrogen peroxide (UV/H_2O_2)	<280	Nonthermal plasma (surface corona discharge)
Water photolysis in vacuum ultraviolet (VUV)	<190	γ-Radiolysis and electron-beam treatment
		Electrochemical oxidation
		Fenton and related processes (Fe^{2+}/H_2O_2)
		Ozonation with hydrogen peroxide (O_3/H_2O_2)
		Alkaline ozonation (O_3/HO^-)

produce sufficient reactive radicals in the water system (Tezcanli-Güyer and Ince, 2004; Naffrechoux et al., 2000) to oxidize the organic compounds.

The introduction of ultrasound waves into the water creates a hot spot thus increasing the temperature of the liquids, which leads to the thermal decomposition of the water solution into the reactive radical species. The chain reaction generated at the gas-liquid interface can be written as (Ince et al., 2000; Beckett and Hua, 2000):

$$H_2O \xrightarrow{\text{Ultrasound waves}} \bullet OH + \bullet H(\text{thermolysis}) \tag{3.1}$$

$$\bullet OH + \bullet H \rightarrow H_2O \tag{3.2}$$

$$\bullet OH + \bullet OH \rightarrow H_2O_2 \tag{3.3}$$

$$2 \bullet OH \rightarrow H_2O + O \tag{3.4}$$

Moreover, some additional radicals may produce if the solution is saturated by oxygen. The reactions can be expressed as:

$$O_2 \xrightarrow{\text{Ultrasound waves}} O + O \tag{3.5}$$

$$O + H_2O \rightarrow \bullet OH + \bullet OH \tag{3.6}$$

$$O_2 + \bullet H \rightarrow \bullet O_2H \tag{3.7}$$

$$\bullet O_2H + \bullet O_2H \rightarrow H_2O_2 + O_2 \tag{3.8}$$

The ultrasonic irradiation of wastewater or industrial containing organic recalcitrant compounds results in thermochemical degradation in liquids or in the gas-liquid interface due to the production of highly reactive hydroxyl radical species. Ma et al. (2006) investigated the degradation of an acidic dye, acid orange 7, from wastewater in the presence of ultrasound and visible light, and the reported removal was higher than 65% (Fig. 3.1). The high degradation rate was due to the synergistic effect to the reduction of the

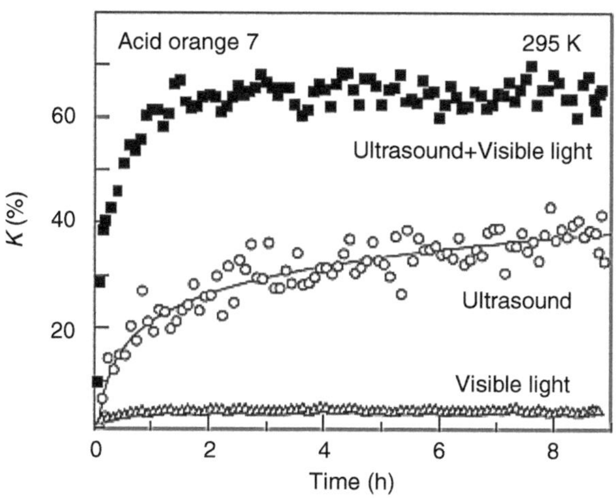

Figure 3.1 ***Graph shows the percentage removal of acid orange 7 in the presence of visible light, ultrasound, and ultrasound + visible light at different time intervals.*** *(Source: Ma et al., 2006; © Elsevier).*

bandgap energy by generation of high temperature and high pressure via the bubble collapse under the ultrasonic irradiation in aqueous solutions. Recently, acid orange 7 dye degradation has been carried out successfully under the presence of ultrasound irradiation, UV light, O_3, and H_2O_2 (Tezcanli-Güyer and Ince, 2004). The high rate of degradation was attributed to O_3 diffusion in the presence of ultrasound irradition and the photolysis of ultrasound-generated •OH radicals from H_2O_2 in aqueous media.

The synergistic effect of ultrasound irradiation on the degradation of methyl orange dye in the presence of TiO_2 was investigated, and higher removal of dye was observed due to the sonoluminescence and generation of hot spots, which leads to the production of reactive •OH radicals (Wang et al., 2006). The solution pH plays an important role, and the higher percentage removal was found in the basic medium (pH 9) under the contact time of 120 min. Table 3.2 shows the sonochemical degradation of hazardous dyes by TiO_2-based catalysts.

3.2.2 Electrochemical Oxidation Processes

Electrochemical oxidation process is an old technique for the degradation and remediation of environmental contaminants present in water by suitable electrode materials. Application of this technique dates back to 1970. During the last few decades efforts have been taken to develop new electrocatalytic electrode materials with high selectivity, increased oxidation-reduction rates, high stability, investigation of the operating parameters, and examination of degradation kinetics and their mechanisms (Chen, 2004;

Table 3.2 Sonochemical degradation of hazardous dyes by TiO_2-based catalysts

Catalysts	Pollutants	Reaction conditions	References
TiO_2, ZnO, TiO_2/ZnO	Acid Red B	0.01 g catalyst dose, 40 kHz	Wang et al. (2009)
CeO_2/TiO_2, SnO_2/TiO_2, ZrO_2/TiO_2	Acid Red B	0.01 g catalyst dose, 40 kHz	Wang et al. (2010)
TiO_2, Au/TiO_2	Orange II	0.01 g catalyst dose, 40 kHz	Wang et al. (2008a)
Fe^{3+}–TiO_2	Fuchsine	0.05 g catalyst dose, 40 kHz	Wang et al. (2008b)
Nanosized TiO_2	Basic Blue 41	0.01 g catalyst dose, 35 kHz	Abbasi and Asl (2008)
Rutile TiO_2	Methyl orange	0.01 g catalyst dose, 40 kHz	Wang et al. (2007)

Sires et al., 2007). The tannery effluent wastewater was treated by electrochemical methods using Ti/Pt and Ti/Pt/Ir as electrode materials, and the removal of NH_4^+ of different strength was investigated (Szpyrkowicz et al., 1995). The Ti/Pt/Ir electrode shows electrocatalytic properties and was more effective in the treatment process of wastewater.

Reactive azo dye containing wastewater from a textile dyeing process was treated by an electrochemical oxidation method using Ti/Pt as anode and stainless steel 304 as cathode (Vlyssides et al., 1999). Due to the strong oxidizing potential of chlorine, oxygen, hydroxyl radicals, and other oxidant reactive species when the wastewater was passed through the electrolytic cell, the organic pollutants were oxidized to carbon dioxide and water. The result shows the 100% reduction in color units.

The reactive azo dye containing wastewater from a textile dyeing process from the total dyeing finishing stage, and wastewater only from the dyeing stage, was treated by an electrochemical oxidation method using Ti/Pt as anode and stainless steel 304 as cathode (Vlyssides et al., 2000). In this degradation method, sodium chloride was used as an electrolyte, and the mixture was passed through an electrolytic cell. A high rate of color removal was obtained due to the production of highly reactive strong oxidizing potentials of chlorine, oxygen, hydroxyl radicals, and other oxidant species. When the wastewater was passed through the electrolytic cell the organic pollutants were oxidized to carbon dioxide and water.

Szpyrkowicz et al. (2005) investigated the decrease of chemical oxygen demand, total nitrogen, ammonia, chromium, and sulfide in an electrochemical reactor equipped with anodes based on noble metals and metal oxides (Ti/Pt–Ir, Ti/PbO_2, Ti/PdO–Co_3O_4, and Ti/RhO_x–TiO_2). The type of anode material and electrochemical parameters significantly affects the removal mechanism of contaminants from tannery wastewater. Several parameters control the removal processes, and better removal was obtained at high current/voltage. The removal kinetics fitted to the pseudo-first order reaction and Ti/Pt–Ir and Ti/PdO–Co_3O_4 were better anode materials in comparison with two other electrodes.

The electrochemical oxidation of oxalic acid at boron-doped diamond or IrO_2–Ta_2O_5 (DSA-O_2) anodes in a continuous batch recirculation reaction system in the presence and in the absence of sodium chloride was investigated (Scialdone et al., 2009). In the presence of chlorides, the anodic oxidation of oxalic acid is partially replaced by an indirect oxidation process, and in the absence of chlorides, high current efficiency is obtained at boron-doped diamond when most of the process is under charge transfer controlled kinetics. An abatement of oxalic acid of about 99 and 39% was observed in acidic and in basic medium respectively at iridium anode in the presence of NaCl of 5 g/L. Interestingly, a low abatement of oxalic acid was observed at pH 9. The suggested possible mechanism was due to the fact that at a pH close to 7–8 the chemical reaction between hypochlorous acid and hypochlorite was expected to be maximized since relevant concentrations of both these compounds were present in the solution.

Using boron-doped diamond anodes, complete mineralization of organic pollutants was carried out, and additionally NH_3–N was also removed completely without adjusting the solution pH (Zhu et al., 2009). Moreover, the total organic carbon and NH_3–N removal rates in boron-doped diamond anode cells were much greater than those in other common anode systems such as SnO_2 and PbO_2 anode cells. The results established improved performance of the boron-doped diamond anode system for the treatment of coking wastewater with great potential.

3.2.3 Advanced Photochemical Oxidation Processes

The photocatalytic activity of TiO_2 is highly dependent on UV light and the rapid recombination response to the generated electron hole pairs. Thus, the homogeneous coating of TiO_2 by graphene may reduce this effect. It was found that the covalent anchoring of graphene oxide on TiO_2 leads to the reduction of electron hole pair recombination, and it further increases the reduction of highly toxic Cr(VI) in aqueous solutions (Zhang et al., 2012).

Recently, the carbon-doped titanium dioxide photocatalysts were prepared and applied for the photocatalytic degradation of 4-chlorophenol under UV and solar irradiation (Kavitha and Devi, 2014). The pretreatment of carbon-doped titanium dioxide with UV light reduced the visible light activity due to the removal of surface carbonaceous species, and it has been postulated that the electron transfer rate to oxygen molecules adsorbed on the TiO_2 surface increases the dioxygen reduction with subsequent formation of free radicals, which accelerates the rate of 4-chlorophenol degradation (Tian et al., 2008).

In another study, ozonation of azo dye containing wastewater was carried out in a semi-batch reactor (Wu and Wang, 2001). It was concluded from the experimental results that the values of sample chemical oxygen demand (COD) and biochemical oxygen demand (BOD) did not change with ozonation time for 30 min. It was revealed that after the ozonation, the COD reduced in the water samples ranges 27 $\pm$ 87% and the increase in BOD_5:COD ratio was 11 $\pm$ 66 times.

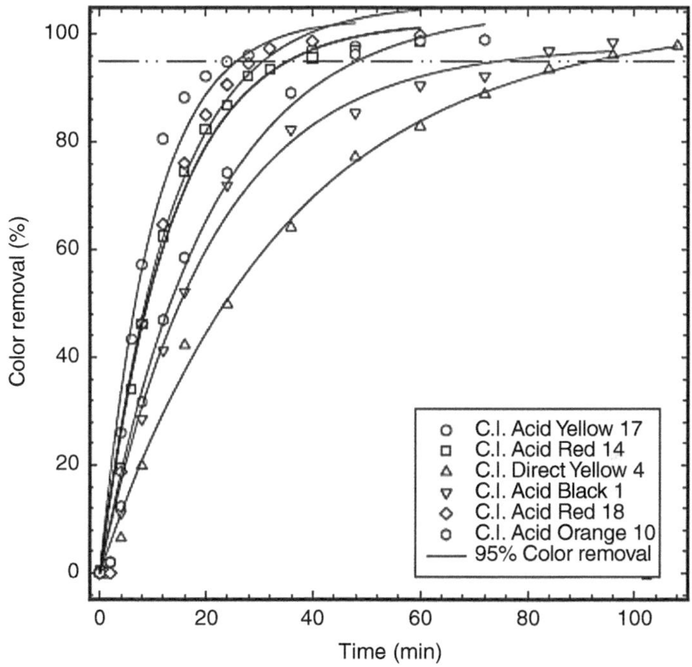

Figure 3.2 ***The effect of contact time on the removal of azo dyes from aqueous solutions by the UV/H_2O_2 process.*** *(Source: Shu and Chang, 2005; © Elsevier).*

The degradation of six azo dyes from wastewater was carried out using the O_3, UV/O_3, and UV/H_2O_2 processes (Shu and Chang, 2005). The solution's acidic medium favors the degradation of azo dyes from wastewater as the pH turns from 5.4 to 3.5. Increased acidity favors the degradation and breakdown of sulfate groups attached on the acidic azo dye molecules. Fig. 3.2 shows the effect of contact time on the removal of azo dyes from aqueous solutions by the UV/H_2O_2 process. More than 95% color removal was obtained with minimum energy utilization.

Different advanced oxidation processes (O_3, O_3/H_2O_2, UV, UV/O_3, UV/H_2O_2, O_3/UV/H_2O_2, Fe^{2+}/H_2O_2, and photocatalysis) have been studied for the catalytic degradation of phenol in aqueous solutions (Esplugas et al., 2002). In comparison to the UV processes (UV, UV/H_2O_2, and photocatalysis), the degradation rate with the UV/H_2O_2 process was almost five times higher than photocatalysis and UV alone. It was suggested that the Fenton's reagent showed better degradation rates at 40 times higher than the UV process and five times higher than ozonation. The suggested photocatalytic process was able to efficiently degrade the phenol in water with high rates and low contact time.

Photocatalytic degradation of acetaminophen on graphene/titanium dioxide nanotubes from aqueous solutions was recently studied by Tao et al. (2015). The photocatalytic degradation rate was higher than 96% for the acetaminophen under UV light irradiation for 3 h and with the 5% graphene/titanium dioxide nanotubes dosage of 0.1 g/L. On the basis of experimental results it was concluded that the electron-hole pairs were the main oxidation species in the photocatalytic process. The obtained results recognized a new prospect for acetaminophen degradation by using high efficiency catalysts.

3.2.4 Heterogeneous Photocatalysis

Recently, degradation of environmental contaminants by heterogeneous catalysts has been recognized as the best type of advanced oxidation process for the treatment of wastewater. The heterogeneous catalysis of environmental pollutants have several advantages, such as complete and fast degradation of contaminants to form carbon dioxide, water, and mineral acids without the formation of polycyclic by-products, high efficient catalysts, less contact time, high percentage removal, low cost, high stability, and environmentally sustainable materials (Neppolian et al., 2007; Gomez et al., 2012).

Water treatment on heterogeneous catalysts involves the irradiation of appropriate semiconductor materials with light energy that exceeds the bandgap, generating charge carrier molecules (electrons and holes) that have the ability to recombine and migrate to the catalysts surface (Nguyen and Juang, 2015). In this way, various redox reactions may take place on the surface with different species, such as oxygen or hydroxyl groups. These reactions on the heterogeneous catalysts can increase the production of highly reactive oxidizing or reducing agents that may react with the environmental contaminants in aqueous media. In degradation reaction, hydroxyl radicals have been well established for organic compound treatment in aqueous solutions (Hoffmann et al., 1995).

As far as the semiconductor material is concerned, the TiO_2 nanoparticles have been widely reported in literature (Gaya and Abdullah, 2008) due to their distinct lone electron features in the outer orbital. This lone electron gets photoexcited to the empty conduction band in femtoseconds, as the TiO_2 receives photon energy greater than bandgap energy, which is usually 3.2 eV. Fig. 3.3 shows the generation of electron-hole pairs in a semiconductor TiO_2 nanoparticle due to the photo-induced mechanism in the presence of environmental pollutants.

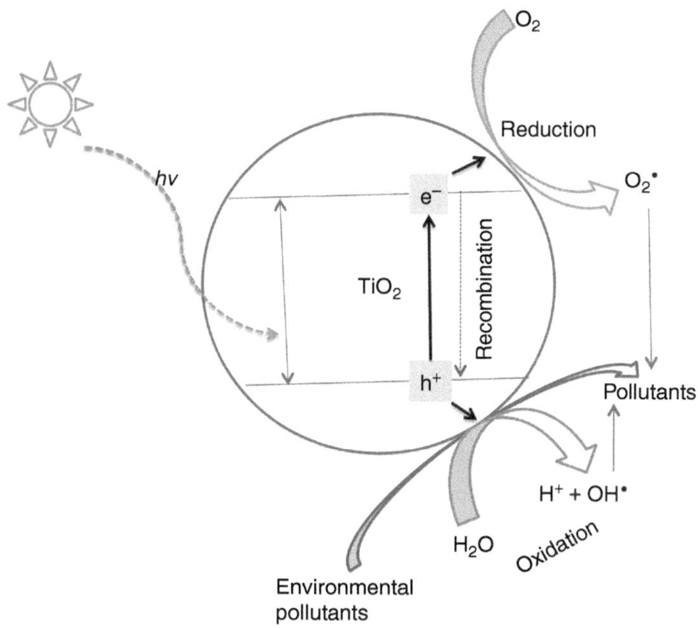

Figure 3.3 ***Schematic diagram for the generation of electron-hole pair in a semiconductor TiO$_2$ nanoparticle due to the photoinduced mechanism in the presence of environmental pollutants.***

The photo induced excitation leaves behind an empty unfilled valence band and thus generates an electron-hole pair (e^--h^+). The reaction mechanism during the photoexcitation and photodegradation processes on TiO_2 can be summarized as follows (Nguyen and Juang, 2015; Chong et al., 2010):

$$TiO_2 \xrightarrow{hv} h^+ + e^- \tag{3.9}$$

$$H_2O + h^+ \rightarrow \bullet OH + H^+ \tag{3.10}$$

$$\bullet O_2^- + H^+ \rightarrow \bullet HO_2 \tag{3.11}$$

$$\bullet HO_2 + \bullet HO_2 \rightarrow H_2O_2 + O_2 \tag{3.12}$$

$$\bullet O_2^- + \bullet HO_2 \rightarrow HO_2^- + O_2 \tag{3.13}$$

$$HO_2^- + H^+ \rightarrow H_2O_2 \tag{3.14}$$

These reactions show that the hydroxyl radicals are formed under photoexcitation when the positive holes react with H_2O and OH^- on the TiO_2

surface. It has been reported that the reaction proceeds in the reverse direction when the pH exceeds the pK_a (4.88) for the HO_2 radical (Nguyen and Juang, 2015).

Chlorophenol and *p*-chlorophenols are toxic and harmful compounds that are generally found in wastewater as persistent pollutants released from paper, pharmaceutical, and dye manufacturing industries. Mostly they are considered nonbiodegradable because of their complex structure and high physicochemical stability, thus they are continuously posing serious risk to the aquatic organism and humans. Recently, Nguyen and Juang (2015) applied the photocatalysts for the degradation of chlophenol and *p*-chlorophenol under the hybrid $H_2O_2/TiO_2/UV$ irradiation. Several important parameters include initial chlorophenol concentration, TiO_2 dose, H_2O_2 concentration, solution pH, and different intensity of UV irradiation. More than 98% degradation of *p*-chlorophenol was obtained within 150 min by using hybrid $H_2O_2/TiO_2/UV$ irradiation. Fig. 3.4 shows the effect of the initial solution pH on p-chlorophenol degradation on the TiO_2/UV at 25°C. The photocatalysis degradation ability of the TiO_2/UV system

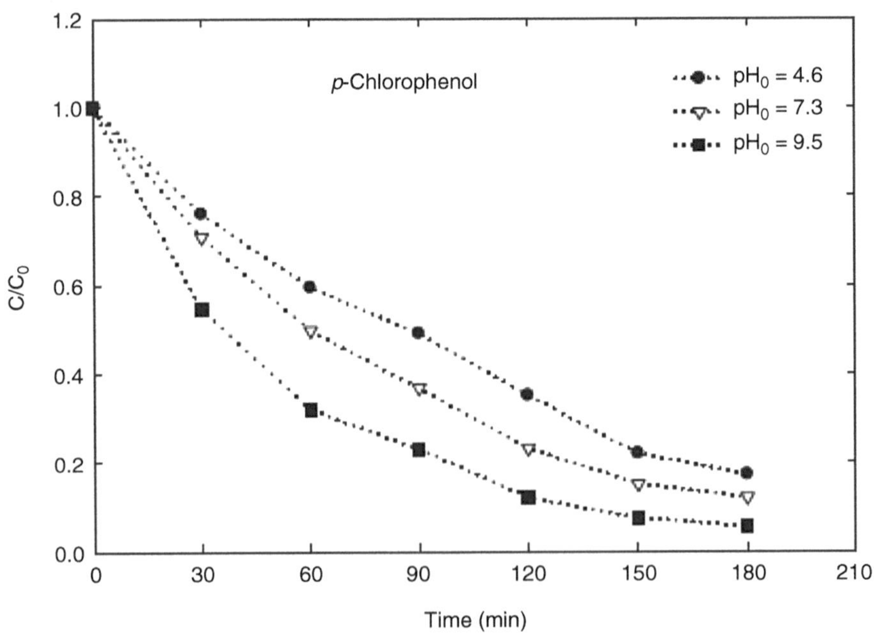

Figure 3.4 ***Effect of the initial solution pH on p-chlorophenol degradation on the TiO_2/UV at 25°C (100-W UV, initial concentration = 0.50 mM, TiO_2 = 1 g/L).*** *(Source: Nguyen and Juang, 2015; © Elsevier).*

expressed no significant effect of solution pH; however, the removal percentage of *p*-chlorophenol was 82%, 88%, and 94% at pH 4.6, 7.3, and 9.5, respectively. Moreover, the significant results were obtained at neutral pH, where *p*-chlorophenol was degraded approximately 93% within 180 min of irradiation.

A novel nanocomposite based on TiO_2/Cu_2O/reduced graphene oxide with enhanced solar-light-driven photocatalytic activity and the individual materials were studied through the photodegradation of methylene blue under solar light irradiation (Almeida et al., 2015). Approximately 95% methylene blue photodegradation was obtained within 5 h of contact time. The highest photochemical degradation due to the synergy of this carbonaceous structure with TiO_2 and Cu_2O, which could absorb a wider portion of the solar spectrum, adsorb higher quantities of methylene blue on the surface, and improve the effective separation of the generated electron–hole pairs. More specifically, the TiO_2 shows the highest removal efficiency in comparison with Cu_2O nanoparticles.

A series of TiO_2–graphene photocatalysts were prepared from titanium dioxide and graphene oxide by a simple hydrothermal method, and the photocatalyst material was used for the photocatalytic degradation of methylene blue in water under visible light irradiation (Rong et al., 2015). The photocatalytic degradation efficiency was higher than 98.8% within 100 min of contact time. The degradation behavior was fitted well to the pseudo-first order kinetic model. It was clearly seen that the characteristic absorption of methylene blue solution at 663 nm gradually diminishes in intensity with the increasing irradiation time (Fig. 3.5). Moreover, a fairly weak absorption peak at 663 nm and no new absorption peak appears, indicating the complete degradation of methylene blue.

3.2.5 Photo-Fenton Oxidation Reaction

In 1890, British chemist H.J.H. Fenton introduced a chemical reaction based on hydrogen peroxide (H_2O_2) and an iron catalyst to generate hydroxyl ions for the degradation of organic compounds present in water or wastewater. When a source of light (such as UV) is introduced to the above system, it is commonly called a photo-Fenton oxidation reaction. Under acidic solution (below pH ~3), the solution leads to the formation of $Fe(OH)^{2+}$ complex. Further, the exposure of UV irradiation produces highly reactive (•OH) radicals and Fe^{2+} (Munter, 2001).

$$Fe^{3+} + H_2O \rightarrow Fe(OH)^{2+} + H^+ \tag{3.15}$$

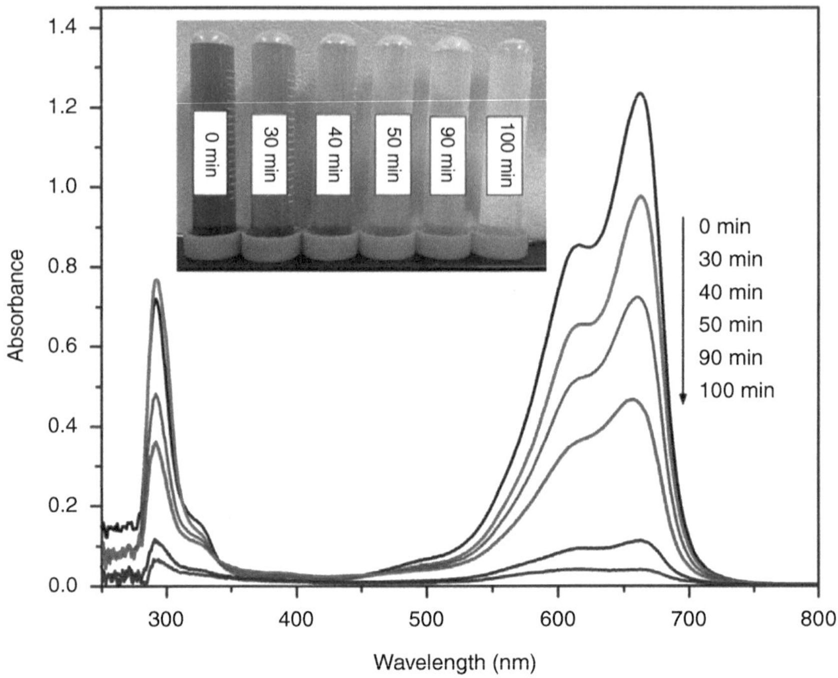

Figure 3.5 ***The evolution absorption spectra of methylene blue solution after treatment with TiO_2–graphene photocatalyst under visible light irradiation at different time intervals.*** *(Source: Rong et al., 2015; © Elsevier).*

$$Fe(OH)^{2+} \rightarrow Fe^{3+} + OH^- \quad (3.16)$$

$$Fe(OH)^{2+} \xrightarrow{hv} Fe^{2+} + \bullet OH \quad (3.17)$$

Malato et al. (2002) studied the photocatalytic degradation of four water-soluble pesticides (diuron, imidacloprid, formetanate, and methomyl) at pilot scale in two well-defined systems of special interest because natural-solar UV light was used as: heterogeneous photocatalysis with titanium dioxide and homogeneous photocatalysis by photo-Fenton. Complete removal of the parent compounds and approximately 90% mineralization was obtained with all pesticides tested; however, photocatalytic degradation of methomyl compound was slightly more difficult with both treatments.

Recently, Módenes et al. (2012) investigated the performance of a photo-Fenton process-based textile effluent treatment using both solar and artificial light sources on the operating parameters like pH, Fe^{2+}, and

H_2O_2. The chemical oxygen demand removal and decolorization was estimated by optimizing the photo-Fenton process. The optimum conditions of the photo-Fenton process were attained at concentration values of 0.05 g Fe^{2+} L^{-1} and 6.0 g H_2O_2 L^{-1} and pH 3 for both solar and artificial light sources. The highest chemical oxygen demand and decolorization was obtained at pH range 2–3. Low amounts of residual peroxide and iron was detected in the treated effluent, which was under the dischargeable prescribed limits. An operational cost of US$6.85 per m^3 of treated effluent was estimated using solar irradiation. The removal of residual organic pollutants was conducted from the cosmetic wastewater by photo-Fenton oxidation processes (Ebrahiem et al., 2013). Several optimizing parameters were investigated for the degradation of dye from wastewater and pH 3, 1 ml/L H_2O_2, 0.70 g/L Fe^{2+} and Fe^{3+}, and 40 min of contact time was suitable for maximum removal. The results show the removal efficiency was higher than 95%.

Three types of iron oxides—nano- and submicro-structured magnetite and nanostructured maghemite—were tested as catalysts for the degradation of paracetamol from aqueous solutions (Velichkova et al., 2013). The paracetamol aqueous solutions (100 mg/L) were treated by heterogeneous Fenton oxidation at acidic pH (2.6), and it was found that the high temperature enhances the paracetamol mineralization due to the radical scavenging effects. The catalytic degradation was achieved within 5 h; however, the complete mineralization was not yet achieved in aqueous media.

3.3 CONCLUSIONS

Advanced oxidation processes have been widely accepted for the degradation of recalcitrant environmental pollutants from wastewater due to high efficiency, stability, excellent performance, feasibility, and ecofriendly nature. A variety of advanced oxidation process-based technologies such as sonochemical treatment, ozonation, heterogeneous catalysis, electrochemical treatment, Fenton's oxidation, and photo-Fenton's processes have been developed and utilized in water remediation.

Till now the TiO_2 nanoparticles have been used as a model catalyst in the oxidation processes due to their low cost and high catalytic performance. However, there is a need to further enhance the catalytic performance of TiO_2 semiconductors. Hence, the TiO_2 nanoparticles have been continuously doped with metals such as Cr, V, Fe, Mn, and nonmetals such as carbon nanotubes, activated carbon, nitrogen, sulfur, and graphene to narrow the bandgap, increase the removal efficiency, and enhance the chemical stability of semiconductors in visible light irradiation.

Fenton oxidation also has been widely utilized for the decontamination of organic pollutants from water; however, the cost of catalysts and oxidizing agents limits its application at the industrial scale. Secondary by-products may impose another problem that may be more harmful than their parent compounds. During ozonation and peroxidation processes, extensive and proper handling should be used, as the release of these compounds may pose harmful effects due to their oxidizing nature.

Photocatalysis of environmental contaminants may be a superior approach to treating the wastewater with small loading of organic compounds as it utilizes low-cost energy from solar or UV irradiation in open atmospheric temperature and pressure. Light-sensitive degradation of organic dye and color compounds on TiO_2 is of particular interest for developing countries located at the equator where solar irradiation is plentiful. There is an urgent need to develop some more cost effective and high-performing semiconductor materials for the photocatalytic degradation of dyes and pigments. Efforts should also be made to use these techniques at pilot plant scale to treat the industrial effluents.

REFERENCES

Abbasi, M., Asl, N.R., 2008. Sonochemical degradation of Basic Blue 41 dye assisted by nano TiO_2 and H_2O_2. J. Hazard. Mater. 153, 942–947.

Almeida, B.M., Melo, Jr., M.A., Bettini, J., Benedetti, J.E., Nogueira, A.F., 2015. A novel nanocomposite based on TiO_2/Cu_2O/reduced graphene oxide with enhanced solar-light-driven photocatalytic activity. App. Surf. Sci. 324, 419–431.

Beckett, M.A., Hua, I., 2000. Elucidation of the 1,4-dioxine decomposition pathway at discrete ultrasonic frequencies. Environ. Sci. Technol. 34, 3944–3953.

Chen, G., 2004. Electrochemical technologies in wastewater treatment. Sep. Purif. Technol. 38, 11–41.

Chong, M.N., Jin, B., Chow, C.W.K., Saint, C., 2010. Recent developments in photocatalytic water treatment technology: a review. Wat. Res. 44, 2997–3027.

Ebrahiem, E.E., Al-Maghrabi, M.N., Mobarki, A.R., 2013. Removal of organic pollutants from industrial wastewater by applying photo-Fenton oxidation technology. Arabian J. Chem., doi: http://dx.doi.org/10.1016/j.arabjc.2013.06.012.

Esplugas, S., Gimenez, J., Contreras, S., Pascual, E., Rodriguez, M., 2002. Comparison of different advanced oxidation processes for phenol degradation. Wat. Res. 36, 1034–1042.

Gaya, U.I., Abdullah, A.H., 2008. Heterogeneous photocatalytic degradation of organic contaminants over titanium dioxide: a review of fundamentals, progress and problems. J. Photochem. Photobiol. C 9, 1–12.

Gomez, C.M., Del Angel, G., Tzompantzi, F., Gomez, R., Torres-Martinez, L.M., 2012. Photocatalytic degradation of p-cresol on Pt/gAl_2O_3-TiO_2 mixed oxides: effect of oxidizing and reducing pretreatments. J. Photochem. Photobiol. A 236, 21–25.

Hoffmann, M.R., Martin, S.T., Choi, W.Y., Bahnemann, D.W., 1995. Environmental applications of semiconductor photocatalysis. Chem. Rev. 95, 69–96.

Ince, N.H., Tezcanli, G., Belen, R., Apikyan, I.G., 2000. Ultrasound as a catalyzer of aqueous reaction systems: the state of the art and environmental applications. Appl. Catal. B 29, 167–176.
Kavitha, R., Devi, L.G., 2014. Synergistic effect between carbon dopant in titania lattice and surface carbonaceous species for enhancing the visible light photocatalysis. J. Environ. Chem. Eng. 2, 857–867.
Litter, M.I., Quici, N., 2010. Photochemical advanced oxidation processes for water and wastewater treatment. Recent Patents on Engineering 4, 217–241.
Ma, C.Y., Xu, J.Y., Liu, X.J., 2006. Decomposition of an azo dye in aqueous solution by combination of ultrasound and visible light. Ultrasonics 44, e375–e378.
Malato, S., Blanco, J., Cáceres, J., Fernández-Alba, A.R., Agüera, A., Rodriguez, A., 2002. Photocatalytic treatment of water-soluble pesticides by photo-Fenton and TiO_2 using solar energy. Catal. Today 76, 209–220.
Módenes, A.N., Espinoza-Quiñones, F.R., Manenti, D.R., Borba, F.H., Palácio, S.M., Colombo, A., 2012. Performance evaluation of a photo-Fenton process applied to pollutant removal from textile effluents in a batch system. J. Environ. Manage. 104, 1–8.
Munter, R., 2001. Advanced oxidation processes—current status and prospect. Proc. Estonian Acad. Sci. Chem. 50 (2), 59–80.
Naffrechoux, E., Chanoux, S., Petrier, C., Suptil, J., 2000. Sonochemical and photochemical oxidation of organic matter. Ultrason. Sonochem. 7 (4), 255–259.
Neppolian, B., Jung, H., Choi, H., 2007. Photocatalytic degradation of 4-chlorophenol using TiO_2 and Pt-TiO_2 nanoparticles prepared by sol-gel method. J. Adv. Oxid. Technol. 10, 1–6.
Nguyen, A.T., Juang, R.S., 2015. Photocatalytic degradation of p-chlorophenol by hybrid H_2O_2 and TiO_2 in aqueous suspensions under UV irradiation. J. Environ. Manage. 147, 271–277.
Rong, X., Qiu, F., Zhang, C., Fu, L., Wang, Y., Yang, D., 2015. Preparation, characterization and photocatalytic application of TiO_2–graphene photocatalyst under visible light irradiation. Ceram. Int. 41, 2502–2511.
Scialdone, O., Randazzo, S., Galia, A., Silvestri, G., 2009. Electrochemical oxidation of organics in water: Role of operative parameters in the absence and in the presence of NaCl. Wat. Res. 43, 2260–2272.
Shu, H.Y., Chang, M.C., 2005. Decolorization effects of six azo dyes by O_3, UV/O_3 and UV/H_2O_2 processes. Dyes Pigm. 65, 25–31.
Sires, I., Centellas, F., Garrido, J.A., Rodriguez, R.M., Arias, C., Cabot, P.L., et al.,2007. Mineralization of clofibric acid by electrochemical advanced oxidation processes using a boron doped diamond anode and Fe^{2+} and UVA light as catalysts. Appl. Catal. B 72, 373–381.
Szpyrkowicz, L., Naumczyk, J., Zilio-Grandi, F., 1995. Electrochemical treatment of tannery wastewater using Ti/Pt, and Ti/Pt/Ir electrodes. Wat. Res. 29, 517–524.
Szpyrkowicz, L., Kaul, S.N., Neti, R.N., Satyanarayan, S., 2005. Influence of anode material on electrochemical oxidation for the treatment of tannery wastewater. Water Res. 39, 1601–1613.
Tao, H., Liang, X., Zhang, Q., Chang, C.T., 2015. Enhanced photoactivity of graphene/titanium dioxide nanotubes for removal of Acetaminophen. Appl. Surf. Sci. 324, 258–264.
Tezcanli-Güyer, G., Ince, N.H., 2004. Individual and combined effects of ultrasound, ozone and UV irradiation: a case study with textile dyes. Ultrasonics 42, 603–609.
Tian, B., Zhang, J., Tong, T., Chen, F., 2008. Preparation of Au/TiO_2 catalysts from Au(I)-thiosulfate complex and study of their photocatalytic activity for the degradation of methyl orange. Appl. Catal. B 79, 394–401.
Velichkova, F., Julcour-Lebigue, C., Koumanova, B., Delmas, H., 2013. Heterogeneous Fenton oxidation of paracetamol using iron oxide (nano)particles. J. Environ. Chem. Eng. 1, 1214–1222.

Vlyssides, A.G., Loizidou, M., Karlis, P.K., Zorpas, A.A., Papaioannou, D., 1999. Electrochemical oxidation of a textile dye wastewater using a Pt/Ti electrode. J. Hazard. Mater. B70, 41–52.

Vlyssides, A.G., Papaioannou, D., Loizidoy, M., Karlis, P.K., Zorpas, A.A., 2000. Testing an electrochemical method for treatment of textile dye wastewater. Waste Manag. 20, 569–574.

Wang, J., Ma, T., Zhang, Z., Zhang, X., Jiang, Y., Pan, Z., Wen, F., Kang, P., Zhang, P., 2006. Investigation on the sonocatalytic degradation of methyl orange in the presence of nanometer anatase and rutile TiO_2 powders and comparison of their sonocatalytic activities. Desalination 195, 294–305.

Wang, J., Ma, T., Zhang, Z., Zhang, X., Jiang, Y., Zhang, G., Zhao, G., Zhao, H., Zhang, P., 2007. Investigation on transition crystal of ordinary rutile TiO_2 powder and its sonocatalytic activity. Ultrason. Sonochem. 14, 246–252.

Wang, Y., Zhao, D., Ma, W., Chen, C., Zhao, J., 2008a. Enhanced sonocatalytic degradation of azo dyes by Au/TiO_2. Environ. Sci. Technol. 42 (16), 6173–6178.

Wang, J., Sun, W., Zhang, Z., Jiang, Z., Wang, X., Xu, R., Li, R., Zhang, X., 2008b. Preparation of Fe-doped mixed crystal TiO_2 catalyst and investigation of its sonocatalytic activity during degradation of azo fuchsine under ultrasonic irradiation. J. Colloid Interface Sci. 320, 202–209.

Wang, J., Jiang, Z., Zhang, L., Kang, P., Xie, Y., Lv, Y., Xu, R., Zhang, X., 2009. Sonocatalytic degradation of some dyestuffs and comparison of catalytic activities of nanosized TiO_2, nano-sized ZnO and composite TiO_2/ZnO powders under ultrasonic irradiation. Ultrason. Sonochem. 16 (2), 225–231.

Wang, J., Lv, Y., Zhang, L., Liu, B., Jiang, R., Han, G., Xu, R., Zhang, X., 2010. Sonocatalytic degradation of organic dyes and comparison of catalytic activities of CeO_2/TiO_2, SnO_2/TiO_2 and ZrO_2/TiO_2 composites under ultrasonic irradiation. Ultrason. Sonochem. 17 (4), 642–648.

Wu, J., Wang, T., 2001. Ozonation of aqueous azo dye in a semi-batch reactor. Wat. Res. 35, 1093–1099.

Zhang, K., Kemp, K.C., Chandra, V., 2012. Homogeneous anchoring of TiO_2 nanoparticles on graphene sheets for waste water treatment. Mater. Lett. 81, 127–130.

Zhu, X., Ni, J., Lai, P., 2009. Advanced treatment of biologically pretreated coking wastewater by electrochemical oxidation using boron-doped diamond electrodes. Wat. Res. 43, 4347–4355.

CHAPTER 4

Graphene-Based Nanocomposites as Nanosorbents

4.1 INTRODUCTION

Contamination of aqueous solutions by heavy metals and dyes is a severe problem worldwide. The negative effects of pollutants on the biodiversity and on human health are diverse and depend on the degree of pollution. Heavy metals and metalloids like Hg, Pb, Cr, Cd, Cu, Co, Zn, and As are among the ubiquitous trace contaminants of aquatic ecosystems. Organic dyes also possess toxicity and reduce the light penetration and oxygen solubilization in aquatic media, thus affecting the rate of photosynthesis. These contaminations raise concerns as small quantities of heavy metals and dyes have been shown to be carcinogenic to humans and animals and can pose a risk to the aquatic biota. Hence, there is an urgent need to treat the wastewater containing heavy metals and dyes before they are discharged into the water bodies. Various physicochemical methods such as coagulation, flocculation, ion exchange, membrane separation, oxidation, and adsorption are available for the treatment of pollutants from aquatic media. The major drawbacks of these methods are high amounts of sludge production, handling and disposal problems, high cost, technical constraints, and tedious design. This necessitates cost-effective and environmentally sound techniques for treatment of wastewater containing heavy metals and organic dyes. This search brought nanomaterials to the foreground of scientific interest as a potential basis for the design of novel wastewater treatment processes. Graphene nanocomposites, composed of both inorganic and organic moieties, have recently been examined as promising platforms for detection and separation applications (Gao et al., 2011). This unique class of nanomaterials can retain not only beneficial features of both the inorganic and organic components, but can also provide the ability to systematically tune the properties of hybrid materials through a combination of appropriate functional components. Graphene nanocomposites have received much attention due to their unique properties, such as extremely small size, high surface-area-to-volume ratio, surface modifiability, multifunctionality, excellent conductivity, low-cost synthesis, and great biocompatibility. These nanocomposites offer a new

Nanomaterials for Wastewater Remediation
http://dx.doi.org/10.1016/B978-0-12-804609-8.00004-2

vista of separation and purification technology for heavy metals and dyes. Graphene and its nanocomposites have been successfully synthesized and applied for the separation and purification of heavy metals and dyes from wastewater effluents.

4.2 GRAPHENE-BASED NANOCOMPOSITES AS NANOSORBENTS

Industrialization, urbanization, and rapid population growth have resulted in the discharge of a number of toxic chemicals of anthropogenic origin in natural surface water bodies (Mishra and Ramaprabhu, 2011). The presence of toxic contaminant poses immense problems for aquatic flora and fauna. Hence, action should be taken to avoid and treat toxic contaminants before their discharge into the water bodies. Therefore, several methods have been applied for the treatment of heavy metals and dyes from aqueous solutions before their discharge into water bodies. The presence of toxic contaminants in the source water influences the efficiency of conventional water treatment plants. Advanced oxidation processes (Oller et al., 2011), membrane filtration techniques (Yun et al., 2006), coagulation (Joo et al., 2007), flocculation (Fang et al., 2010), and adsorption (Gautam et al., 2013, 2014) have been used for the treatment of wastewater. In all of them, adsorption exhibits superior performance in the treatment and removal of heavy metal and dye contaminants over conventional systems. Still, uses of other methods are limited, owing to their maintenance, energy supply, and high operating cost. Among all the approaches proposed for toxic metal impurities removal, adsorption is found to be economical and efficient on solid phase adsorbents (Hosamani et al., 2010; Nguyen et al., 2009; Thomas et al., 2007).

Several low-cost adsorbents such as clays (Weng et al., 2007; Sprynskyy et al., 2006; Erdem et al., 2004), bagasse fly ash (Gupta et al., 1998, 2003; Gupta and Ali, 2000), coal fly ash (Wang et al., 2005, 2006; Mohan et al., 2002), silica gel (Kushwaha and Chattopadhyaya, 2014; Kushwaha et al., 2013), hydroxyapatite (Gupta et al., 2011), agricultural wastes products (Rawat et al., 2013; Banerjee et al., 2014a), medicinal plant wastes (Rao et al., 2010), saw dust (Banerjee et al., 2013, 2014b), and biomass-derived adsorbents (Gautam et al., 2013, 2014; Gautam and Chattopadhyaya, 2013; Gupta et al., 2012) have been used for the adsorption and removal of heavy metals and dyes from aqueous solutions.

These adsorbents tested have achieved tremendous attention, but their performance at industrial scale limits their application because of the high cost of maintenance, energy requirements, and high sludge generation. High sludge generation creates further problems in industries concerning their disposal, which can create a secondary environmental pollution. The proper disposal of industrial sludge is a burden in the developing industrial sector in most developing countries.

Therefore, the immense focus of scientific community is going on to develop low cost adsorbents which have high surface area to volume ratio, surface tunable properties, well dispersible in liquids, and easy regeneration, and recycling properties and abilities in harsh environment. Hence, carbon nanotubes have been applied for the purpose of adsorbents to remove and recover the toxic metal ions and dyes from liquids (Stafiej and Pyrzynska, 2007; Rao et al., 2007; Wu, 2007; Li et al., 2003).

4.2.1 Adsorption of Heavy Metals

Mishra and Ramaprabhu (2011) synthesized the graphene sheets by hydrogen-induced exfoliation of graphitic oxide followed by functionalization. These functionalized graphene sheets were used for simultaneous removal of high concentration of inorganic species of arsenic and sodium from aqueous solution using supercapacitor–based water filters. In addition, these functionalized graphene sheet–based water filters were used for desalination of seawater. Maximum adsorption capacities, using the Langmuir isotherm, for arsenate, arsenite, and sodium were found to be nearly 142, 139, and 122 mg/g, respectively.

Lee and Yang (2012) synthesized self-assembled flowerlike TiO_2 on exfoliated graphite oxide (GO–TiO_2) for heavy metal removal from aqueous solutions. The surface-attached oxygenated functional groups of exfoliated graphite oxide showed a high removal capacity of heavy metals. The flowerlike TiO_2 on GO structure significantly improved the removal efficiency of heavy metals. The GO–TiO_2 hybrid adsorption capacities of heavy metal ions, after 6 and 12 h of hydrothermal treatment at 100°C, were respectively 44.8 ± 3.4 and 88.9 ± 3.3 mg/g for removing Zn^{2+}, 65.1 ± 4.4, and 72.8 ± 1.6 mg/g for removing Cd^{2+}, and 45.0 ± 3.8 and 65.6 ± 2.7 mg/g for removing Pb^{2+} at pH 5.6. Comparatively, colloidal GO under identical conditions showed removal capacities of 30.1 ± 2.5, 14.9 ± 1.5, and 35.6 ± 1.3 mg/g for Zn^{2+}, Cd^{2+}, and Pb^{2+}, respectively. TiO_2 blossoms markedly formed upon GO as the hydrothermal treatment time at 100°C increased from 6 to 12 h. Longer treatment times resulted in an increase in the surface area of GO–TiO_2 hybrid and thus increased removal capacity of heavy metals from aqueous solutions.

4.2.2 Graphene Oxide

Recently, Wang et al. (2013a) studied the adsorption of Zn(II) on graphene oxide (GO) as a function of pH, adsorbent dosage, foreign ions, contact time, and temperature from aqueous solutions using the batch technique. They observed that the suitable pH for Zn(II) removal was about 7.0, and the optimal dosage was 2 mg. The adsorption equilibrium of Zn(II) onto GO increased sharply within 20 min. The pseudo-second-order rate model for the adsorption process was found to be more suitable. Equilibrium studies indicated that the data of Zn(II) adsorption followed the Langmuir model, with maximum adsorption capacity for Zn(II) up to 246 mg/g. The thermodynamic parameters suggested that Zn(II) sorption on GO was an exothermic and spontaneous process in nature. The possibility of maximum Zn(II) recovery was achieved with hydrochloric acid.

Zhao et al. (2011a) synthesized a few layered GO nanosheets and used them as sorbents for the removal of Cd(II) and Co(II) ions from aqueous solutions. The effects of operating parameters such as pH, ionic strength, and humic acid on Cd(II) and Co(II) sorption were investigated. Their results indicated that Cd(II) and Co(II) sorption on GO nanosheets were strongly dependent on pH and weakly dependent on ionic strength. The maximum adsorption capacities of Cd(II) and Co(II) on GO nanosheets at pH 6.0 and $T = 303$ K were about 106.3 and 68.2 mg/g, respectively. Thermodynamic results showed that the adsorption process was spontaneous and endothermic in nature. The abundance of oxygen-containing functional groups on the surface of graphene nanosheets played a major role in adsorption of cations from aqueous solutions. Zhao et al.'s (2011a) results indicated that the oxygen-bearing functional groups on graphene nanosheets can be successfully utilized for the removal of toxic Cd(II) ions from large volumes of water.

4.2.3 Graphene-Oxide-Based Chitosan–Gelatin Composites

Zhang et al. (2011) synthesized ordered porous chitosan–gelatin/graphene oxide (CGGO) monoliths with more than 97% porosity by a unidirectional freeze-drying method and used as adsorbents for metal ions sequestration from aqueous solutions. The adsorption behavior of the CGGO monoliths and the influencing factors such as solution pH, GO concentration, and metal ion concentration as well as the effect of ethylenediaminetetraacetic acid (EDTA) were investigated. The incorporation of GO significantly increased the compressive strength of the CGGO monoliths in both their wet and dry states and changed their porous structure. They exhibited an extremely high adsorbing ability for metal ions, which decreased at low pH but increased from 20 to 88% upon the addition of EDTA at low pH. The CGGO monoliths have good stability and were recycled several times, with only a slight loss in adsorption ability. The adsorption ability of the porous CGGO monoliths for Cu^{2+} was greatly influenced by the pH of the Cu^{2+} solution. The effect of pH on the adsorption ability for Pb^{2+} was similar with that of the Cu^{2+} ions. The adsorption percentage for Cu^{2+} was dramatically increased from 15.6 to 88.6% as pH changed from 2.0 to 6.0. The CGGO monolith was doped with only 0.5 wt.% GO, and the maximum adsorption capacity of CGGO for Pb^{2+} increased from 68.5 to 75.5 mg/g. As the GO concentration was increased further, the maximum adsorption capacity for Pb^{2+} gradually increased to a maximum value of about 100 mg/g, whereas the changes in the maximum adsorption capacity for Cu^{2+} was very minute.

4.2.4 Graphene Oxide Aerogels

Graphene oxide aerogels were prepared by a unidirectional freeze-drying method, their structure was observed, and their adsorption performance for Cu^{2+} in aqueous solution was evaluated by Mi et al. (2012). The results showed that GO aerogels were unidirectional porous structures and had good adsorption ability for Cu^{2+}. The adsorption system was dependent on the initial Cu^{2+} ion concentration and follows a pseudo-second-order kinetic model. An adsorbent dose of 0.06 g of GO aerogels was sufficient to remove maximum Cu^{2+} from 100 mL aqueous solution within 15 min of equilibrium contact time. The adsorption isotherms were simulated by the Langmuir, Freundlich, and Temkin models and fit well with classical Langmuir adsorption isotherm with a maximum capacity of 29.59 mg/g at 40°C. The adsorption of Cu^{2+} on GO aerogels was strongly dependent on pH, indicating an ion exchange mechanism. The obtained results by Mi et al. (2012) have demonstrated that the GO aerogels can be used as an effective adsorbent for Cu^{2+} removal from water. However, GO can be aggregated by Cu^{2+} interaction in aqueous solution with a huge Cu^{2+} absorption capacity (Yang et al., 2010a). The Cu^{2+} causes GO sheets to be folded and also to form large aggregates. The folding/aggregation was most likely triggered by the coordination between GO and Cu^{2+} in aqueous solutions. The average distance between the folded layers was around 2 nm, which was higher than the typical single layer thickness of GO (around 0.9 nm). The equilibrium Cu^{2+} concentrations and equilibrium absorption capacity of GO were measured to estimate the maximum absorption capacity of GO for Cu^{2+} using Langmuir and Freundlich isotherm models. The adsorption of Cu^{2+} by GO fit well to the Langmuir isotherm model with a maximum adsorption capacity of 46.6 mg/g. Graphene oxide has a huge absorption capacity for Cu^{2+}, which is around 10 times that of active carbon.

4.2.5 Graphene-Oxide-Based Ferric Hydroxide Nanocomposites

A series of novel composites based on GO cross-linked with ferric hydroxide was developed for effective removal of arsenate from contaminated drinking water (Zhang et al., 2010a). Graphene oxide, which was used as a supporting matrix, was first treated with ferrous sulfate. Then, the ferrous compound cross-linked with GO was in situ oxidized to ferric compound by hydrogen peroxide, followed by treatment with ammonium hydroxide. These composites were evaluated as absorbents for arsenate removal from drinking water. More than 95% of arsenate was absorbed by composite with an absorption capacity of 23.78 mg/g of composite for the water, with arsenate concentration at 51.14 ppm. Effective arsenate removal occurred in a wide range of pH from 4 to 9.

The adsorption properties of Pb(II) and Cd(II) on functionalized graphene prepared by the electrolysis method have been estimated by Deng et al. (2010). The adsorption capacity for the removal of Pb(II) or Cd(II) ions from water were 406.6 mg/g (pH = 5.1) for Pb(II) and 73.42 mg/g (pH = 6.2) for Cd(II), respectively. Pseudo-second-order model was found to be the most suitable for the adsorption of Pb(II) and Cd(II) on functionalized graphene.

Reduced graphene oxide–magnetite and graphene oxide–ferric hydroxide composites were used for the adsorptive removal of arsenic from aqueous media (Chandra et al., 2010). Iron-based oxides and hydroxides doped with activated carbon are known to remove arsenic from drinking water (Zhang et al., 2010b). The reports show that reduced graphene oxide (RGO) and graphene oxide–supported materials have higher binding capacity compared to free nanoparticles (Chandra et al., 2010). Chandra et al. (2010) have synthesized the magnetite graphene hybrids via a chemical reaction with a magnetite particle size of ~10 nm. The composites were superparamagnetic at room temperature and were separated by an external magnetic field. As compared to bare magnetite particles, the hybrids show a high binding capacity for As(III) and As(V) from drinking water. Their high binding capacity was due to the increased adsorption sites in the M-RGO composite, which occurs by reducing the aggregation of bare magnetite.

Since the composites showed near complete (over 99.9%) arsenic removal within 1 ppb, they are practically usable for arsenic separation from water. The wide scan XPS spectra of the M-RGO showed photoelectron lines at a binding energy of about 285, 530, and 711 eV attributed to C1s, O1s, and Fe2p, respectively. In the spectrum of Fe2p, the peaks $Fe2p_{3/2}$ and $Fe2p_{1/2}$ were located at 711.29 and 724.82 eV, not at 710.35 and 724.0 eV, which were for γ-Fe_2O_3. In addition, there was no satellite peak at ~719.0 eV, characteristic of γ-Fe_2O_3, which was indicative of formation of the Fe_3O_4 phase in the RGO matrix. It was known that at least three types of oxygen species may contribute to the O1s peak, that is, the contribution of the anionic oxygen in Fe_3O_4 at about 530.2 eV, the oxygen-containing functional groups at around 531.8 eV, and water at higher binding energies. The peak around 530.4 eV was due to the oxygen in the M-RGO, while the oxygen in graphene oxide was around 532.6 eV. Deconvolution of the C1s peak of graphene oxide showed the presence of different oxygen-containing functional groups of (a) the nonoxygenated C at 284.8 eV, (b) the carbon in C-O at 286.2 eV, (c) the carbonyl carbon (CAO) at 287.9 eV, and (d) the carboxylate carbon (O-CAO) at 289.0 eV. The C1s spectra of M-RGO shows mainly the nonoxygenated carbon (284.8 eV) and the carbon in C-O (286.2 eV). Arsenic ion removal on the M-RGO surface was due to the electrostatic attraction between the positively charged surface of M-RGO and the negatively charged arsenic/arsenous acid. Under most pH conditions, As(V) was present in negative ionic form ($H_2AsO_3^-$), whereas As(III) was in a nonionic form (H_3AsO_4). The pH-value dependence of arsenic ions adsorption onto M-RGO was explained by point of zero charge (pH_{PZC}) of the adsorbent. At pH lower than pH_{PZC}, the M-RGO surface was positively charged, whereas at a pH greater than pH_{PZC}, the M-RGO surface was negatively charged. Due to the net positive surface charge of M-RGO at a pH lower than pH_{PZC}, it attracts As(V) anions, resulting in large adsorption at a low pH-value.

4.2.6 Reduced Graphene Oxide–Metal Oxide Composites

Recently, a wide range of reduced graphene oxide–metal/metal oxide composites was synthesized by Sreeprasad et al. (2011), and their application in water purification was carried out. Reduced GO is progressively oxidized primarily to GO and the formed metal nanoparticles are anchored onto the carbon sheets. Metal ion scavenging applications of reduced GO–MnO_2 and reduced GO–Ag were demonstrated by taking Hg(II) as the model pollutant. Reduced GO and the composites give a high distribution coefficient (K_d), greater than 10 L/g for Hg(II) uptake.

The Ni (II) removal potential of graphene nanosheet/δ-MnO_2 (GNS/MnO_2) from solution was investigated by Ren et al. (2011). GNS/MnO_2 composite was synthesized by a microwave-assisted method. The adsorption optimization parameters had been examined with respect to solution pH effect, removal of kinetic data, equilibrium isotherms, and thermodynamic data in batch experiments. The pseudo-second-order kinetic model was best fitted to the adsorption data, showing that a chemical sorption process plays a significant role in Ni (II) ions adsorption on GNS/MnO_2. The Langmuir model was most suited to describe the equilibrium isotherm with adsorption capacity of 46.6 mg/g at room temperature, which was 1.5 and 15 times higher than those of pure δ-MnO_2 and GNS, respectively. Maximum adsorption capacity 66.01 mg/g was obtained for Ni(II) removal at 318 K with a high value of correlation coefficient 0.999. The positive values of both ΔH and ΔS suggested an endothermic reaction and increase in randomness at the solid–liquid interface during the Ni (II) adsorption process. The negative ΔG values indicated a spontaneous adsorption process, and GNS/MnO_2 was reused five times with a high recovery rate of 91%.

4.2.7 Magnetic Chitosan/Graphene Oxide

Fan et al. (2013) have fabricated magnetic chitosan/GO through a facile and fast process, and their application as excellent adsorbents for metal ions was also demonstrated. The magnetic chitosan/GO was used as a sorbent for the removal of Pb(II) ions from aqueous solutions. The effects of optimization parameters such as pH, contact time, and concentration on Pb(II) ion sorption were investigated. The results indicated that Pb(II) ion sorption on magnetic chitosan/GO was strongly dependent on pH. The abundant functional groups on the surfaces of magnetic chitosan/GO played an important role on Pb(II) sorption. Equilibrium studies showed that the data of Pb(II) adsorption followed the Langmuir model. The maximum adsorption capacity for Pb(II) was estimated to be 76.94 mg/g.

Adsorbents for chromium removal based on GO functionalized with magnetic cyclodextrin–chitosan (CCGO) was synthesized by Li et al. (2013). The adsorption equilibrium of CCGO for Cr(VI) fitted well to Langmuir isotherm with maximum adsorption capacity of 67.66 mg/g, and the novel adsorbent exhibited better Cr(VI) removal efficiency in solutions with pH 2.0. It was found that the Cr(VI) adsorption performance of CCGO strongly depends on their surface charge concentration of hydroxyl and amino groups and specific surface area. It was found that the decrease in the adsorption amount of Cr(VI) was less than 5% after five adsorption and desorption cycles.

Li et al. (2010) synthesized graphene nanoflakes with superior specific surface area and conductivity and used them as electrosorption electrodes for ferric chloride ($FeCl_3$). It was found that the electrosorptive capacity of graphene nano-flakes was 0.5 mg/g for an initial $FeCl_3$ concentration of 20 mg/L. The maximum equilibrium electrosorption capacity and rate constant for $FeCl_3$ at 2.0V were 0.88 mg/g and 0.27 min^{-1}, respectively. These data were validated by both Langmuir isotherm and pseudo-first order adsorption.

4.2.8 Sulfonated Magnetic Graphene Oxide Composite

A sulfonated magnetic GO composite was synthesized from GO and was used as an adsorbent for removal of Cu(II) ions from aqueous solutions (Hu et al., 2013). The effects of operating parameters such as pH, Cu(II) concentration, and temperature on the Cu(II) adsorption were investigated. The optimum adsorption Cu(II) uptake of 62.73 mg/g was achieved at pH 4.68, Cu(II) concentrations 73.71 mg/L, and temperature 50°C. The pseudo-second-order kinetic model was best fitted to the adsorption process. The experimental data of isotherm followed the Langmuir isotherm model with a maximum adsorption capacity of 63.67 mg/g at 323.15 K. Both maximum adsorption and Langmuir constant of the Langmuir model increased with increase in temperature and showed the bonding between Cu(II) and active sites of the adsorbent weakened at lower temperature, and the adsorption process was endothermic.

Magnetic graphene/iron oxide composite (Fe_3O_4/GO), which was synthesized from graphene using a chemical reaction approach, had been employed as a novel adsorbent for the preconcentration and solidification of U(VI) ions from aqueous solutions (Zong et al., 2013). The sorption behavior of U(VI) on the surface of Fe_3O_4/GO was carried out under ambient conditions such as contact time, pH, and ionic strength. The U(VI) sorption on Fe_3O_4/GO demonstrated that the sorption mechanism of U(VI) was pH-dependent and ionic strength-independent inner-sphere surface complexation at low pH values, whereas the removal of U(VI) was achieved by simultaneous precipitation and inner-sphere surface complexation at high pH values. The maximum sorption capacity of U(VI) on Fe_3O_4/GO at 293 K and pH = 5.5 was about 69.49 mg/g.

4.2.9 Graphene-Layered Double Hydroxide Nanocomposites

Graphene/MgAl-layered double hydroxide (G–MgAl-LDH) nanocomposites were prepared by urea hydrolyzed hydrothermal reaction of $Al(NO_3)_3{\cdot}9H_2O$, $Mg(NO_3)_2{\cdot}6H_2O$, and GO (Yuan et al., 2013). The calcined G–MgAl-LDH was easily prepared by heating G–MgAl-LDH at higher temperatures and used as a potential nanoadsorbent for the removal of Cr(VI) from the aqueous solution. Higher adsorption capacity was exhibited by calcined G–MgAl-LDH in comparison to virgin MgAl-LDH with lower dosage when removing Cr(VI) from aqueous solutions. Maximum adsorption capacity of 172.55 mg/g with 1.0 g/L adsorbent dosage of calcined G–MgAl-LDH was achieved. The equilibrium data was modeled with the classical Freundlich and Langmuir isotherm models. The higher R^2 value of 0.9953 for the Freundlich model indicated that the Freundlich model fitted the experimental data better than the Langmuir model (R^2 0.9573), which showed the adsorption of Cr(VI) onto calcined G–MgAl-LDH took place in a heterogeneous manner. The results about the adsorption kinetics, isotherms, and thermodynamics indicated that the adsorption mechanism of Cr(VI) involved the surface adsorption of graphene, and the memory effect of calcined LDH and synergistic contribution from each component in calcined G–MgAl-LDH enhanced Cr(VI) adsorption capacity.

Gao et al. (2012) synthesized AlOOH-reduced GO via a hydrothermal method directly from the GO and $Al(NO_3)_3{\cdot}9H_2O$, and applied for the removal of Cd(II) and Pb(II) from drinking water. Due to the strong affinity of AlOOH to cations and the fast electron transfer properties of graphene, the combination of solid-phase extraction and stripping voltammetric analysis allowed fast and sensitive determination of Cd(II) and Pb(II) in drinking water, making these new nanocomposites promising candidates for practical applications in the fields of detecting heavy metal ions.

4.2.10 Simultaneous Removal of Heavy Metals and Dyes

Several heavy metals and ionic dyes commonly coexist in aqueous media of industrial effluents making a dangerous source of environmental pollution (Ay et al., 2012). Therefore, there is an urgent need to find an efficient, cost-effective and ecofriendly solution to simultaneously remove these coexisting pollutants. In an effort to remove heavy metals and ionic dyes simultaneously, magnetic graphene oxide (MGO) was synthesized and used as an adsorbent for simultaneous removal of Cd(II) and ionic dyes, including methylene blue (MB) and orange G (OG) (Deng et al., 2013). Kinetic data followed a pseudo-second-order model for the adsorption process. Equilibrium data were well fitted by the Langmuir model, with the maximum sorption capacities in ultrapure water for Cd(II), MB, and OG being 91.29 mg/g, 64.23 mg/g, and 20.85 mg/g, respectively. Increase in the MB concentration suppressed the sorption capacity of Cd(II), and MB removal was not affected with increasing Cd(II) concentration in the Cd(II)–MB binary system. Nevertheless, increased Cd(II) concentration enhanced the sorption capacity of OG and for Cd(II) was independent of the concentration of OG in Cd(II)–OG binary system.

Koo et al. (2011) synthesized the graphene-based multifunctional iron oxide nanosheets with tunable properties for the removal of As(V) and Cr(VI) from aqueous solutions. BET surface area of the IO/GNSs showed great enhancement up to 1460 m^2/g after the synthesis of IO needles, due to the densely packed tip-like surface structures in comparison with the bare GNSs, where BET surface area was found to be 375 m^2/g. Rapid removal of 97 and 94% for As(V) and Cr(VI) ions, respectively, were obtained from the water within 5 min. Within 20 min, most of the metal ions from the water were removed and the removal capacities were found to be 218 and 190 mg/g for As(V) and Cr(VI), respectively.

4.2.11 Graphene-Based Polymer Composites

The higher stability and nontoxicity of polypyrrole (PPy) and high surface area of RGO motivated synthesis of the PPy–RGO composite for removal of Hg^{2+} from aqueous solutions (Chandra and Kim, 2011). The isotherm data was modeled with the Freundlich and Langmuir isotherm models and fitted well to Langmuir model with a maximum capacity of 980 mg/g. The adsorption kinetics of mercury on the PPy–RGO composite was very fast and the equilibrium was attained within 20 min. The kinetic data were fitted to the pseudo-second-order kinetics.

EDTA have been effectively utilized as chelating agents toward the separation of metal ions and functionalization of nanocomposites and their analogues in wide dimensions because of their high ability to bind the metal ions in aqueous media (Repo et al., 2011). Seeking the extra advantage of this chelating agent, several scientists worldwide have functionalized the nanoadsorbents to increase their metal separation capacity (Yang et al., 2010b; Zhang et al., 2010c). Dupont et al. (2014) have utilized EDTA functionalized magnetic and nonmagnetic nanoparticles for the selective separation of rare earth elements from aqueous solutions.

Recently, this chelating agent was successfully linked to GO surfaces through a silanization reaction between *N*-(trimethoxysilylpropyl) ethylenediamine triacetic acid (EDTA-silane) and hydroxyl groups on GO surface (Madadrang et al., 2012). EDTA–GO was found to be an ideal adsorbent for Pb(II) removal from aqueous media with a higher adsorption capacity of 479 mg/g at pH 6.8, within the 20 min of equilibrium time. The experimental data agreed well with the Langmuir adsorption isotherm model. EDTA modification enhances the adsorption capacity of GO because of the chelating ability of EDTA. The desorption process was carried out on Pb(II) adsorbed on EDTA–GO adsorbents; the preloaded-Pb, EDTA–GO adsorbents were kept in 100 mL of HCl solution with pH adjusted at 2.0, 3.0, 4.0, 5.0, 6.0, and 7.0, respectively. After 4 h of treatment, Pb(II) adsorbed onto EDTA-GO surface was desorbed and dissolved in HCl solution. The filtrate was collected carefully to determine the concentration of Pb(II) ions, and the EDTA–GO adsorbent morphology was studied by SEM EDAX. The surface element ratio result reflected the surface element enrichment in the recycled adsorbent.

Polyvinylpyrrolidone-RGO was synthesized and used as adsorbent for removal of Cu ions from aqueous solutions (Zhang et al., 2014). The effects of contact time and Cu ion concentration on adsorption capacity were examined, and the maximum adsorption capacity of 1689 mg/g was observed. It was described that there was strong interaction between functional groups of polyvinylpyrrolidone and metal ions, which acted as adsorption sites for removal of Cu ions.

Chelating agent 1-(2-pyridylazo)-2-naphthol (PAN) forms stable complexes with a number of metals and has found numerous applications in trace element separation and preconcentration methods (Narin and Soylak, 2003). Graphene-based magnetic nanoparticles (G-Fe_3O_4) were prepared and used as an effective adsorbent for the solid phase extraction of trace quantities of cadmium in the presence of PAN from water and vegetable samples (Alvand and Shemirani, 2014). The effects of process parameters, such as pH, the amount of G–Fe_3O_4, extraction time, type and volume of eluant, desorption time, and interfering ions on the extraction efficiency, were optimized. It was reported that the removal of cadmium (II), increased as the pH in the aqueous solution, was increased up to 6.0, and then after that the extraction recovery started to decrease.

Amino group–modified graphene nanosheets were used for the extraction of cadmium and lead ions (Sayar et al., 2014). The effect of pH, metal concentration, flow rate, and influences of other metal ions were studied. The effect of pH was investigated on extraction efficiency of graphene nanosheets, and the pH of sample solutions containing 1 mg/L of Pb(II) or Cd(II) ions was adjusted to the range of 2–8. It was found that the optimum pH for adsorption of both Pb(II) and Cd(II) ions was pH = 7. This was attributed to the fact that, given an acidic pH, the amino group was protonated and could not coordinate to ions, and given an alkaline pH, the ions were precipitated as hydroxyl forms, causing a reduction in recovery.

Monitoring trace amounts of cadmium(II), copper(II), nickel(II), zinc(II), and lead(II) in different food and water samples by flame atomic absorption spectrophotometer after preconcentration by amine-functionalized graphene nanosheet were carried out by Behbahani et al. (2014). Amine-functionalized graphene nanosheets were prepared by adding GO (0.05 g) and DMF (5.0 mL) in a round flask under ultrasonic pulse for 30 min, and then 3-aminopropyltriethoxysilane was added to the solution under nitrogen atmosphere. The reaction mixture was heated to 80–110°C and stirred for 40 h. For coupling reaction, 20 μL of triethylamine was typically added to the reaction mixture to promote the graft reaction. Then, the mixture was subjected to centrifugation and thoroughly washed with toluene and THF at least five times. The product was carefully collected and dried under vacuum at 60°C until constant weight. Amine-functionalized graphene was obtained as a black powder. For the adsorption experiments, 100 mg of amine-functionalized graphene was packed in a glass column with dimensions of 120 mm in length and 20 mm in diameter and blocked by two polypropylene filters at the ends to prevent loss of the material during sample loading. Thereafter, 25 mL of standard solution containing 2 mg/L Cd, Cu, Ni, and Pb ions was passed through the column at a flow rate of 18 mL/min using a peristaltic pump, and the pH of the sample was adjusted to 6.0. Afterward, the analytes were determined by flame atomic absorption spectrophotometer. Fig. 4.1 provides a schematic representation of extraction procedure. In any adsorption experiments the solution pH plays an important role. It has been illustrated that among the tested variables, pH was found to be the most critical parameter for the adsorption of metals on the modified graphene. To evaluate the effect of pH on the retention efficiency, the pH of 25 mL of sample solution containing 2 mg/L of Cd, Cu, Ni, Zn, and Pb ions was adjusted to fit in the range of 2–9. The adsorption of Cd, Cu, Ni, Zn, and Pb ions on the amine-functionalized graphene was obtained at the pH of 6.0 (Fig. 4.2).

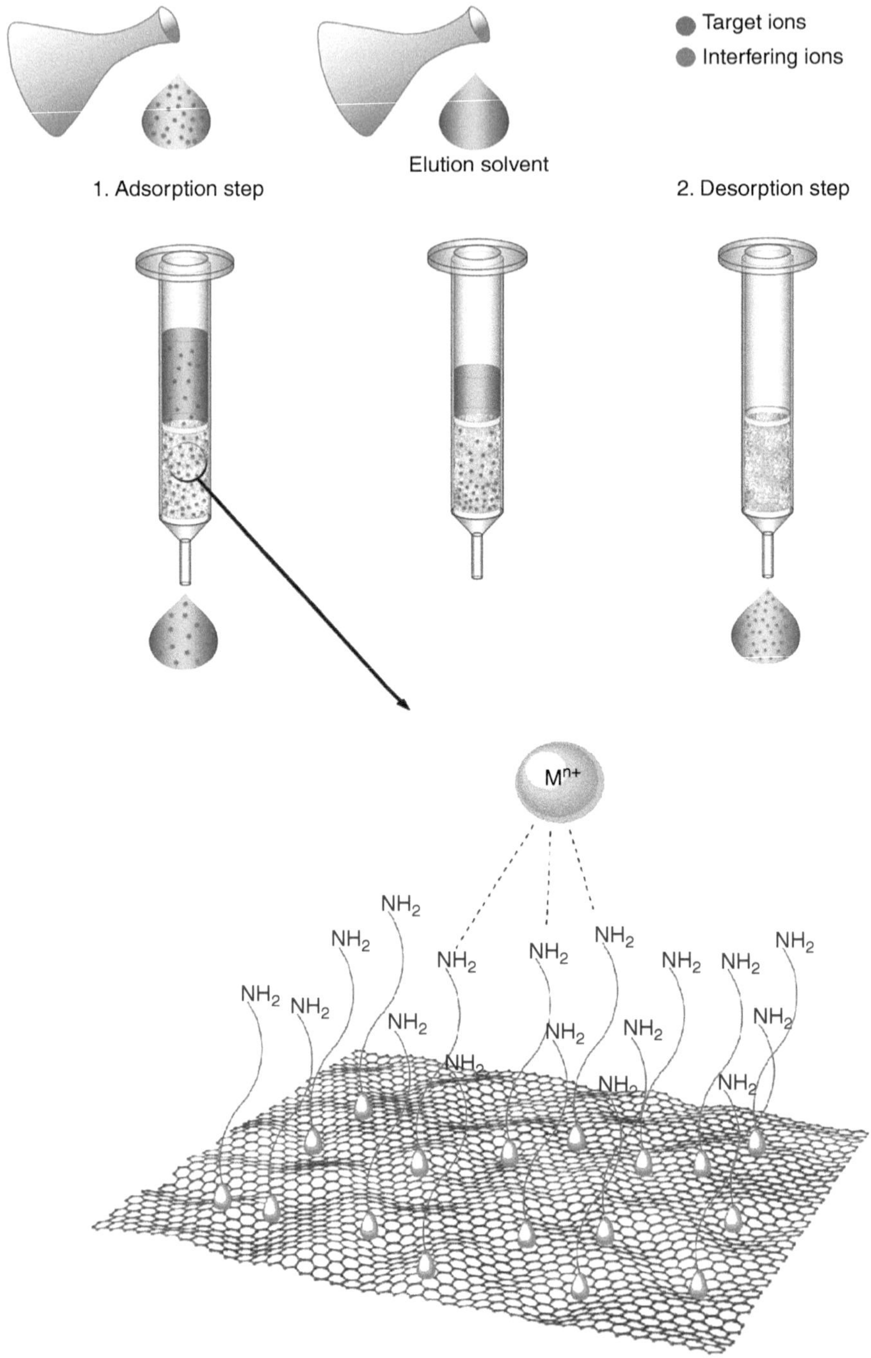

Figure 4.1 ***A schematic illustration of the extraction procedure of heavy metals by amine-functionalized graphene.*** *(Source: Behbahani et al., 2014; © Reproduced with permission from Springer).*

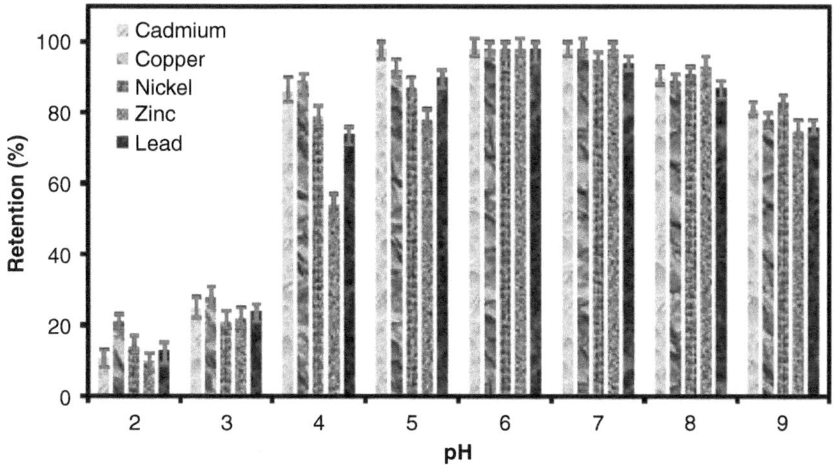

Figure 4.2 ***The effect of pH of sample solutions on retention of lead ions by amine-functionalized graphene (conditions: sample concentration, 2 mg/L; sample volume, 25 mL).*** *(Source: Behbahani et al., 2014; © Reproduced with permission from Springer).*

4.2.12 Graphene-Based Magnetic Nanocomposites

Graphene oxide/Fe_3O_4 (GO/Fe_3O_4) composites were synthesized and applied for the removal of Cu(II) and a natural organic macromolecule fulvic acid by GO/Fe_3O_4 from aqueous solutions (Li et al., 2012). Their finding indicated that Cu(II) sorption on GO/Fe_3O_4 were strongly dependent on pH and independent of ionic strength, indicating that the sorption was mainly dominated by inner-sphere surface complexation rather than outer-sphere surface complexation or ion exchange. The addition of fulvic acid enhanced the Cu(II) sorption at low pH and decreased at high pH, whereas the presence of Cu(II) led to an increase in fulvic acid sorption. In the GO/Fe_3O_4 adsorbed fulvic acid, the sorption capacity for Cu(II) was enhanced at pH 5.3, and the sorption capacity for fulvic acid was also enhanced after Cu(II) sorption on GO/Fe_3O_4. The experimental data was modeled with Langmuir and Freundlich isotherm models and fitted well to the Langmuir isotherm equation with a maximum adsorption capacity of 25.57 mg/g for Cu(II) at 333 K. These results are important for estimating and optimizing the removal of metal ions and organic substances by GO/Fe_3O_4 composites.

Recently, Lei et al. (2014) synthesized a novel three-dimensional (3D) GO foam/Fe_3O_4 nanocomposite (GOF/Fe_3O_4) and evaluated its adsorption performance for Cr(IV) removal from water. The GOF/Fe_3O_4 revealed a very large surface area of 574.2 m^2/g, a high saturation magnetization of

40.2 emu/g, and a maximum absorption capacity of 258.6 mg/g for Cr(IV) removal. It was found that the adsorption capacity of GOF/Fe_3O_4 increased with increasing the concentration of Cr(VI) ions, and finally equilibrium was established. The experimental isotherm data were analyzed by the Langmuir and Freundlich models. The maximum Langmuir adsorption capacity of Cr(VI) onto GOF/Fe_3O_4 was 258.6 mg/g. However, the Freundlich model seems to be slightly better than the Langmuir model, as suggested by the heterogeneous surfaces of the GOF/Fe_3O_4. The heterogeneous surfaces might be attributed to the heterogeneous porous structure of the GOF/Fe_3O_4. The effects of adsorption time and pH on the adsorption capacity of the GOF/Fe_3O_4 have been investigated. The adsorption processes was very fast during the initial time, and the equilibrium was established in 20 min. The effect of pH on the adsorption capacity of GOF/Fe_3O_4 for Cr(VI) have been studied in the pH range of 1–6 and the pH 2 was found to be most suitable for the removal of Cr(VI) from aqueous solutions.

The adsorption mechanism was further illustrated by implying the XPS to analyze the existing forms of Cr(VI) on GOF/Fe_3O_4 and their interactions after the adsorption process. Fig. 4.3a shows the wide-scan XPS spectrum of GOF/Fe_3O_4 after Cr(VI) adsorption. The high-resolution Cr 2p spectrum (Fig. 4.3b) can be deconvoluted into two peaks of Cr $2p_{3/2}$ (577.3 eV) and Cr $2p_{1/2}$ (586.8 eV), which corresponded to the typical XPS characteristic of Cr(III) oxidation state (Peterson et al., 1997). The chemical interactions of Cr(VI) ions with GOF/Fe_3O_4 originates the presence of Cr(III) in the XPS spectra. Fig. 4.3c shows the high resolution O 1s spectrum of GOF/Fe_3O_4 after Cr(VI) adsorption. It has been concluded that the increase in the intensity of the −COOH peak is attributed to the reduction of the adsorbed Cr(VI) ions by unprotected ferrous ions (Gollavelli et al., 2013; Cao and Zhang, 2006), with a simultaneous formation of additional carboxylic groups on the surface of GOF/Fe_3O_4, which can act as binding sites for Cr(III). In addition, the decrease in the intensity of the —OH peak indicates the loss of hydroxyl groups during adsorption, suggesting that ion exchange occurs between the hydroxyl groups and $Cr_2O_7^{2-}$/$HCrO^{4-}$.

A graphene-based magnetic nanocomposite was synthesized and used as an adsorbent for the removal of a dye from aqueous solutions (Wang et al., 2011). The as-prepared graphene-based magnetic nanocomposite adsorbents were examined using an organic dye fuchsine as the adsorbate. The effect of various optimization parameters such as solution pH, adsorbent dose, adsorption capacity, and contact time were investigated in batch mode. The Freundlich model and Langmuir model were used to study the

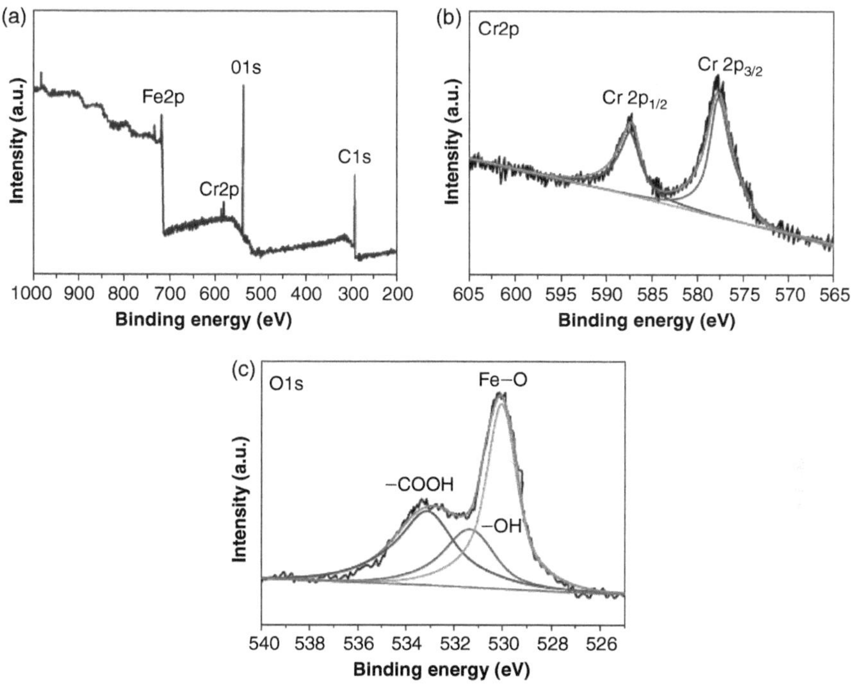

Figure 4.3 ***XPS spectra of GOF/Fe_3O_4 after absorption of Cr(IV).*** (a) Wind scan; (b) Cr 2p; and (c) O 1s. *(Source: Lei et al., 2014; © Reproduced with permission from Springer).*

adsorption isotherms, and the experimental data showed that the Langmuir model fitted with the adsorption system. The maximum adsorption capacity and Langmuir constant values for the adsorption of fuchsine by the magnetic graphene nanocomposites were 89.4 mg/g and 1.175 mg/g, respectively. The adsorption kinetic data were well fitted by a pseudo-second-order model. The graphene magnetic nanocomposites proved to be a highly efficient adsorbent and may easily be used for separation purposes.

4.2.13 Effects of Coexisting Ions, Contact Time, and Temperature

During the adsorption process several parameters such as effect of polymer matrix, nanoparticle content, pH of solution, temperature, contact time, metal ion concentration, adsorbent dose, and coexisting ions played an important role. Metal ion speciation has been well recognized for the adsorption of toxic forms of heavy metals, such as chromium, and have mostly Cr(VI) and Cr(III) oxidation states of which Cr(VI) is well known for its carcinogenic properties in biological systems. Thus, the speciation of Cr as Cr(VI) exists in the pH range of 2 to 3. So the role of pH for any medium from which we have to remove Cr(VI) should be maintained at pH 2 to 3. The adsorption behaviors of chitosan/magnetite nanocomposite beads for Pb^{2+} and Ni^{2+} ion removal were investigated at pH 4–6 and at room temperature by Tran et al. (2010). It was observed that there was an increase in metal uptake along with an increase in pH up to 6. Pb^{2+} was more strongly attracted to the adsorbent than Ni^{2+}, with higher removal percentage than Ni^{2+} at the same pH. Moreover, the sorption behavior also can be explained by the hydrolysis of metal salts. For instance, the main hydrolyzed copper species in the pH range of 3–6 appear to be Cu^{2+}, $Cu(OH)^{+}$, and $Cu(OH)_2$. Therefore, less adsorption of Cu^{2+} that takes place on some nanocomposites at lower pH can be explained by the fact that at these pH values the H^{+} concentration is high, which competes with Cu^{2+} for the same adsorption sites.

Many polymer matrix–coated nanoparticles also played important roles in the adsorption of toxic metal ions from aqueous solutions. Polymer matrices have several functional groups, such as carboxylic acids, hydroxyl acids, animo groups, thiolic compounds, phenol groups, and ester compounds, and are well reported for the adsorption of metal ions. The role of anionic carboxylic and hydroxyl acid is well reported for the removal of cations such as Cd, Cu, Co, Zn, Ni, and Pb from aqueous solutions at various operating parameters.

The contact time between the adsorbate and adsorbent is an important parameter for designed adsorption process in order to achieve an economical wastewater treatment process. The efficient studies of contact time have been widely reported for the adsorption process because it is very important to know the effect of time on the adsorption system to optimize the reactor. After a certain time the adsorption process becomes steady and equilibrium gets established, afterward a constant adsorption capacity obtained for a given adsorption is processed at optimized conditions.

Temperature is known to have a profound effect on the various chemical processes and affects the rate by altering the molecular interactions and the solubility of the adsorbate. Several studies investigated the effect of temperature on the adsorption of metal ions (Badruddoza et al., 2011, 2013). Most of the laboratory-based experiments have been carried out in the temperature range 20 to 50°C to optimize the wastewater treatment plants in batch reactors (Setshedi et al., 2013). It has been observed that the adsorption of Cr(VI) increases with an increase in temperature, indicating the endothermic nature of the process. For all the studied temperatures, the isotherms display analogous features, that is, very high uptake values in the low concentration range followed by a plateau, which showed an established equilibrium. It is possible that increasing the temperature could have increased the mobility and the driving force of the Cr(VI) ions toward the active sites of the exfoliated polypyrrole–organically modified montmorillonite clay nanocomposite, thus a higher Cr(VI) uptake was obtained at high temperatures (Setshedi et al., 2013).

Thermodynamic parameters such as change in Gibbs free energy (ΔG°) and enthalpy (ΔH°) and entropy (ΔS°) for the adsorption of metals onto the adsorbents can be calculated by using the following relationships.

$$\ln\left(\frac{q_e}{C_e}\right) = \frac{\Delta S^\circ}{R} + \frac{-\Delta H^\circ}{RT}$$

$$\Delta G = -RT\ln\left(\frac{q_e}{C_e}\right)$$

where *R* (J/mol/K) is the gas constant and *T* is the temperature in K. ΔH° and ΔS° were obtained from the slope and intercept of the linear Van't Hoff plots of $\ln(q_e/C_e)$ versus 1/T.

Industrial wastewater discharges may contain a large number of other ions, which may have competitive effect on a particular metal ion in the adsorption process. Thus, the experiment should be carried out in batch mode and continuous flow system to check the effective competition between the target metal ions and other coexisting cations and anions, where a suitable amount of adsorbent or optimized adsorbent dose of the sorption media can be contacted with a certain amount of aqueous solution containing metal ions and known quantities of commonly occurring ions (Ni^{2+}, Co^{2+}, Cu^{2+}, NO^{3-}, Cl^-) in wastewater. Nitrates, phosphates, and sulfates commonly occur in wastewater and also cause eutrophication and mostly compete with target anions such as adsorption of Cr and As from aqueous systems. Competitive effect of Ni^{2+}, Co^{2+}, Cu^{2+}, NO^{3-}, and Cl^- with Cr(VI) adsorption from aqueous solutions is widely studied on clay-based nanocomposites (Setshedi et al., 2013). It has been observed that the adsorption of Cr(VI) nanocomposites was not affected by the presence of other occurring ions, which showed that the nanocomposites were selective for Cr(VI). One possible reason for Cr(VI) selectivity in the binary system with cations may be that, at lower pH, the nanocomposite was positively charged, hence giving a repulsive force to the approaching positively charged cations, and further adsorbing Cr(VI) without competition. In case of the Cr(VI)/NO^{3-} and Cr/Cl^- binary systems, the selectivity of Cr(VI) can be explained by the behavior of NO^{3-} and Cl^- in adsorption processes. Nitrate (NO^{3-}) and chloride (Cl^-) are low affinity ligands which are well known to form outer-sphere complexes with binding surfaces. Their low affinity nature thus implies weaker adsorption mechanisms via outer-sphere complexation, and consequently, their competitive influence during Cr(VI) adsorption was negligible.

4.2.14 Polydopamine-Functionalized Graphene Hydrogel

Gao et al. (2013) presented a one-step approach to polydopamine-modified graphene hydrogel (PDA-GH), with dopamine serving as both reductant and surface functionalization agents. The preparation method was based on the spontaneous polymerization of dopamine and the self-assembly of graphene nanosheets into porous hydrogel structures. The compositional change of GO before and after reduction by dopamine or hydrothermal methods was examined by X-ray photoelectron spectroscopy (XPS). The peaks of C 1s and O 1s are observed in GO and hydrothermally synthesized graphene hydrogel (HT-GH), while a new peak of N 1s emerges in PDA-GH, which should arise from PDA on the graphene hydrogel. Deconvolution of the C 1s signal in GO sample shows the presence of C═C/C—C (~284.6 eV), C—O (hydroxyl and epoxy, ~286.5 eV), C═O (carbonyl, ~288.3 eV), and O—C═O (carboxyl, ~290.3 eV) groups. After hydrothermal reduction, the intensities of the peaks decrease significantly for the heavily oxygenated carbon species, and the peak associated with C═C/C—C (~284.6 eV) becomes dominant. Although the C1s spectrum of PDA-GH also exhibits the same functionalities of GO, the majority of oxygen-containing groups are removed and some of them may be from PDA. The O 1s spectrum of PDA-GH peaks is deconvoluted into three components of C═O at 531.1 eV, O—C═O at 532.3 eV, and C—O at 533.3 eV, which are consistent with those of C 1s. The three components of N 1s peak correspond to amine groups (R—NH2, 401.7 eV), substituted amines (R-NH-R or indole groups, 399.8 eV), and imino groups (=N-R, 398.5 eV). Benefiting from the abundant functional groups of polydopamine and the high specific surface areas of graphene hydrogel with three-dimensional interconnected pores, the prepared material exhibited high adsorption capacities toward a wide spectrum of contaminants, including heavy metals, synthetic dyes, and aromatic pollutants. Importantly, the free-standing graphene hydrogel was easily removed from water after adsorption process, and was regenerated by altering the pH values of the solution for adsorbed heavy metals or using low-cost alcohols for synthetic dyes and aromatic molecules. The maximum adsorption capacities of Pb(II) and Cd(II) onto PDA-GH (1:2, w:w) were 336.32 and 145.48 mg/g (ie, 1.62 and 1.29 mmol/g), showing a significant improvement relative to that of HT-GH, PDA, and PDA-GH (0.5:2, w:w) with lower degree of functionalization. When the mass ratio of PDA:GO increased to 2:2, the strong capping and stabilizing effect of polydopamine led to the formation of a black rGO dispersion rather than hydrogels. The relatively higher adsorption capacity of Pb(II) should possibly arise from the higher binding affinity of phenolic and carboxylic groups to Pb (II) compared with Cd(II).

4.2.15 Thiol-Modified Graphene Oxide

Retaining the inherent hydrophilic character of graphite oxide nanosheets, sp^2 domains on graphite oxide are covalently modified with thiol groups by diazonium chemistry (Zhao et al., 2011a). The surface-modified graphite oxide adsorbed sixfold higher concentration of aqueous mercuric ions than the unmodified graphite oxide. "Core-shell" adsorbent granules, readily useable in filtration columns, were synthesized by assembling aqueous graphite oxide over sand granules. The nanostructured graphite oxide-coated sand retained at least fivefold higher concentration of heavy metal and organic dye than pure sand. It has been suggested that the research results could open avenues for developing low-cost water purification materials for the developing economies.

4.3 GRAPHENE OXIDE FOR REMOVAL OF PHENOL AND NAPHTHOL

Persistent aromatic pollutants are widely found in the industrial effluents of the electrical, pharmaceutical, petrochemical, dyestuff, pesticide, and manufacturing industries. Since their high solubility in water, they transport into the environment widely and do harm to humans and animals. Various technologies have been developed for the efficient removal of organic pollutants from aqueous solutions such as photocatalysis, thermal degradation, electrolysis, adsorption, and bioremediation. Among these techniques adsorption has been widely applied for the removal of persistent organic pollutants from aqueous medium due to its simplicity and cost effectiveness. However, the adsorption capacities of the available adsorbents are not enough and still suffer for selectivity, and limited options of adsorbents. Hence it becomes important to develop novel adsorbents with high adsorption capacity and selectivity for persistent organic pollutants from its aqueous solutions.

Zhang et al. (2013a) reported the removal of tetrabromobisphenol A (TBBPA) from its aqueous solution by GO in batch mode. The effects of various optimization parameters such as contact time, pH, and temperature, and the presence of coexisting anions as well as humic acid were studied. The adsorption capacity decreased with an increase in temperature as well as solution pH. The presence of coexisting anions such as NO_3^-, SO_4^{2-}, HPO_4^{2-}, and HCO_3^{2-} reduced TBBPA adsorption in the order of $NO_3^- < SO_4^{2-} < HPO_4^{2-} < HCO_3^{2-}$. The presence of humic acid also appreciably decreased the adsorption of TBBPA onto GO. The adsorption kinetics followed well to the pseudo-second-order model, and the adsorption isotherm was well described by the Langmuir model with a maximum adsorption capacity of 115.77 mg/g at 298 K. Thermodynamic studies indicated that the adsorption of TBBPA onto GO was feasible, spontaneous, and exothermic in nature at all temperatures. Both π–π stacking interactions and hydrogen bonding were proposed to be responsible for the adsorption of TBBPA by GO.

Sun et al. (2012) discovered the chemically RGO, with an ID/IG >1.4, which was able to effectively activate peroxymonosulfate (PMS) to produce active sulfate radicals ($SO_4^{\bullet-}$) for decomposition of various aqueous contaminants. Graphene demonstrated a higher activity than several carbon allotropes, such as activated carbon, graphite powder, GO, and multiwall carbon nanotubes. Kinetic study of graphene-catalyzed activation of PMS was carried out on various environmental pollutants. It was shown that graphene catalysis was superior to that on transition metal oxide in degradation of phenol, 2,4-dichlorophenol, and methylene blue dye in water.

GO and RGO were both decorated with iron oxide nanoparticles and was applied for the adsorption of Pb(II), 1-naphthol, and 1-naphthylamine, as representatives of inorganic and organic pollutants, on GO-iron oxides and RGO-iron oxides (Yang et al., 2012). The results showed that the GO-iron oxide material was a good adsorbent for Pb(II) but not for 1-naphthol and 1-naphthylamine due to oxygen-containing groups on the surface, whereas the RGO–iron oxide material was a good adsorbent for 1-naphthol and 1-naphthylamine but not for Pb(II). The adsorption of 1-naphthol and 1-naphthylamine on RGO-iron oxides was an endothermic and spontaneous process. Both materials were easily separated by magnetic separation.

Zhao et al. (2011b) introduced a kind of sulfonated graphene with high dispersion properties in aqueous solution capable of absorbing naphthalene and 1-naphthol aromatic pollutants from aqueous solutions. The adsorption capability of the prepared sulfonated graphene nanomaterials approaches ~2.3–2.4 mmol/g for naphthalene and 1-naphthol. Graphene oxides were prepared from graphite by using a modified Hummers method, and then the prereduction of GO was performed with sodium borohydride at 80°C for 1 h to remove the majority of the oxygen functionality. The sulfonation of GO was then carried out with the aryl diazonium salt of sulfanilic acid in an ice bath for 2 h, followed by postreduction with hydrazine to remove any remaining oxygen functionality at 100°C for 24 h.

X-ray photoelectron spectroscopy (XPS) was used to analyze the samples of GO and sulfonated graphene sheets. The C1s XPS spectrum of GO indicated a considerable degree of oxidation with different functional groups: the nonoxygenated ring C (284.4 eV), the C atom in the C—O bond (286.2 eV), the carbonyl C (287.8 eV), and the carboxylate carbon (O—C=O) (289.0 eV). While in the C1s XPS spectrum of the sulfonated graphene, the intensities for the peaks corresponded to the epoxy/ether group (286.2 eV), the carbonyl C (287.8 eV), and the carboxylate carbon (O—C=O) (289.0 eV) exhibited much smaller relative contents, indicating that most of the oxygen-containing functional groups were successfully removed by the reduction process. The effect of the contents of sulfonated graphene sheets on the adsorption of naphthalene and 1-naphthol suggested that the adsorption percentages of naphthalene and 1-naphthol enhanced with the increase of sulfonated graphene content in the suspension. When the content of sulfonated graphene sheets was higher than 0.15 g/L, more than 95% of 1-naphthol and naphthalene were adsorbed on the sulfonated graphene. The adsorption isotherms of naphthalene and 1-naphthol on sulfonated graphene were applied to the Langmuir model to describe the adsorption characteristics of the sulfonated graphene sheets. The maximum adsorption capacities of sulfonated graphene were 2.326 mmol/g for naphthalene and 2.407 mmol/g for 1-naphthol, which was attributed to the strong π–π interaction in the form of face-to-face stacking. The effect of pH on the adsorption of naphthalene and 1-naphthol to sulfonated graphene was also investigated. It was found that at low pH, the adsorption of both naphthalene and 1-naphthol on sulfonated graphene sheets decreased with the increase of pH. At high pH, adsorption of naphthalene did not change accordingly with the increase of pH while that of 1-naphthol increased again at pH > 7.

4.4 GRAPHENE OXIDE FOR REMOVAL OF ALGAL TOXINS

Recently, Pavagadhi et al. (2013) have shown that GO can also be utilized as an adsorbent for algal toxins—microcystin-LR (MC-LR) and microcystin-RR (MC-RR). The adsorption performance of GO was compared to that of commercially available activated carbon. Graphene oxide showed a very high adsorption capacity of 1700 μg/g for microcystin-LR and 1878 μg/g for microcystin-RR, whereas the maximum adsorption capacity obtained with commercial-activated carbon was 1481.7 and 1034.1 μg/g for microcystin-LR and microcystin-RR, respectively. The adsorption kinetic experiments revealed that more than 90% removal of both microcystin-LR and microcystin-RR was achieved within 5 min for all the doses studied from aqueous media. Graphene oxide could also be reused as an adsorbent following 10 cycles of adsorption/desorption with no significant loss in its adsorption capacity.

4.5 GRAPHENE FOR REMOVAL OF PERSISTENT ORGANIC POLLUTANTS

A magnetic microsphere-confined graphene adsorbent (Fe_3O_4@SiO_2-G) was fabricated and used for the extraction of five polycyclic aromatic hydrocarbons (fluorene, anthracene, phenanthrene, fluoranthene, and pyrene) from aqueous solutions prior to high performance liquid chromatography with fluorescence detection (Wang et al., 2013). Various optimization parameters such as the amount of Fe_3O_4@SiO_2-G, the solution pH, ionic strength, the extraction time, and desorption conditions were investigated. Due to the high surface area and excellent adsorption capacity of the Fe_3O_4@SiO_2-G, satisfactory extraction can be achieved with only 15 mg of the adsorbent within 5 min of extraction time from a 250 mL solution. Under the optimum conditions, a linear response was observed in the concentration range of 5–1500 ng/L for fluorene; 2.5–1500 ng/L for anthracene; and 15–1500 ng/L for phenanthrene, fluoranthene, and pyrene, with high correlation coefficients ranging from 0.9897 to 0.9961. The results indicated that graphene-based magnetic nanocomposites had a great adsorptive ability toward the five polycyclic aromatic hydrocarbons from aqueous solutions.

An electrostatic self-assembly method was employed to prepare Fe_3O_4/GO nanocomposites, and their application in the magnetic solid-phase extraction of polycyclic aromatic hydrocarbons from aqueous solution was investigated (Han et al., 2012). Five polycyclic aromatic hydrocarbons were selected as model analytes to validate the extraction performance of the Fe_3O_4/GO nanocomposites. The excellent adsorption capacity was attributed to the dominant roles of π–π stacking interaction and hydrophobic interaction between hydrocarbons and magnetic nanocomposites. After optimizing the various experimental conditions, the results indicated that the recoveries of these compounds were in the range of 76.8–103.2%, with relative standard deviations ranging between 1.7 and 11.7%; the limits of detection were in the range of 0.09–0.19 ng/mL. This result showed that the Fe_3O_4/GO nanocomposites can be utilized for the extraction of polycyclic aromatic hydrocarbons from aqueous solution.

A new kind of functional GO with fine stability in water was fabricated by mixing GO and brilliant blue (BB) with a certain weight ratio (Zhang et al., 2013b). The adsorption performance of this mixture of BB and GO (BBGO) to polycyclic aromatic hydrocarbons anthracenemethanol and fluoranthene was investigated. The results indicated that the BBGO possessed an adsorption capacity of 1.676 and 2.212 mmol/g for anthracenemethanol and fluoranthene, respectively. The removal efficiency was 72.7 and 93.2% for anthracenemethanol and fluoranthene, respectively. When the adsorption equilibrium was achieved, the pH and temperature-sensitive coagulation method was used to remove the anthracenemethanol/BBGO or fluoranthene/BBGO complex and proved to be an effective approach to flocculate the anthracenemethanol/BBGO or fluoranthene/BBGO complex into large flocs, which tended to be removed from the aqueous solution.

4.6 CONCLUSIONS

Graphene based nanomaterials offer a tremendous potential for the removal and recovery of environmental pollutants due to their large surface area to volume ratio, high physicochemical stability, enhanced functionality, biocompatibility, and strong adsorption capability. Graphene-based nanocomposites have been widely used for the separation of heavy metals, dyes, and aromatic organic pollutants from aqueous media. Surface functionalized graphene by thiol, amine, EDTA, and various polymers have established their potential in the extraction of environmental hazardous materials from water. The functionalized graphene enhanced the adsorption capacity of nanocomposites for heavy metals, dyes, and aromatic pollutants due to increased sorption sites on the nanocomposite surface. Graphene-based inorganic transition materials have been used for the photocatalytic degradation of environmental pollutants from water. Graphene-doped CdS, ZnO, MnO_2, and TiO_2 have been popularly used for elimination of hazardous dyes from aqueous solutions. Graphene-based magnetic nanocomposites have been widely used for the adsorption of metals and dyes from water and wastewater. Various operating parameters such as solution pH, contact time, pollutant concentration, ionic strength, and adsorbent dose played a major role in the adsorption and separation of pollutants from aquatic medium. The separations of single pollutants from aquatic media have been well established in batch mode under various optimization conditions. However, there is urgent need to perform experiments in column mode to use their application in real wastewater systems. There is also need to perform the toxicity tests before their application in real wastewater systems.

REFERENCES

Alvand, M., Shemirani, F., 2014. Preconcentration of trace cadmium ion using magnetic graphene nanoparticles as an efficient adsorbent. Microchim Acta 181, 181–188.

Ay, C.O., Ozcan, A.S., Erdogan, Y., Ozcan, A., 2012. Characterization of *Punica granatum* L. peels and quantitatively determination of its biosorption behavior towards lead (II) ions and acid blue 40, Colloids Surf. B-Biointerfaces 100, 197–204.

Badruddoza, A.Z.M., Tay, A.S.H., Tan, P.Y., Hidajat, K., Uddin, M.S., 2011. Carboxymethyl-β-cyclodextrin conjugated magnetic nanoparticles as nano-adsorbents for removal of copper ions: synthesis and adsorption studies. J. Hazard. Mater. 185, 1177–1186.

Badruddoza, A.Z.M., Shawon, Z.B.Z., Tay, W.J.D., Hidajat, K., Uddin, M.S., 2013. Fe_3O_4/cyclodextrin polymer nanocomposites for selective heavy metals removal from industrial wastewater. Carbohydr. Polym. 91, 322–332.

Banerjee, S., Nigam, V., Gautam, R.K., Haider, S., Chattopadhyaya, M.C., 2013. Removal of an anionic dye, Orange G from aqueous solutions by adsorption on unmodified sheesham (Dalbergia sisso) saw dust. J. Indian Chem. Soc. 90 (5), 555–563.

Banerjee, S., Chattopadhyaya, M.C., Uma, Sharma, Y.C., 2014a. Adsorption characteristics of modified wheat husk for the removal of a toxic dye, methylene blue, from aqueous solutions. J. Hazard. Toxic Radioact. Waste 18 (1), 56–63.

Banerjee, S., Chattopadhyaya, M.C., Srivastava, V., Sharma, Y.C., 2014b. Adsorption studies of methylene blue onto activated saw dust: kinetics, equilibrium, and thermodynamic studies. Environ Prog. Sustain. Energy 33 (3), 790–799.

Behbahani, M., Tapeh, N.A.G., Mahyari, M., Pourali, A.R., Amin, B.G., Shaabani, A., 2014. Monitoring of trace amounts of heavy metals in different food and water samples by flame atomic absorption spectrophotometer after preconcentration by amine-functionalized graphene nanosheet. Environ. Monit. Assess. 186, 7245–7257.

Cao, J.S., Zhang, W.X., 2006. Stabilization of chromium ore processing residue (COPR) with nanoscale iron particles. J. Hazard. Mater. 132, 213–219.

Chandra, V., Kim, K.S., 2011. Highly selective adsorption of Hg^{2+} by a polypyrrole–reduced graphene oxide composite. Chem. Commun. 47, 3942–3944.

Chandra, V., Park, J., Chun, Y., Lee, J.W., Hwang, I.-C., Kim, K.S., 2010. Water-dispersible magnetite-reduced graphene oxide composites for arsenic removal. ACS Nano 4, 3979–3986.

Deng, X., Lu, L., Li, H., Luo, F., 2010. The adsorption properties of Pb(II) and Cd(II) on functionalized graphene prepared by electrolysis method. J. Hazard. Mater. 183, 923–930.

Deng, J.H., Zhang, X.R., Zeng, G.M., Gong, J.L., Niu, Q.Y., Liang, J., 2013. Simultaneous removal of Cd(II) and ionic dyes from aqueous solution using magnetic graphene oxide nanocomposite as an adsorbent. Chem. Eng. J. 226, 189–200.

Dupont, D., Brullot, W., Bloemen, M., Verbiest, K., Binnemans, 2014. Selective uptake of rare earths from aqueous solutions by EDTA functionalized magnetic and nonmagnetic nanoparticles. ACS Appl. Mater. Interfaces 6, 4980–4988.

Erdem, E., Karapinar, N., Donat, R., 2004. The removal of heavy metal cations by natural zeolites. J. Colloid Interface Sci. 280, 309–314.

Fan, L., Luo, C., Sun, M., Li, X., Qiu, H., 2013. Highly selective adsorption of lead ions by water-dispersible magnetic chitosan/graphene oxide composites. Colloids Surf. B 103, 523–529.

Fang, R., Cheng, X., Xu, X., 2010. Synthesis of lignin-base cationic flocculant and its application in removing anionic azo-dyes from simulated wastewater. Bioresour. Technol. 101, 7323–7329.

Gao, W., Majumder, M., Alemany, L.B., Narayanan, T.N., Ibarra, M.A., Pradhan, B.K., Ajayan, P.M., 2011. Engineered graphite oxide materials for application in water purification. ACS Appl. Mater. Interfaces 3, 1821–1826.

Gao, C., Yu, X.Y., Xu, R.X., Liu, J.H., Huang, X.J., 2012. AlOOH-reduced graphene oxide nanocomposites: one-pot hydrothermal synthesis and their enhanced electrochemical activity for heavy metal ions. ACS Appl. Mater. Interfaces 4, 4672–4682.

Gao, H., Sun, Y., Zhou, J., Xu, R., Duan, H., 2013. Mussel-inspired synthesis of polydopamine-functionalized graphene hydrogel as reusable adsorbents for water purification. ACS Appl. Mater. Interfaces 5, 425–432.

Gautam, R.K., Chattopadhyaya, M.C., 2013. Biosorption of heavy metals: recent trends and challenges. Sharma, S.K., Sanghi, R. (Eds.), Wastewater Reuse and Management, 2013, London, Springer, pp. 305–322.

Gautam, R.K., Mudhoo, A., Chattopadhyaya, M.C., 2013. Kinetic, equilibrium, thermodynamic studies and spectroscopic analysis of Alizarin Red S removal by mustard husk. J. Environ. Chem. Eng. 1 (4), 1283–1291.

Gautam, R.K., Mudhoo, A., Lofrano, G., Chattopadhyaya, M.C., 2014. Biomass-derived biosorbents for metal ions sequestration: adsorbent modification and activation methods and adsorbent regeneration. J. Environ. Chem. Eng. 2 (1), 239–259.

Gollavelli, G., Chang, C.C., Ling, Y.C., 2013. Facile synthesis of smart magnetic graphene for safe drinking water: heavy metal removal and disinfection control. ACS Sustain Chem. Eng. 1, 462–472.

Gupta, V.K., Ali, I., 2000. Utilization of bagasse fly ash (a sugar industry waste) for the removal of copper and zinc from wastewater. Sep. Purif. Technol. 18, 131–140.

Gupta, V.K., Mohan, D., Sharma, S., 1998. Removal of lead from wastewater using bagasse fly ash-a sugar industry waste material. Sep. Sci. Technol. 33, 1331–1343.

Gupta, V.K., Jain, C.K., Ali, I., Sharma, M., Saini, V.K., 2003. Removal of cadmium and nickel from wastewater using bagasse fly ash—a sugar industry waste. Water Res. 37, 4038–4044.

Gupta, N., Kushwaha, A.K., Chattopadhyaya, M.C., 2011. Adsorption of cobalt(II) from aqueous solution onto hydroxyapatite/zeolite composite. Adv. Mat. Lett. 2 (4), 309–312.

Gupta, N., Kushwaha, A.K., Chattopadhyaya, M.C., 2012. Adsorption studies of cationic dyes onto Ashoka (*Saraca asoca*) leaf powder. J. Taiwan Inst. Chem. E. 43 (4), 604–613.

Han, Q., Wang, Z., Xia, J., Chen, S., Zhang, X., Ding. F M., 2012. Facile and tunable fabrication of Fe_3O_4/graphene oxide nanocomposites and their application in the magnetic solid-phase extraction of polycyclic aromatic hydrocarbons from environmental water samples. Talanta 101, 388–395.

Hosamani, K.M., Keri, R.S., Nataraj, S.K., Aminabhavi, T.M., Harisha, R.S., 2010. Arsenic removal from drinking water using thin film composite nanofiltration membrane. Desalination 252, 75–80.

Hu, X.J., Liu, Y.G., Wang, H., Chen, A.W., Zeng, G.M., Liu, S.M., Guo, Y.M., Hu, X., Li, T.T., Wang, Y., Zhou, L., Liu, S.H., 2013. Removal of Cu(II) ions from aqueous solution using sulfonated magnetic graphene oxide composite. Sep. Purif. Technol. 108, 189–195.

Joo, D.J., Shin, W.S., Choi, J.H., Choi, S.J., Kim, M.C., Han, M.H., Ha, T.W., Kim, Y.H., 2007. Decolorization of reactive dyes using inorganic coagulants and synthetic polymer. Dyes Pigments 73, 59–64.

Koo, H.Y., Lee, H.J., Go, H.A., Lee, Y.B., Bae, T.S., Kim, J.K., Choi.F W.S., 2011. Graphene-based multifunctional iron oxide nanosheets with tunable properties. Chem. Eur. J. 17, 1214–1219.

Kushwaha, A.K., Chattopadhyaya, M.C., 2014. Surface modification of silica gel for adsorptive removal of Ni^{2+} and Cd^{2+} from water. Desalin. Water Treat. 54, 1642–1650.

Kushwaha, A.K., Gupta, N., Chattopadhyaya, M.C., 2013. Enhanced adsorption of methylene blue on modified silica gel: equilibrium, kinetic, and thermodynamic studies. Desalin. Water Treat. 52, 4527–4537.

Lee, Y.C., Yang, J.W., 2012. Self-assembled flower-like TiO_2 on exfoliated graphite oxide for heavy metal removal. J. Ind. Eng. Chem. 18, 1178–1185.

Lei, Y., Chen, F., Luo, Y., Zhang, L., 2014. Three-dimensional magnetic graphene oxide foam/Fe_3O_4 nanocomposite as an efficient absorbent for Cr(VI) removal. J. Mater. Sci. 49, 4236–4245.

Li, Y.H., Ding, J., Luan, Z., Di, Z., Zhu, Y., Xu, C., Wu, D., Wei, B., 2003. Competitive adsorption of Pb, Cu and Cd ions from aqueous solutions by multiwalled carbon nanotubes. Carbon 41, 2787–2792.

Li, H., Zou, L., Pan, L., Sun, Z., 2010. Using graphene nano-flakes as electrodes to remove ferric ions by capacitive deionization. Sep. Purif. Technol. 75, 8–14.

Li, J., Zhang, S., Chen, C., Zhao, G., Yang, X., Li, J., Wang, X., 2012. Removal of Cu(II) and fulvic acid by graphene oxide nanosheets decorated with Fe_3O_4 nanoparticles. ACS Appl. Mater. Interfaces 4, 4991–5000.

Li, L., Fan, L., Sun, M., Qiu, H., Li, X., Duan, H., Luo, C., 2013. Adsorbent for chromium removal based on graphene oxide functionalized with magnetic cyclodextrin–chitosan. Colloids Surf. B 107, 76–83.

Madadrang, C.J., Kim, H.Y., Gao, G., Wang, N., Zhu, J., Feng, H., Gorring, M., Kasner, M.L., Hou, S., 2012. Adsorption behavior of EDTA-graphene oxide for Pb (II) removal. ACS Appl. Mater. Interfaces 4, 1186–1193.

Mi, X., Huang, G., Xie, W., Wang, W., Liu, Y., Gao, J., 2012. Preparation of graphene oxide aerogel and its adsorption for Cu^{2+} ions. Carbon 50, 4856–4864.

Mishra, A.K., Ramaprabhu, S., 2011. Functionalized graphene sheets for arsenic removal and desalination of sea water. Desalination 282, 39–45.

Mohan, D., Singh, K.P., Singh, G., Kumar, K., 2002. Removal of dyes from wastewater using flyash, a low-cost adsorbent. Ind. Eng. Chem. Res. 41, 3688–3695.

Narin, I., Soylak, M., 2003. The uses of 1-(2-pyridylazo)-2-naphthol (PAN) impregnated Ambersorb 563 resin on the solid phase extraction of traces heavy metal ions and their determinations by atomic absorption spectrometry. Talanta 60, 215–221.

Nguyen, C.M., Bang, S., Cho, J., Kim, K.W., 2009. Performance and mechanism of arsenic removal from water by a nanofiltration membrane. Desalination 245, 82–94.

Oller, I., Malato, S., Sánchez-Pérez, J.A., 2011. Combination of advanced oxidation processes and biological treatments for wastewater decontamination-A review. Sci. Total Environ. 409, 4141–4166.

Pavagadhi, S., Tang, A.L.L., Sathishkumar, M., Loh, K.P., Balasubramanian, R., 2013. Removal of microcystin-LR and microcystin-RR by graphene oxide: adsorption and kinetic experiments. Water Res. 47, 4621–4629.

Peterson, M.L., Brown, G.E., Parks, G.A., Stein, C.L., 1997. Differential redox and sorption of Cr(III/VI) on natural silicate and oxide minerals: eXAFS and XANES results. Geochim. Cosmochim. Ac. 61, 3399–3412.

Rao, G.P., Lu, C., Su, F., 2007. Sorption of divalent metal ions from aqueous solution by carbon nanotubes: a review. Sep. Purif. Technol. 58, 224–231.
Rao, R.A.K., Khan, M.A., Rehman, F., 2010. Utilization of Fennel biomass (*Foeniculum vulgari*) a medicinal herb for the biosorption of Cd(II) from aqueous phase. Chem. Eng. J. 156, 106–113.
Rawat, V., Rai, P., Gautam, R.K., Chattopadhyaya, M.C., 2013. Kinetic and equilibrium isotherm studies for the adsorptive removal of Brilliant Green Dye from aqueous solution by Oplismenus frumentaceus husk. J. Indian Chem. Soc. 90 (5), 577–583.
Ren, Y., Yan, N., Wen, Q., Fan, Z., Wei, T., Zhang, M., Ma, J., 2011. Graphene/δ-MnO_2 composite as adsorbent for the removal of nickel ions from wastewater. Chem. Eng. J. 175, 1–7.
Repo, E., Warchol, J.K., Bhatnagar, A., Sillanpaa, M., 2011. Heavy metals adsorption by novel EDTA-modified chitosan–silica hybrid materials. J. Colloid Interface Sci. 358, 261–267.
Sayar, O., Mehrani, K., Hoseinzadeh, F., Mehrani, A., Sadeghi, O., 2014. Comparison of the performance of different modified graphene oxide nanosheets for the extraction of Pb(II) and Cd(II) from natural samples. Microchim Acta 181, 313–320.
Setshedi, K.Z., Bhaumik, M., Songwane, S., Onyango, M.S., Maity, A., 2013. Exfoliated polypyrrole-organically modified montmorillonite clay nanocomposite as a potential adsorbent for Cr(VI) removal. Chem. Eng. J. 222, 186–197.
Sprynskyy, M., Buszewski, B., Terzyk, A.P., Namiesnik, J., 2006. Study of the selection mechanism of heavy metal (Pb^{2+}, Cu^{2+}, Ni^{2+}, and Cd^{2+}) adsorption on clinoptilolite. J. Colloid Interface Sci. 304, 21–28.
Sreeprasad, T.S., Shihabudheen, M., Maliyekkal, Lisha, K.P., Pradeep, T., 2011. Reduced graphene oxide–metal/metal oxide composites: facile synthesis and application in water purification. J. Hazard. Mater. 186, 921–931.
Stafiej, A., Pyrzynska, K., 2007. Adsorption of heavy metal ions with carbon nanotubes. Sep. Purif. Technol. 58, 49–52.
Sun, H., Liu, S., Zhou, G., Ang, H.M., Tade, M.O., Wang, S., 2012. Reduced graphene oxide for catalytic oxidation of aqueous organic pollutants. ACS Appl. Mater. Interfaces 4, 5466–5471.
Thomas, S.Y., Choong, T.G., Chuah, Y., Robiah, F.L., Gregory Koay, I., 2007. Azni Arsenic toxicity, health hazards and removal techniques from water: an overview. Desalination 217, 139–166.
Tran, H.V., Tran, L.D., Nguyen, T.N., 2010. Preparation of chitosan/magnetite composite beads and their application for removal of Pb(II) and Ni(II) from aqueous solution. Mater. Sci. Eng. C 30, 304–310.
Wang, S., Boyjoo, Y., Choueib, A., Zhu, J., 2005. Removal of dyes from aqueous solution using fly ash and red mud. Water Res. 39, 129–138.
Wang, S., Soudi, M., Li, L., Zhu, Z.H., 2006. Coal ash conversion into effective adsorbents for removal of heavy metals and dyes from wastewater. J. Hazard. Mater. B133, 243–251.
Wang, C., Feng, C., Gao, Y., Ma, X., Wu, Q., Wang, Z., 2011. Preparation of a graphene-based magnetic nanocomposite for the removal of an organic dye from aqueous solution. Chem. Eng. J. 173, 92–97.
Wang, H., Yuan, X., Wu, Y., Huang, H., Zeng, G., Liu, Y., Wang, X., Lin, N., Qi.F Y., 2013a. Adsorption characteristics and behaviors of graphene oxide for Zn(II) removal from aqueous solution. Appl. Surf. Sci. 279, 432–440.
Wang, W., Ma, R., Wu, Q., Wang, C., Wang, Z., 2013b. Magnetic microsphere-confined graphene for the extraction of polycyclic aromatic hydrocarbons from environmental water samples coupled with high performance liquid chromatography–fluorescence analysis. J. Chromatogr. A 1293, 20–27.
Weng, C.H., Tsai, C.Z., Chu, S.H., Sharma, Y.C., 2007. Adsorption characteristics of copper(II) onto spent activated clay. Sep. Purif. Technol. 54, 187–197.
Wu, C.H., 2007. Adsorption of reactive dye onto carbon nanotubes: equilibrium, kinetics and thermodynamics. J. Hazard. Mater. 144, 93–100.

Yang, S.T., Chang, Y., Wang, H., Liu, G., Chen, S., Wang, Y., Liu, Y., Cao, A., 2010a. Folding/aggregation of graphene oxide and its application in Cu^{2+} removal. J. Colloid Interface Sci. 351, 122–127.
Yang, L.Q., Li, Y.F., Wang, L.Y., Zhang, Y., Ma, X.J., Ye, Z.F., 2010b. Preparation and adsorption performance of a novel bipolar PS-EDTA resin in aqueous phase. J. Hazard. Mater. 180, 98–105.
Yang, X., Chen, C., Li, J., Zhao, G., Ren, X., Wang, X., 2012. Graphene oxide-iron oxide and reduced graphene oxide-iron oxide hybrid materials for the removal of organic and inorganic pollutants. RSC Advances 2, 8821–8826.
Yuan, X., Wang, Y., Wang, J., Zhou, C., Tang, Q., Rao, X., 2013. Calcined graphene/MgAl-layered double hydroxides for enhanced Cr(VI) removal. Chem. Eng. J. 221, 204–213.
Yun, M.A., Yeon, K.M., Park, J.S., Lee, C.H., Chun, J., Lim, D.J., 2006. Characterization of biofilm structure and its effect on membrane permeability in MBR for dye wastewater treatment. Water Res. 40, 45–52.
Zhang, K., Dwivedi, V., Chi, C., Wu. F J., 2010a. Graphene oxide/ferric hydroxide composites for efficient arsenate removal from drinking water. J. Hazard. Mater. 182, 162–168.
Zhang, S., Li, X.-Y., Chen, J.P., 2010b. Preparation and evaluation of a magnetite-doped activated carbon fiber for enhanced arsenic removal. Carbon 48, 60–67.
Zhang, Y., Li, Y.F., Yang, L.Q., Ma, X.J., Wang, L.Y., Ye, Z.F., 2010c. Characterization and adsorption mechanism of Zn^{2+} removal by PVA/EDTA resin in polluted water. J. Hazard. Mater. 178, 1046–1054.
Zhang, N., Qiu, H., Si, Y., Wang, W., Gao, J., 2011. Fabrication of highly porous biodegradable monoliths strengthened by graphene oxide and their adsorption of metal ions. Carbon 49, 827–837.
Zhang, Y., Tang, Y., Li, S., Yu, S., 2013a. Sorption and removal of tetrabromobisphenol A from solution by graphene oxide. Chem. Eng. J. 222, 94–100.
Zhang, C., Wu, L., Cai, D., Zhang, C., Wang, N., Zhang, J., Wu. F Z., 2013b. Adsorption of polycyclic aromatic hydrocarbons (fluoranthene and anthracenemethanol) by functional graphene oxide and removal by pH and temperature-sensitive coagulation. ACS Appl. Mater. Interfaces 5, 4783–4790.
Zhang, Y., Chi, H., Zhang, W., Sun, Y., Liang, Q., Gu, Y., Jing, R., 2014. Highly efficient adsorption of copper ions by a PVP-reduced graphene oxide based on a new adsorptions mechanism. Nano-Micro Lett. 6 (1), 80–87.
Zhao, G., Li, J., Ren, X., Chen, C., Wang. F X., 2011a. Few-layered graphene oxide nanosheets as superior sorbents for heavy metal ion pollution management. Environ. Sci. Technol. 45, 10454–10462.
Zhao, G., Jiang, L., He, Y., Li, J., Dong, H., Wang, X., Hu, W., 2011b. Sulfonated graphene for persistent aromatic pollutant management. Adv. Mater. 23, 3959–3963.
Zong, P., Wang, S., Zhao, Y., Wang, H., Pan, H., He, C., 2013. Synthesis and application of magnetic graphene/iron oxides composite for the removal of U(VI) from aqueous solutions. Chem. Eng. J. 220, 45–52.

CHAPTER 5

Kinetics and Equilibrium Isotherm Modeling: Graphene-Based Nanomaterials for the Removal of Heavy Metals From Water

5.1 INTRODUCTION

Continuous discharge of pollutants in the environment calls for the development and compilation of a variety of information on one platform to understand their behavioral characteristics so that the appropriate techniques can be adopted to prevent, minimize, or remove their toxic effect on biological systems and environmental biota. When toxic heavy metals and dyes became major concerns, it also became necessary to understand their role in the environment through their speciation and degradation into different matrices of environmental systems. When chemical contaminants enter into the environment, they are usually not confined to a specific location, but they are dynamic in nature and tend to move in media or in adjacent media. The tendency of pollutants to move and distribute themselves between a medium is determined by its physicochemical properties, environmental variables, and biological factors. The quantity of a given heavy metal in a given environmental matrix and the state of metal speciation existence directly governs the toxicity and their environmental impacts. Hence, it becomes important to understand the factors that drive a contaminant from one medium to another and the manner and extent that a pollutant associates with the different environmental media.

Thus, the study of kinetics, adsorption equilibrium, and isotherm modeling is essential in supplying the basic information required for the design and operation of adsorption processes for wastewater treatment plants. The area of chemical science that is concerned with the speeds or rates of reactions is called chemical kinetics. The study of chemical kinetics of pollutants' adsorption is a subject of broad importance in environmental analysis. It relates, for example, to how quickly an adsorbate is adsorbed on an adsorbent's surface. Chemical kinetics directly governs the chemical reaction in

Nanomaterials for Wastewater Remediation
http://dx.doi.org/10.1016/B978-0-12-804609-8.00005-4

the adsorption–desorption processes. Much of our study in the adsorption system is concerned with the adsorption and separation of pollutants from aqueous media. However, it is equally important to understand how rapidly these chemical reactions occur. The rates of chemical reaction periods have a wide range, from those that are complete within fractions of a second, for example, adsorption of certain molecules onto super-adsorbents by pore diffusion. Until now, several kinetic models have been developed and applied to understanding the mechanism of adsorption of pollutants from aquatic media. However, the most classic adsorption kinetic models, such as pseudo-first-order, pseudo-second-order, and diffusion models, have been widely used and accepted for adsorption systems in aqueous solutions.

Adsorption processes of pollutants on adsorbents are usually studied through graphs known as adsorption isotherms. Adsorption isotherms describe the equilibrium of the sorption of a material on the surface of adsorbent at constant temperature. Adsorption isotherms are essential for the description of how pollutants' concentration will interact with adsorbent surfaces and are useful to optimize the use of adsorbents for the removal of pollutants from aqueous solutions (Emmanuel and Rao, 2008). The equilibrium adsorption isotherms are some of the most important data for understanding the mechanism of the sorption. Several isotherm equations are available and few important isotherms, which have applications in adsorption studies, are namely the Langmuir, Freundlich, Temkin, Redlich–Peterson, and Dubinin–Radushkevich (D–R) isotherms.

Our goal in this chapter is not only to understand how to determine the rates at which reactions occur, but also to consider the factors that control these rates, for example, effect of time, effect of dose, effect of pollutants' concentration, and effects of other chemical constituents in solution media.

5.2 KINETIC STUDIES AND MODELS

Adsorption equilibrium studies are important to conclude the efficacy of an adsorbent in any adsorption system. In spite of this, it is also necessary to identify the adsorption mechanism type in a given system. Kinetic models have been exploited to evaluate the mechanisms of adsorption of pollutants and their potential rate-controlling steps that include mass transport and chemical reaction processes (Aksu, 2001). In addition, information on the kinetics of pollutant uptake is required to select the optimum condition for full-scale batch removal processes. Predicting the rate of adsorption for a given system is among the most important factors in adsorption system design, as the system's kinetics determines adsorbate residence time and the reactor dimensions (Febrianto et al., 2009). As previously noted, various factors govern adsorption capacity, that is, initial pollutant concentration, temperature, pH of solution, adsorbent particle size, pollutant's nature, and contact time; a kinetic model is only concerned with the effect of observable parameters on the overall rate.

Several adsorption kinetic models have been established to understand the adsorption kinetics and rate-limiting step. These include pseudo-first and pseudo-second-order rate model, the Weber and Morris sorption kinetic model, the Adam–Bohart–Thomas relation, the first-order reversible-reaction model, the external mass transfer model, the first-order equation of Bhattacharya and Venkobachar, Elovich's model, and Ritchie's equation. The pseudo-first and pseudo-second-order kinetic models are the most celebrated model to study the adsorption kinetics of pollutants and quantify the extent of uptake in adsorption kinetics. Sorption kinetic models can be divided into two main types: reaction-based models and diffusion-based models (Al-Degs et al., 2006; Ho et al., 2000).

5.2.1 Reaction-Based Models

5.2.1.1 Pseudo-First-Order Model

The Lagergren first-order rate expression based on solid capacity is generally expressed as follows (Lagergren, 1898):

$$\frac{dq}{dt} = k_1(q_e - q) \tag{5.1}$$

where q and q_e are amounts of adsorbate adsorbed (mg/g) at time t (min) and at equilibrium, respectively; k_1 is the rate constant of adsorption (l/min). Integration of the aforementioned equation with the boundary conditions: $t = 0, q = 0$, and $t = t, q = q$ gives

$$\ln(q_e - q) = \ln q_e - k_1 t \tag{5.2}$$

which can be represented as follows:

$$q = q_e\left[1 - \exp(-k_1 t)\right] \tag{5.3}$$

or it can be expressed as:

$$\log(q_e - q_t) = \log(q_e) - \frac{k_1}{2.303}t \tag{5.4}$$

The equation applicable to experimental results generally differs from a true first-order equation mainly in two ways (Gautam et al., 2014; Gerente et al., 2007).

1. The parameter $k_1(q_e - q_t)$ does not represent the number of available sites.
2. The parameter $log(q_e)$ is an adjustable parameter which is often not found equal to the intercept of a plot of $log(q_e - q_t)$ against t, whereas in a true first-order sorption reaction $log(q_e)$ should be equal to the intercept of a plot of $log(qe - qt)$ against t.

In order to fit the aforementioned equation to experimental data, the equilibrium sorption capacity, q_e, must be known. In many cases q_e is unknown and as chemisorption tends to become unmeasurably slow; the amount sorbed is still significantly smaller than the equilibrium amount. In most cases in the literature, the pseudo-first-order equation of Lagergren does not fit well for the whole range of contact time and is generally applicable over the initial 20 to 30 min of the sorption process (Gerente et al., 2007). Furthermore, one has to find some means of extrapolating the experimental data to $t = \infty$ or treat q_e as an adjustable parameter to be determined by trial and error. For this reason, it is necessary to use trial and error to obtain the equilibrium sorption capacity, q_e, in order to analyze the pseudo-first-order model kinetics.

The pseudo-first-order kinetic model has been used extensively to describe the sorption of metal ions onto a variety of adsorbents (Hanif et al., 2007). The main disadvantages of this model are (1) that the linear equation does not give theoretical q_e values that agree with experimental q_e values, and (2) that the plots are only linear over the first 30 min, approximately. Beyond this initial 30 min period the experimental and theoretical data do not correlate well. Various authors have made proposals for these deviations. One suggestion for the differences in the q_e values is that there is a time lag, possibly due to a boundary layer or external resistance controlling the beginning of the sorption process (McKay and Allen, 1980). This time lag is difficult to quantify and does little to help diminish the differences in experimental and theoretical q_e values. In terms of the short time correlation between experimental and theoretical data for the pseudo-first-order model, it has been proposed that another first-order reaction supersedes the first, or another reaction of another order becomes predominant (Sarkar and Chattoraj, 1993; Singh et al., 1996; Varshney et al., 1996). No mechanisms have been proposed for this stage by any authors. Several researchers have tested the Lagergren pseudo-first-order reaction model for pollutants' adsorption from aqueous solutions (Wu et al., 2000, 2001).

5.2.1.2 Pseudo-Second-Order Model

Pseudo-second-order model is derived on the basis of the sorption capacity of the solid phase, which may be expressed as (Ho and McKay, 1999):

$$\frac{dq}{dt} = k_2(q_e - q)^2 \tag{5.5}$$

Integration of the aforementioned equation with the boundary conditions $t = 0$, $q = 0$, and at $t = t$, $q = q$, results in

$$\frac{1}{q_e - q} = \frac{1}{q_e} + k_2 t \tag{5.6}$$

This equation can be expressed as

$$\frac{t}{q} = \frac{t}{q_e} + \frac{1}{k_2 q_e^2} \tag{5.7}$$

where k_2 is the equilibrium rate constant of pseudo-second-order adsorption (g/mg/min). The pseudo-second-order rate constants can be determined experimentally by plotting t/q versus t. As such, in comparison to pseudo-first-order kinetics this model is considered more appropriate for representing the kinetic data in adsorption systems. The pseudo-first and pseudo-second-order rate expressions have been and still are in wide use for studying the adsorption of pollutants from aqueous solutions. In the pseudo-second-order model the chemisorption process is superior to the pseudo-first-order model, as it takes into account the interaction of adsorbent–adsorbate through their valency forces (Febrianto et al., 2009).

The pseudo-second-order kinetic model has been the most widely celebrated model for the adsorption of pollutants from aqueous solutions. There are several examples where the pseudo-second-order model was best fitted for the adsorption of metal ions and dyes onto nanosorbents, that is, copper (Hu et al., 2013; Mi et al., 2012), nickel (Ren et al., 2011), lead (Fan et al., 2013), and methylene blue (Ai et al., 2011).

5.2.2 Diffusion-Based Models

5.2.2.1 External Diffusion Model

If external diffusion of pollutant (within the diffuse layers outside the sorbent) is the rate-limiting step, then it has been shown that Eq. (5.8) can be fitted into sorption data with some success (Lee et al., 1999):

$$\ln\frac{C_t}{C_0} = -k_f \frac{A}{V} t \tag{5.8}$$

where C_0, C_t, A/V, and t are the initial pollutant concentration, concentration at time t, the external sorption area to the total solution volume, and sorption time, respectively. The external diffusion coefficient k_f (cm/s) can be calculated from the slope of the straight line obtained from Eq. (5.8).

However, when the transport of the solute molecules from the liquid phase up to the solid phase boundary plays the most significant role in adsorption, the liquid film diffusion model may be applied (Gupta and Bhattacharyya, 2006):

$$\ln(1-F) = -k_{fd}t \tag{5.9}$$

where F is the fractional attainment of equilibrium ($F = q_t/q_e$) and k_{fd} is the adsorption rate constant. A linear plot of $\ln(1 - F)$ versus t with zero intercept would suggest that the kinetics of the adsorption process is controlled by diffusion through the liquid film surrounding the solid adsorbents.

5.2.2.2 Internal Diffusion Model (Weber and Morris Sorption Kinetic Model)

When the diffusion (internal surface and pore diffusion) of pollutant inside the sorbent is the rate-limiting step, then sorption data can be presented by the internal diffusion model. Originally this model was employed by Weber and Morris (1963), hence it is also known as the Weber and Morris sorption kinetic model. The equation can be expressed as:

$$q = K_{WM}\sqrt{t} + C \tag{5.10}$$

where q and K_{WM} are the amount adsorbed at time t, and Weber and Morris intraparticle diffusion rate constant, respectively. The $\sqrt{t}$ represents the square root of time t. C is the intercept. Pavasant et al. (2006) employed this model for the biosorption by *Caulerpa lentillifera* biomass for Cu(II), Cd(II), Pb(II), and Zn(II) and concluded that the adsorption was governed by two main mechanisms, that is, intraparticle diffusion and external mass transfer. The intraparticle diffusion can be estimated with

$$D = \frac{\pi}{8640}\left(\frac{d_p K_{WM}}{q_e}\right)^2 \tag{5.11}$$

where D and d_p is the intraparticle diffusion coefficient and mean particle diameter, respectively, and K_{WM} and q_e is the amount adsorbed at time t and amount of solute adsorbed at equilibrium condition; respectively.

The external mass transfer process was determined by

$$\frac{dq}{dt} = K_L' A\left(C - C_s^i\right) \tag{5.12}$$

where C and C_s^i is the liquid phase concentration of sorbate in the bulk solution at time t and concentration of sorbate in the inner pore of sorbent respectively, and K_L and A is the liquid–solid mass transfer coefficient and specific surface area of biomass respectively.

5.3 OTHER KINETIC MODELS

5.3.1 Boyd Kinetic Model

Due to the double nature of intraparticle diffusion (both film and pore diffusion), and in order to determine the actual rate-controlling step involved in the solute adsorption, the kinetic data can be analyzed by using the Boyd model. The Boyd kinetic equation is represented as (Boyd et al., 1947):

$$F = \frac{6}{\pi^2}\exp(-Bt) \tag{5.13}$$

and

$$F = q/q_e$$

where q_e is the amount of solute adsorbed at equilibrium (mg/g) and q represents the amount of solute adsorbed at any time t (min), F represents the fraction of solute adsorbed at any time t, and Bt is the mathematical function of F. The Eq. (5.13) may be rearranged as:

$$Bt = -0.4977 - \ln(1 - F) \tag{5.14}$$

The plot of [$-0.4977 \ln (1 - F)$] against t can be employed to test the linearity of the experimental data. The linearity of these plots can be employed to distinguish between external transport and intraparticle transport controlled rates of adsorption of solute.

A straight line passing through the origin is indicative of adsorption processes governed by particle diffusion mechanisms; otherwise, they are governed by film diffusion (Mohan and Singh, 2002).

The diffusion coefficient D_i (m^2/g) can be calculated from the B values using the relationship:

$$B = \frac{\pi^2 D_i}{r^2} \tag{5.15}$$

where D_i is the effective diffusion coefficient of solute in the adsorbent phase and r is the radius of the adsorbent particles.

5.3.2 Bhattacharya and Venkobachar Model

The linearized form of the Bhattacharya and Venkobachar kinetic equation can be written as:

$$\log[1 - U(T)] = -(k/2.303)t \tag{5.16}$$

where $U(T) = [(C_0 - C_t)/(C_0 - C_e)]$, and k is the Bhattacharya and Venkobachar rate constant (min^{-1}), which can be calculated from the slope of the plot $\log[1 - U(T)]$ versus t. C_0 and C_e are the initial and equilibrium concentration of pollutants in aqueous media, respectively. C_t is the concentration of pollutant at any time t.

5.3.3 Elovich Kinetic Model

Elovich equation has been used successfully to describe adsorption kinetics, assuming that the actual solid surfaces are energetically heterogeneous (Thomas and Thomas, 1997). The linear form may be written as:

$$q_t = \beta \ln(\alpha\beta) + \beta \ln t \tag{5.17}$$

where α and β, known as the Elovich coefficients, represent the initial sorption rate [mg/(g min)] and the desorption constant (g/mg), respectively. The Elovich coefficients could be computed from the plots of q_t versus ln t.

5.4 MODELING OF EQUILIBRIUM ADSORPTION PROCESSES

Judgment of a solid-solution sorption system is usually based on two types of investigations: equilibrium batch sorption experiments and dynamic continuous-flow sorption studies. Equilibrium isotherm model equations such as Langmuir (1918), Freundlich (1906), Redlich–Peterson (Redlich and Peterson, 1959), and Dubinin–Raduskevich (Dubinin, 1960) are used to describe experimental adsorption data in batch mode. It is important to find best-fitting isotherms to evaluate the efficacy of the prepared adsorbent to develop suitable industrial adsorption system designs.

5.4.1 Equilibrium Adsorption Isotherms

Equilibrium isotherm models are generally classified into empirical equations and mechanistic models. The mechanistic models are based on the mechanism of pollutants' adsorption, which are able not only to represent but also to explain and predict the experimental behavior. Table 5.1 represents some empirical models for single solute systems. The Langmuir and Freundlich models are the most widely accepted and used in literature (Afkhami and Moosavi, 2010; Afkhami et al., 2009; Chatterjee et al., 2009; Parham et al., 2012; Tian et al., 2011). The Brunauer–Emmett–Teller (BET) model (Brunauer et al., 1938) describes the multilayer adsorption at the surface of adsorbents and assumes that the Langmuir isotherm applies to each layer. These models can provide information on pollutant uptake capacity and the difference in pollutant uptake between various species.

Langmuir adsorption isotherms are widely used to describe the relationship between the amount of adsorbate adsorbed on the adsorbent and its equilibrium concentration in aqueous solution. The Langmuir equation relates the coverage of molecules on a solid surface to concentration of a medium above the solid surface at a fixed temperature. Langmuir isotherms are based on three assumptions: (1) the surface of the adsorbent is in contact with a solution containing an adsorbate, which is strongly attracted to the surface; (2) the surface has a specific number of sites where the solute molecules can be adsorbed; and (3) the adsorption involves the attachment of only one layer of molecules to the surface, that is, monolayer adsorption (Gautam et al., 2014).

At lower concentrations, an alternate isotherm developed by Herbert F. Freundlich frequently describes the data better. The Freundlich isotherm is an empirical equation. This equation is one among the most widely used isotherms for the explanation of adsorption equilibrium. The Freundlich isotherm is capable of describing the adsorption of pollutants on a wide variety of nanosorbents. On average, a favorable adsorption tends to have a Freundlich constant n between 1 and 10. A larger value of n (smaller value of $1/n$) implies stronger interaction between adsorbent and pollutant, while $1/n$ equal to 1 indicates linear adsorption leading to identical adsorption energies for all sites (Febrianto et al., 2009). Sometimes, these empirical models do not reflect any mechanisms of sorbate uptake and hardly have a meaningful physical interpretation for adsorption. It has been mentioned

Table 5.1 Summary of widely used isotherms for adsorption systems with their advantages and disadvantages

Isotherm	Functional form	Linear form	Plot	Advantages	Disadvantages
Langmuir	$q_e = \frac{q_m K_L C_e}{1 + K_L C_e}$	$\frac{C_e}{q_e} = \frac{1}{K_L q_m} + \frac{1}{q_m} C_e$	$\frac{C_e}{q_e} VS. C_e$	Has Henry law and finite saturation limit so valid over a wide range of concentrations	Based on monolayer assumption
Freundlich	$q_e = K_F C_e^{1/n}$	$\ln q_e = \ln K_F + \frac{1}{n} \ln C_e$	$\ln q_e VS. \ln C_e$	Simple expression and has parameter for surface heterogeneity	Does not have Henry law and no saturation limit, not structured, not applicable over wide range of concentrations
Temkin	$q_e = \frac{RT}{b} \ln(K_T C_e)$	$q_e = \frac{RT}{b} \ln K_T + \frac{RT}{b} \ln C_e$	$q_e VS. \ln C_e$	Simple expression	Same as Freundlich; does not have correct Henry law limit and finite saturation limit, not applicable over wide range of concentrations
Redlich–Peterson (R–P)	$q_e = \frac{K_{RP} C_e}{1 + a_{RP} C_e^{\beta}}$	$\ln\left[\left(\frac{K_{RP} C_e}{q_e}\right) - 1\right] = \ln a_{RP} + \beta \ln C_e$	$\ln\left[\left(\frac{K_{RP} C_e}{q_e}\right) - 1\right] VS. \ln C_e$	Approaches Freundlich at high concentration	No special advantages

that the results from empirical models cannot be extrapolated and no predictive conclusions can be made for adsorption of pollutants from aqueous solutions operating under different environmental conditions (Wang and Chen, 2006). Both simple basic models (Langmuir and Freundlich models) also do not incorporate the effects of any external variable environmental factors, although they are capable of describing many adsorption isotherms in most cases. The mechanistic conclusions from the good fit of the models alone should be avoided.

In some cases the isotherm model such as Langmuir or Freundlich fails to describe the adsorption behavior of pollutants from aqueous solutions onto the nanosorbent, and in most cases, more than one model has been applied to illustrate the adsorption mechanism. Some of these equations contain two fitting parameters, such as the Temkin isotherm (Temkin and Pyzhev, 1940), Flory-Huggins (Foo and Hameed, 2010), and the Dubinin-Raduskevich equations (Dubinin, 1960), whereas others can have more than two parameters, such as Redlich–Paterson (Redlich and Peterson, 1959) and Sips isotherms (Sips, 1948).

5.4.1.1 Langmuir Isotherm

The Langmuir sorption isotherm is valid for the adsorption of a solute from solution as monolayer adsorption on a surface containing a finite number of identical sites of uniform strategies of adsorption with no transmigration of adsorbate in the plane of surface (Gautam et al., 2014). The nonlinear equation of the Langmuir isotherm model can be expressed as (Langmuir, 1918):

$$q_e = \frac{Q_0 K_L C_e}{(1 + K_L C_e)} \tag{5.18}$$

where C_e and q_e are the equilibrium liquid-phase concentrations of adsorbate (mg/L) and the amount of adsorbate adsorbed onto the adsorbent (mg/g), respectively. Q_0 is the maximum adsorption capacity reflected on a complete monolayer (mg/g); K_L is the adsorption equilibrium constant (L/mg) that is related to the apparent energy of sorption. The Langmuir isotherm parameters can be obtained from its linearized form as shown in Eq. (5.19):

$$\frac{C_e}{q_e} = \frac{1}{(K_L Q_0)} + \frac{C_e}{Q_0} \tag{5.19}$$

A plot of C_e/q_e versus C_e should indicate a straight line of slope $1/Q_0$ and an intercept of $1/(K_L\ Q_0)$.

Essential characteristics of the Langmuir isotherms can be described by a separation factor (R_L), which is defined by the following equation (Karagoz et al., 2008):

$$R_L = \frac{1}{(1 + K_L C_0)} \tag{5.20}$$

where C_o is the initial concentration of the adsorbate (mg/L) and K_L is the Langmuir adsorption constant (L/mg). The value of the separation factor (R_L) indicates the nature of the adsorption process.

5.4.1.2 Freundlich Isotherm

The Freundlich isotherm is an empirical equation based on adsorption on a heterogeneous surface and assumes that the adsorption occurs at sites with varying energy of adsorption. The equation is commonly represented by Eq. (5.21) (Freundlich, 1906):

$$\ln q_e = \ln K_F + \left(\frac{1}{n}\right) \ln C_e \tag{5.21}$$

where K_F (mg/g) and n are the Freundlich constants' characteristics of the system, indicating the adsorption capacity and the adsorption intensity, respectively. The n is an empirical parameter that varies with the degree of heterogeneity and is related to the distribution of bonded ions on the sorbent surface. In general, $n > 1$ illustrates that the adsorbate is favorably adsorbed on an adsorbent, and the higher the n value the stronger the adsorption intensity (Chakraborty et al., 2011). In particular, the value of n was significantly higher than unity at all the temperatures and concentrations studied.

5.4.1.3 Temkin Isotherm

The linear form of Temkin isotherm can be expressed as follows (Temkin and Pyzhev, 1940):

$$q_e = \frac{RT}{b} \ln AC_e \tag{5.22}$$

$$q_e = B \ln A + B \ln C_e \tag{5.23}$$

where B is the Temkin constant related to the heat of adsorption, A is the equilibrium binding constant corresponding to the maximum binding energy (L/mg), and R is the gas constant (8.314 J/mol/K). The constants A and B can be determined from the intercept and the slope of the linear plot of data q_e versus ln C_e.

5.4.1.4 Redlich–Peterson Isotherm

The Redlich–Peterson equation (Redlich and Peterson, 1959) is widely used as a compromise between the Langmuir and Freundlich isotherm systems. This model has three parameters and incorporates the advantageous significance of both models. The Redlich–Peterson model can be represented as follows:

$$q_e = \frac{K_{\mathrm{RP}} C_e}{1 + (\alpha C_e)^{\beta}} \tag{5.24}$$

where K_{RP} (l/g) and α (l/mg)$^{\beta}$ are Redlich–Peterson isotherm constants and β is the exponent that lies between 0 and 1. The Redlich–Peterson isotherm model has two limiting cases, when $\beta = 1$, the Langmuir equation results, whereas when $\beta = 0$, the Redlich–Peterson isotherm equation transforms to Henry's law equation.

5.4.1.5 Dubinin and Radushkevich Isotherm

Dubinin–Radushkevich (D–R) isotherm model assumes that the characteristics of sorption curves are related to the porosity of the adsorbent. The linear form of this model is expressed by (Dubinin, 1960):

$$\ln(q_e) = \ln(q_m) - \beta\varepsilon^2 \tag{5.25}$$

where q_e is the amount of adsorbate adsorbed per unit dosage of the adsorbent (mg/g), q_m is the theoretical monolayer saturation capacity (mg/g), β is the activity coefficient (mol^2/kJ^2) related to the mean sorption energy, and ε is the Polanyi potential described as (Chen et al., 2011):

$$\varepsilon = RT \ln\left[1 + \left(\frac{1}{C_e}\right)\right] \tag{5.26}$$

where R is the gas constant in kJ/mol/K, T is the temperature in Kelvin, and C_e is as defined earlier. From the plot of ln q_e versus ε^2 the values of β and q_m were determined by the slope and intercept of the linear plot.

5.5 THERMODYNAMIC ANALYSES

Gibbs free energy (ΔG), change in enthalpy (ΔH), and change in entropy (ΔS) are the general thermodynamic parameters which have been widely used to evaluate an adsorbate–adsorbent system. Negative ΔG indicates the spontaneity of the adsorption process. ΔH is used to identify the nature of adsorption. A positive value of ΔH indicates the reaction is endothermic, and the negative value of ΔH shows that the reaction is exothermic (Gautam et al., 2014). A positive value of ΔS indicates increased randomness of adsorbate molecules on the solid surface than in the solution. The free energy of adsorption (ΔG) can be related with the Langmuir equilibrium constant by the following expression (Gautam et al., 2015):

$$\Delta G = -RT \ln K_L \tag{5.27}$$

Enthalpy and entropy changes are also related to the Langmuir equilibrium constant by the following expression:

$$\ln K_L = \frac{\Delta S}{R} - \frac{\Delta H}{RT} \tag{5.28}$$

Thus, a plot of ln K_L versus $1/T$ should be a straight line. ΔH and ΔS values could be obtained from the slope and intercept of this plot. The thermodynamic parameters such as changes in standard free energy (ΔG), enthalpy (ΔH), and entropy (ΔS) were determined by using the following equation (Gautam et al., 2015):

$$\Delta G = \Delta H - T\Delta S \tag{5.29}$$

The slope and the intercept of the plot of ΔG versus T were used to determine the ΔS and ΔH values.

5.6 ADSORPTION OF HEAVY METALS

Recently, Lee and Yang (2012) developed a flowerlike TiO_2–graphene oxide (GO–TiO_2) hybrid through hydrothermal process by stirring a titanium oxide precursor in isopropyl alcohol with graphene oxide colloidal solution. The GO–TiO_2 hybrid material was applied for the removal of heavy metal ions from aqueous solutions. The flowerlike TiO2 on GO structure significantly improved the removal efficiency of heavy metals due to the oxygenated functional groups of exfoliated graphene oxide. The GO–TiO_2 hybrid adsorption capacities of heavy metal ions, after 6 and 12 h of hydrothermal treatment at 100°C, were respectively 44.8 ± 3.4 and 88.9 ± 3.3 mg/g for removing Zn^{2+}, 65.1 ± 4.4 and 72.8 ± 1.6 mg/g for removing Cd^{2+}, and 45.0 ± 3.8 and 65.6 ± 2.7 mg/g for removing Pb^{2+} at pH 5.6.

Mishra and Ramaprabhu (2011) synthesized graphene sheets by hydrogen-induced exfoliation of graphitic oxide followed by functionalization and utilized the functionalized graphene sheets for simultaneous removal of a high concentration of arsenic and sodium from aqueous solution using a supercapacitor-based water filter. Maximum adsorption capacities, using the Langmuir isotherm for arsenate, arsenite, and sodium were found to be nearly 142, 139, and 122 mg/g, respectively. Their results showed that the high adsorption capacity for both inorganic species of arsenic and sodium along with the desalination ability of a graphene-based supercapacitor provides a solution for commercially feasible water filters. Lead(II) adsorption from an aqueous solution onto a graphene layer (Cπ electrons) was investigated by Machida et al. (2006). The basic sites for Pb(II) adsorption was, therefore, supposed to be on a graphene layer; Cπ-cation interactions were operative for Pb(II) adsorption through the electrostatic interactions.

Mi et al. (2012) successfully prepared graphene oxide aerogels with oriented pore structures using unidirectional freezing technology. The graphene oxide aerogels were used as an adsorbent to adsorb Cu^{2+} from aqueous solutions. The adsorption rate was very fast and reaches equilibrium within 15 min. The kinetic of the adsorption system of Cu^{2+} adsorption followed the pseudo-second-order kinetics, and the equilibrium data were well fitted with the Langmuir isotherms. The adsorption of Cu^{2+} on graphene oxide aerogels increased with the increase in pH, which indicated an ion-exchange mechanism. The Langmuir maximum adsorption capacity was reported as 29.59 mg/g at 40°C within 30 min of contact time.

Few-layered graphene oxide (FGO) was synthesized from graphite by using the modified Hummers method and was used to remove Pb(II) ions from aqueous solutions. The plentiful oxygen-containing groups on the surfaces of FGO played an important role in Pb(II) ion adsorption on FGO. The adsorption of Pb(II) ions on FGO was dependent on pH values and independent of ionic strength. The adsorption of Pb(II) ions on FGO was mainly dominated by strong surface complexation. The Langmuir maximum adsorption capacities of Pb(II) ions on FGO were about 842, 1150, and 1850 mg/g at 293, 313, and 333 K, respectively.

The aggregation behavior of graphene oxide in aqueous solutions containing Cu^{2+} ions was investigated by Yang et al. (2010). The folding/aggregation was most likely triggered by the coordination between graphene oxide and Cu^{2+}. The equilibrium Cu^{2+} concentrations and equilibrium absorption capacity of graphene oxide were measured to estimate the maximum absorption capacity of graphene oxide for Cu^{2+} and the absorption model. Graphene oxide has a huge absorption capacity for Cu^{2+}, which is around 10 times that of active carbon. The Langmuir model was found most suitable for the adsorption of Cu^{2+}, which suggested that the Cu^{2+} were absorbed by specific sites of graphene oxide and formed monolayers. The Langmuir monolayer adsorption capacity was 46.6 mg/g.

Magnetic chitosan/graphene oxide (MCGO) composite materials were fabricated and applied as excellent adsorbents for Pb(II) ion sequestration from aqueous solutions (Fan et al., 2013). The effects of pH, contact time, and concentration on Pb(II) ion sorption were investigated. It has been reported that the Pb(II) ion sorption on MCGO was strongly dependent on pH and was maximum at pH 5.0. This could be explained by the fact that, at a lower pH, the amine groups on surface of MCGO can be easily protonated, inducing an electrostatic repulsion of lead ions. That is because the functional groups of inserted chitosan moieties, which are the most effective ones, on the biomass surfaces offers more metal adsorption sites. The slight decrease of the uptake in the acidic media may be attributed to the protonation of the lone pairs of nitrogen that hinder the complex formation according to the following reactions: $R-NH_2+H^+ \rightarrow R-NH_3^+$, $R-NH_3^+ + Pb^{2+} \rightarrow R-NH_3^+ \cdots Pb^{2+}$. In addition, the COO^- groups were free from the protonation: $2R-CH_2COO^- + Pb^{2+} \rightarrow (R-CH_2COO^-)_2Pb^{2+}$. Thus, pH of 5.0 was selected as the optimum pH value of Pb^{2+} solution for the adsorption experiment. The abundant functional groups on the surfaces of MCGO played an important role on Pb(II) sorption. Equilibrium studies showed that the data of Pb(II) adsorption followed the Langmuir model, with maximum adsorption capacity for Pb(II) estimated to be 76.94 mg/g. The desorption result showed that the solution with pH 1 and 2 were able to efficiently remove the adsorbed metal ions, and the desorption efficiency was 90.3% and 90.1%, respectively.

Recently, a simple chemical bonding method to synthesize magnetic cyclodextrin–chitosan/graphene oxide (CCGO) was reported by Li et al. (2013), and the adsorption behaviors of Cr(VI) in aqueous solution on CCGO were investigated. Due to the advantage of high surface area, abundant hydroxyl and amino groups of CCGO, and the magnetic property of Fe_3O_4, the Cr(VI) was easily and rapidly extracted from the water by magnetic attraction. The adsorption equilibrium of CCGO for Cr(VI) fitted well to the Langmuir isotherm with maximum adsorption capacity of 67.66 mg/g, and the novel adsorbent exhibited better Cr(VI) removal efficiency in solutions with low pH. It has been found that the Cr(VI) adsorption performance of CCGO strongly depends on their surface charge concentration and specific surface area.

Hu et al. (2013) synthesized a sulfonated magnetic graphene oxide composite (SMGO) from graphene oxide and was used as an adsorbent for removing Cu(II) ions from aqueous solutions. The effects of operating parameters such as pH, Cu(II) concentration, and temperature on the Cu(II) adsorption were investigated by using a response surface methodology. Optimum Cu(II) uptake of 62.73 mg/g was achieved at pH 4.68, Cu(II) concentrations 73.71 mg/L, and temperature 50°C. The adsorption process was well described by the pseudo-second-order kinetic model ($k_2 = 1.63 \times 10^{-3}$). The experimental data of isotherms followed the Langmuir isotherm model. Moreover, the thermodynamic parameters calculated from the temperature-dependent isotherms indicated that the adsorption reaction was an endothermic and spontaneous process. All the findings indicated that the SMGO may be a promising adsorbent for the efficient removal of copper ions from wastewater.

A type of graphene nanosheet/δ-MnO_2 (GNS/MnO_2) composite was synthesized by a microwave-assisted method and applied for the removal of Ni(II) from aqueous solutions (Ren et al., 2011). Microstructure measurements showed that the nanoscale δ-MnO_2 particles deposit well onto the surfaces of GNS. The adsorption characteristics were examined with respect to pH effect, removal kinetic data, equilibrium isotherms, and thermodynamic data in batch experiments. The equilibrium data were fitted well by the pseudo-second-order kinetics, which suggested that a chemical sorption process played a significant role in Ni(II) ion adsorption. Adsorption isotherms followed the Langmuir model, with saturation adsorption capacity of Ni(II) onto GNS/MnO_2 at 46.6 mg/g at room temperature, which was 1.5 and 15 times higher than those of pure δ-MnO_2 and GNS, respectively. The positive values of both ΔH and ΔS suggested an endothermic reaction and increase in randomness at the solid–liquid interface during the Ni(II) adsorption process. The negative ΔG values indicated a spontaneous adsorption process, and GNS/MnO_2 were reused five times with a high recovery rate of 91%.

A magnetic graphene/iron oxide composite (Fe_3O_4/GO) was synthesized from graphene using a chemical reaction approach and has been employed as a novel adsorbent for the preconcentration and solidification of

U(VI) ions from aqueous solutions (Zong et al., 2013). The sorption behavior of U(VI) on the surface of Fe_3O_4/GO was carried out under ambient conditions such as contact time, pH, and ionic strength. The Langmuir and Freundlich models were adopted to simulate sorption isotherms of U(VI) at three different temperatures relying on the concentration of U(VI); the experimental results suggested that the sorption reaction was favored at higher temperatures. The pH-dependent and ionic strength-independent U(VI) sorption on Fe_3O_4/GO demonstrated that the sorption mechanism of U(VI) was inner-sphere surface complexation at low pH values, whereas the removal of U(VI) was achieved by simultaneous precipitation and inner-sphere surface complexation at high pH values. The maximum sorption capacity of U(VI) was about 69.49 mg/g on Fe_3O_4/GO at $T = 293$ K and $pH = 5.5 \pm 0.1$ higher than the majority of materials.

Graphene/MgAl-layered double hydroxide (G–MgAl-LDH) nanocomposites were prepared by urea hydrolyzed hydrothermal reaction (Yuan et al., 2013). Urea could effectively reduce GO and simultaneously induce the nucleation and in situ growth of LDH crystallites on the nanosheets of graphene. Both LDH and graphene layers were exfoliated in the nanocomposite. The calcined G–MgAl-LDH was easily prepared by heating G–MgAl-LDH at higher temperatures and used as a potential nanoadsorbent to remove Cr(VI) from the aqueous solution. Compared to virgin MgAl-LDH, calcined G–MgAl-LDH exhibited higher adsorption efficiency with lower dosages in removing Cr(VI) from aqueous solution. Maximum adsorption capacity of approximately 172.55 mg/g with 1.0 g/L dosage of calcined G–MgAl-LDH was achieved. It has been reported about the adsorption kinetics, isotherms, and thermodynamics that adsorption mechanisms of Cr(VI) involved the surface adsorption of graphene and memory effects of calcined LDH, and synergistic contributions from each component in calcined G–MgAl-LDH enhanced Cr(VI) adsorption capacity.

3D interconnected porous graphene oxide foam/Fe_3O_4 (GOF/Fe_3O_4) nanocomposites have been successfully fabricated by the combination of microwave plasma chemical vapor deposition technique and coprecipitation method (Lei et al., 2014). Due to the large specific surface area (574.2 m^2/g) of the 3D interconnected porous structure, the GOF/Fe_3O_4 showed an excellent absorption performance for the removal of Cr(VI) ions with maximum absorption capacity (258.6 mg/g) and fast adsorption rate at solution pH 2, which was significantly higher than the reported 2D graphene-based adsorbents and other conventional adsorbents. The adsorption isotherms agree well with the Freundlich model and revealed the multilayer adsorption of Cr(VI) taking place at the adsorbent. The adsorption mechanism of GOF/Fe_3O_4 for Cr(VI) was the synergetic contribution of the reduction of Cr(VI) to Cr(III) and the ion exchange between the hydroxyl groups and CrO_7^{2-}/$HCrO_4^-$. The excellent adsorption performance of the GOF/Fe_3O_4 may offer a way to prepare a range of 3D magnetic graphene-based nanocomposites for applications in effective removal of heavy metal ions.

The removal of Th^{4+} ions from aqueous solutions was investigated using single-layer graphene oxide as an adsorbent through batch adsorption experiments at room temperature (Pan et al., 2013). Structural characterizations of the sorbent were also investigated. The influences of sorption conditions such as solution pH, contact time, sorbent dose, ionic strength, initial metal ion concentration, and temperature on the adsorption of Th^{4+} were also investigated. Experimental results indicated that the adsorption of Th^{4+} was dependent on the pH of the medium and independent on the ionic strength. The graphene oxide sorbent provided significant Th^{4+} removal (>98.7%) at pH 3.0, and the adsorption equilibrium was achieved within 10 min of contact time. The Langmuir adsorption isotherm fit well to the adsorption system and indicated that a maximum adsorption capacity of 411 mg/g for Th^{4+} ions after 2 h. The thermodynamic parameters showed that adsorption process was endothermic and spontaneous in nature, which means higher adsorptive removal can be obtained at higher temperature. Moreover, the desorption level of 84.2 ± 1.2% for Th^{4+} from graphene oxide was obtained using 0.1 M H_2SO_4 as a stripping agent.

Sahoo et al. (2014) synthesized the CdS/graphene and ZnS/graphene nanocomposites by hydrothermal reaction of graphene oxide and morpholine-4-carbodithioate complex of Cd and Zn, respectively. The morphologies of CdS, ZnS, and their graphene nanocomposites were studied by FESEM. The FESEM images clearly showed that the nanoparticles of CdS and ZnS were in irregular size and formed with its dimension, ranging from 30 to 40 and 20 to 30 nm, respectively. Fig. 5.1 shows the FESEM images of CdS/graphene nanocomposites corresponding to different weight ratios of the Cd–MCDT precursor and GO. It was clearly evident that CdS nanoparticles tend to be uniformly distributed on the graphene platelets with increasing weight ratio of Cd–MCDT: GO from 0.25:1.00 to 1:1, respectively. Similar observations are also noted in the case of ZnS/graphene nanocomposite (Fig. 5.2). The corresponding nanocomposites were investigated separately as photocatalysts in the decomposition of methylene blue in the presence of UV light and also as adsorbents in the removal of Cd(II)

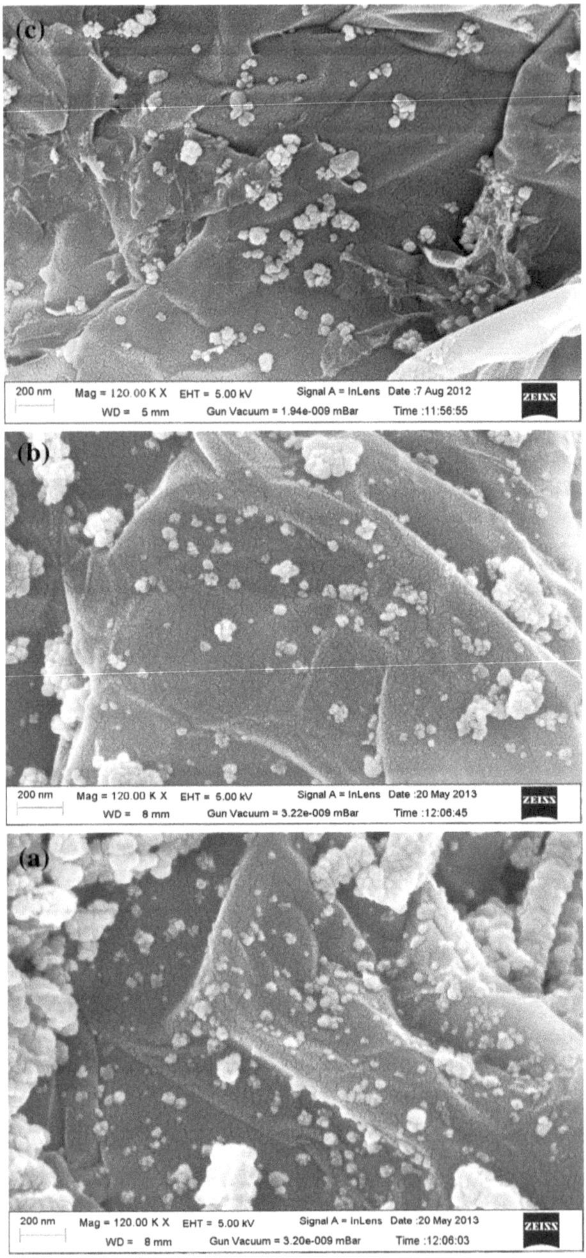

Figure 5.1 FESEM image of hydrothermal reaction product of Cd-MCDT and GO in weight ratio of (a) 1:1, (b) 0.5:1.0, and (c) 0.25:1.00. *(Source: Sahoo et al., 2014; © Reproduced with permission from Springer).*

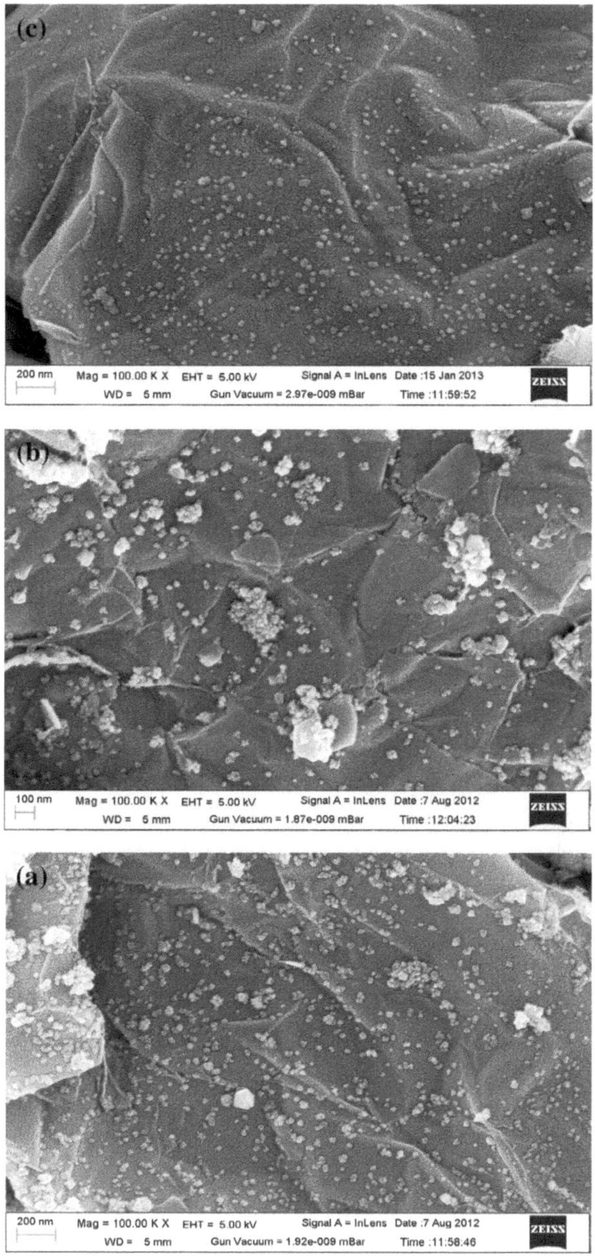

Figure 5.2 FESEM image of hydrothermal reaction product of Zn-MCDT and GO in weight ratio of (a) 1:1, (b) 0.5:1.0, and (c) 0.25:1.00. *(Source: Sahoo et al., 2014; © Reproduced with permission from Springer).*

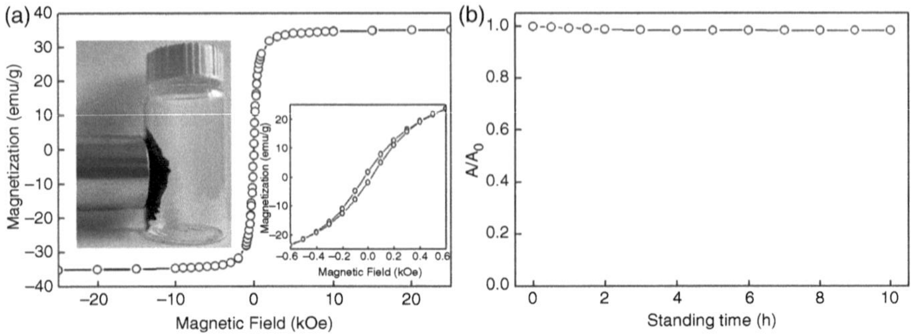

Figure 5.3 Magnetization curve (a) and UV-Vis absorbance at 230 nm as a function of standing time (b) of Fe_3O_4/GO. *(Source: Wang and Yu, 2014; © Reproduced with permission from Springer).*

and Pb(II) ions in aqueous solutions. The results have established that CdS/graphene and ZnS/graphene are effective photocatalysts as well as effective adsorbents in the removal of Cd(II) and Pb(II) ions to an extent of 97 and 99% by ZnS/graphene and CdS/graphene nanocomposite, respectively, under 1 g/L of adsorption dose and at pH ~7. The Langmuir model fit well for the adsorption of Pb(II) and Cd(II) ions with the maximum adsorption capacities of 3.10 mg/g and 3.62 mg/g, respectively.

The composite of Fe_3O_4 and graphene oxide (Fe_3O_4/GO) was prepared by chemical coprecipitation method with excellent adsorption capacity of Cs(I) and Sr(II) (Wang and Yu, 2014). The characterization results of the as-prepared materials confirmed the successful preparation of Fe_3O_4/GO. The saturation magnetization of Fe_3O_4/GO was 35.2 emu/g, indicating excellent magnetic property of Fe_3O_4/GO. Moreover, as can be seen from the magnetic separation Fe_3O_4/GO in Fig. 5.3a, Fe_3O_4/GO particles were effectively concentrated and separated by a simple permanent magnet. The dispersion property of Fe_3O_4/GO was investigated by using UV–Vis spectrophotometer and shown in Fig. 5.3b, which indicates that Fe_3O_4/GO has a very good dispersion property in aqueous solutions. The as-prepared Fe_3O_4/GO was applied as adsorbent for the separation of Cs(I) and Sr(II) from aqueous solutions. The maximum adsorption capacities of Cs(I) and Sr(II) on Fe_3O_4/GO was found to be 15.8 and 38.4 mg/g, respectively, at pH 5.0 and 293 K. Adsorptions of Cs(I) and Sr(II) on Fe_3O_4/GO surface were an endothermic and spontaneous process and were enhanced at high pH and low ionic strength (Fig. 5.4). The desorption of Cs(I) and Sr(II) was performed by washing Fe_3O_4/GO with 0.1 mol/L EDTA solution and deionized water thoroughly, and the regenerated Fe_3O_4/GO was obtained and

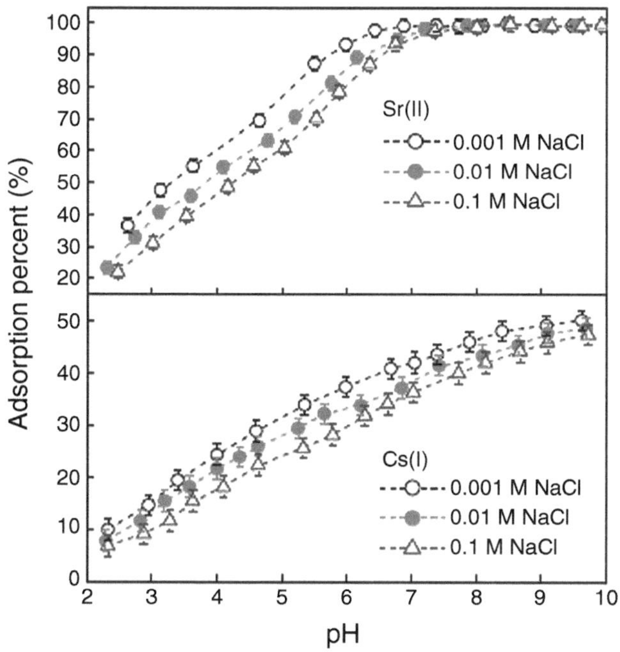

Figure 5.4 ***Effects of pH and ionic strength on the adsorption of Cs(I) and Sr(II) on Fe_3O_4/GO.*** Contact time = 24 h, temperature = 20 ± 1°C, m/V = 0.2 g/L, concentration Cs(I) initial = 10.0 mg/L, concentration Sr(II) initial = 10.0 mg/L. *(Source: Wang and Yu, 2014; © Reproduced with permission from Springer).*

reused in adsorption experiments up to six adsorption–desorption cycles (Fig. 5.5).

Magnetic manganese-incorporated iron(III) oxide ($Mn_x^{2+}Fe_{2-x}^{3+}O_4^{2-}$) (IMBO)–graphene nanocomposites were synthesized and applied for the removal of As(III) (Nandi et al., 2012). BET surface area and pore volume of magnetic nanocomposite were 280 m^2/g and 0.3362 cm^3/g, respectively. The composite showed almost complete (>99.9%) removal of As(III) (≤10 μg/L) from aqueous solutions. An external magnetic field of 0.3 T efficiently separated the water dispersed magnetic composite (0.01 g/10 mL) at room temperature (300 K). The adsorption of As(III) onto the magnetic nanocomposites followed the pseudo-second-order kinetic, and the equilibrium isotherm was better explained by the Langmuir isotherm with the maximum adsorption capacity of 14 mg/g. The As(III)-sorption mechanism modeling indicated the physical nature of the attachment with the composite material via outer surface complexation.

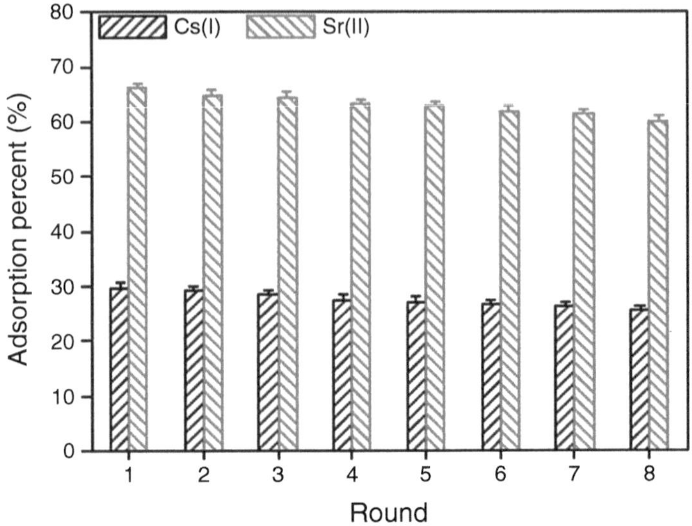

Figure 5.5 ***Recycling of Fe_3O_4/GO in the enrichment of Cs(I) and Sr(II).*** Contact time: 24 h, concentration Cs(I) initial = 10.0 mg/L, concentration Sr(II) initial = 10.0 mg/L, m/V = 0.2 g/L, I = 0.01 mol/L NaCl, pH = 5.0 ± 0.1. *(Source: Wang and Yu, 2014; © Reproduced with permission from Springer).*

The sorption of Th(IV) on graphene oxide was investigated as a function of contact time, solid-to-liquid ratio, solution pH, ionic strength, and in the presence of fulvic acid and humic acid (Li et al., 2014). The sorption percentage of Th(IV) on graphene oxide decreased with increasing ionic strength and decreasing solid-to-liquid ratio. The sorption edge of Th(IV) in the presence of fulvic acid/humic acid is much lower than that in the absence of fulvic acid/humic acid. The removal processes of Th(IV) followed the pseudo-second-order rate model. The equilibrium isotherm fit well to the Langmuir model, and the maximum sorption capacities of Th(IV) were about 5.80×10^{-4} to 7.7×10^{-4} mol/L at 298.15 and 338.15 K, respectively. Adsorption of Th(IV) on graphene oxide was spontaneous and endothermic in nature. The sorption of Th(IV) on graphene oxide was investigated with different background electrolyte solutions. The foreign anions drastically affect Th(IV) sorption on graphene oxide. The sorption of Th(IV) was lowest in 0.1 mol/L $NaClO_4$ and highest in 0.1 mol/L NaCl, although the difference between the three electrolyte solutions was small. The sorption of Th(IV) on graphene oxide in 0.1 mol/L LiCl, NaCl, and KCl solutions was also performed. As the order of the hydrated radii of these monovalent cations is $Li^+ > Na^+ > K^+ >$ (K^+: 2.32 Å, Na^+: 2.76 Å, Li^+: 3.4 Å). Therefore, the hydrated K^+ cation more easily interacts with the functional groups on

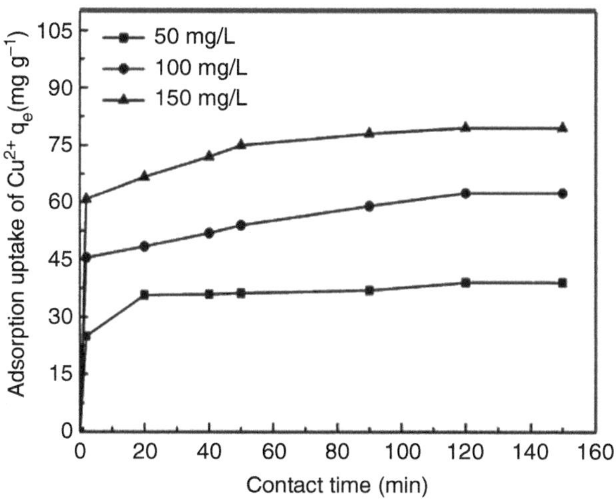

Figure 5.6 ***Effect of contact time on adsorption of Cu(II) (initial Cu(II) concentration 50, 100, and 150 mg L^{-1}, dosage of graphene oxide 1.0 mg mL^{-1}, initial pH of solution 5.3).*** *(Source: Wu et al., 2013; © Reproduced with permission from Springer).*

the surface of the graphene oxide than the hydrated Na^+ and Li^+ cations. The rapid sorption rate and high sorption capacity suggest that graphene oxide can be a promising adsorbent for Th(IV).

The oxygenous functional groups on the surface of graphene oxide are primarily responsible for the sorption of metal ions. The graphene oxide have been used for the removal of Cu(II) ions from water, and the effects of the parameters such as pH value, contact time, Cu(II) concentration, and adsorbent dosage on adsorption were examined (Wu et al., 2013). As the solution pH value increases, the negative charge increases on the adsorbent surface. It could offer electrostatic attractions which were favorable for Cu(II) reacting with functional groups ($—COO^-$ or $—O^-$) on the graphene oxide surface to form a metal ion complex. The effect of the contact time on the adsorption capacity of graphene oxide for Cu(II) with known initial concentrations, the dosage of 1.0 mg m/L, and the pH of 5.3 are described in Fig. 5.6. Obviously, the adsorption equilibrium was attained after stirring for 120 min with initial copper ion concentrations of 50, 100, and 150 mg/L. The sorption process conformed to the Freundlich isotherm, and the maximum sorption capacity of 117.5 mg/g was observed at an initial pH value of 5.3 within 150 min of contact time. It was also found that Cu-pretreated graphene oxide could be desorbed by HCl, and the reusability of graphene oxide could still maintain above 90% of its initial capability after 10 cycles. Table 5.2 represents the adsorption capacities for

Table 5.2 Adsorption capacities for heavy metals and metalloids onto various types of graphene-based nanocomposites from aqueous solutions

Graphene-based composite materials	Metals or metalloids	Maximum adsorption capacity (mg/g)	References
Graphene oxide	Cd(II)	106.3 at 303 K	Zhao et al. (2011b)
	Cd(II)	153.6 at 313 K	Zhao et al. (2011b)
	Cd(II)	167.5 at 333 K	Zhao et al. (2011b)
Functionalized graphene	Cd(II)	73.42	Deng et al. (2010)
Graphene oxide	Co(II)	68.2 at 303 K	Zhao et al. (2011b)
	Co(II)	69.4 at 313 K	Zhao et al. (2011b)
	Co(II)	79.8 at 333 K	Zhao et al. (2011b)
Graphene oxide	Eu(III)	175.44	Sun et al. (2012)
Graphene oxide	U(VI)	97.5	Zhao et al. (2012)
Functionalized graphene sheet–based electrodes	As (V)	~142	Mishra and Ramaprabhu (2011)
	As(III)	~139	Mishra and Ramaprabhu (2011)
Magnetite-reduced graphitic oxide	As (V)	~5.83	Chandra et al. (2010)
	As(III)	~13.10	Chandra et al. (2010)
Magnetite-graphene/hydroxides	As(V)	73.14	Wu et al. (2011)
Few-layered graphene oxide nanosheets	Pb(II)	842 at 293 K	Zhao et al. (2011a)
	Pb(II)	1150 at 313 K	Zhao et al. (2011a)
	Pb(II)	1850 at 333 K	Zhao et al. (2011a)
Graphene	Pb(II)	22.42	Huang et al. (2011)
Functionalized graphene	Pb(II)	406.6	Deng et al. (2010)
EDTA-graphene oxide	Pb(II)	479 ± 46	Madadrang et al. (2012)
Chitosan gelatin/graphene oxide	Pb(II)	100	Zhang et al. (2011)
Polypyrrole/graphene oxide	Cr(VI)	293.4	Li et al. (2012)
Graphene oxide	Cu(II)	46.6	Yang et al. (2010)
Graphene oxide aerogels	Cu(II)	29.59	Mi et al. (2012)
δ-MnO_2-graphene composite	Ni(II)	46.6	Ren et al. (2011)
Graphene oxide	Zn(II)	246	Wang et al. (2013)

heavy metals and metalloids onto various types of graphene-based nanocomposites from aqueous solutions.

5.7 CONCLUSIONS

Nowadays, considerable attention is being paid to separation and removal of heavy metals and metalloids from aquatic media on a variety of adsorbent materials. Graphene-based nanocomposites offer an excellent alternative to the conventional methods for the removal of heavy metals from the liquid phase. The separation and preconcentration of the substance from large volumes of solution can be highly time-consuming, and it is in this field where the use of magnetic adsorbents called magnetic solid-phase extraction gains significance throughout the world. Graphene-based magnetic nanocomposites have been synthesized and applied for the removal of heavy metals from the liquid phase, and the removal kinetics and equilibrium isotherms were modeled by various mathematical equations given by different eminent scientists and researchers. Several kinetic models such as pseudo-first-order, pseudo-second-order, and intraparticle diffusion have been widely tested to the validity of experimental data of the adsorption system. The equilibrium isotherms were modeled with the Langmuir, Freundlich, Redlich–Peterson, and Dubinin–Radushkevich isotherm equation to find out the suitable adsorption isotherms. However, the Langmuir isotherm equation is the most widely used isotherm model to fit the experimental data for various adsorption systems. The bare graphene oxides have low adsorption capacity toward a variety of adsorbents, so it has been widely functionalized with inorganic or organic moieties to enhance the sorption capacity for metal ions in aqueous media. Therefore, the graphene oxide has been modified or functionalized with CdS, ZnO, TiO_2, AlOOH, and MnO_2. The graphene doped with magnetic nanoparticles have enhanced utility, as it can be easily recycled from the treated water systems and can be reused in another cycle. There are enough experimental data on the adsorption experiment in batch mode for the removal of pollutants from water. However, there is an urgent need to perform the experiment for multimetal removal in batch mode, and further it should be explored on column studies for realistic applications at the industrial scale.

REFERENCES

Afkhami, A., Moosavi, R., 2010. Adsorptive removal of Congo red, a carcinogenic textile dye, from aqueous solutions by maghemite nanoparticles. J. Hazard. Mater. 174, 398–403.

Afkhami, A., Madrakian, T., Amini, A., 2009. Mo(VI) and W(VI) removal from water samples by acid-treated high area carbon cloth. Desalination 243, 258–264.
Ai, L., Zhang, C., Chen, Z., 2011. Removal of methylene blue from aqueous solution by a solvothermal synthesized graphene/magnetite composite. J. Hazard. Mater. 192, 1515–1524.
Aksu, Z., 2001. Equilibrium and kinetic modelling of cadmium(II) biosorption by *C. vulgaris* in a batch system: effect of temperature. Sep. Purif. Technol. 21, 285–294.
Al-Degs, Y.S., El-Barghouthi, Musa I., Issa, A.A., Khraisheh, M.A., Walker, G.M., 2006. Sorption of Zn(II), Pb(II), and Co(II) using natural sorbents: equilibrium and kinetic studies. Water Res. 40, 2645–2658.
Boyd, G.E., Adamson, A.W., Myers, L.S., 1947. The exchange adsorption of ions from aqueous solutions by organic zeolites, II, Kinetics. J. Am. Chem. Soc. 69, 2836–2848.
Brunauer, S., Emmett, P.H., Teller, E., 1938. Adsorption of gases in multimolecular layers. J. Am. Chem. Soc. 60, 309–316.
Chakraborty, S., Chowdhury, S., Saha, P.D., 2011. Adsorption of Crystal Violet from aqueous solution onto NaOH-modified rice husk. Carbohydr. Polym. 86, 1533–1541.
Chandra, V., Park, J., Chun, Y., Lee, J.W., Hwang, I.C., Kim, K.S., 2010. Water-dispersible magnetite-reduced graphene oxide composites for arsenic removal. ACS Nano 4, 3979–3986.
Chatterjee, S., Lee, D.S., Lee, M.W., Woo, S.H., 2009. Enhanced adsorption of congo red from aqueous solutions by chitosan hydrogel beads impregnated with cetyl trimethyl ammonium bromide. Bioresour. Technol. 100, 2803–2809.
Chen, L., Lu, L., Shao, W., Luo, F., 2011. Kinetics and equilibria of Cd(II) adsorption onto a chemically modified lawny grass with H[BTMPP]. J. Chem. Eng. Data 56, 1059–1068.
Deng, X., LüL, Li, H., Luo, F., 2010. The adsorption properties of Pb(II) and Cd(II) on functionalized graphene prepared by electrolysis method. J. Hazard. Mater. 183, 923–930.
Dubinin, M.M., 1960. The potential theory of adsorption of gases and vapors for adsorbents with energetically non uniform surface. Chem. Rev. 60, 235–266.
Emmanuel, K.A., Rao, A.V., 2008. Adsorption of Mn(II) from aqueous solutions using *Pithacelobium dulce* carbon. Rasayan J. Chem. 1, 840–852.
Fan, L., Luo, C., Sun, M., Li, X., Qiu, H., 2013. Highly selective adsorption of lead ions by water-dispersible magnetic chitosan/graphene oxide composites. Colloids Surf. B 103, 523–529.
Febrianto, J., Kosasih, A.N., Sunarso, J., Ju, Y.H., Indraswati, N., Ismadji, S., 2009. Equilibrium and kinetic studies in adsorption of heavy metals using biosorbent: a summary of recent studies. J. Hazard. Mater. 162, 616–645.
Foo, K.Y., Hameed, B.H., 2010. Insights into the modeling of adsorption isotherm systems. Chem. Eng. J. 156, 2–10.
Freundlich, H.M.F., 1906. Uber die adsorption in lasugen. J. Phys. Chem 57, 385–470.
Gautam, R.K., Mudhoo, A., Lofrano, G., Chattopadhyaya, M.C., 2014. Biomass-derived biosorbents for metal ions sequestration: adsorbent modification and activation methods and adsorbent regeneration. J. Environ. Chem. Eng. 2, 239–259.
Gautam, R.K., Rawat, V., Banerjee, S., Sanroman, M.A., Soni, S., Singh, S.K., Chattopadhyaya, M.C., 2015. Synthesis of bimetallic Fe-Zn nanoparticles and its application towards adsorptive removal of carcinogenic dye malachite green and Congo red in water. J. Molec. Liquids 212, 227–236.
Gerente, C., Lee, V.K.C., Le Cloirec, P., McKay, G., 2007. Application of chitosan for the removal of metals from wastewaters by adsorption—mechanisms and models review. Crit. Rev. Environ. Sci. Technol. 37, 41–127.
Gupta, S.S., Bhattacharyya, K.G., 2006. Adsorption of Ni(II) on clays. J. Colloid Interface Sci. 295, 21–32.
Hanif, M.A., Nadeem, R., Bhatti, H.N., Ahmad, N.R., Ansari, T.M., 2007. Ni(II) biosorption by *Cassia fistula* (Golden Shower) biomass. J. Hazard. Mater. B139, 345–355.

Ho, Y.S., McKay, G., 1999. Pseudo-second order model for sorption processes. Process Biochem. 34, 451–465.
Ho, Y.S., Ng, J.C., McKay, G., 2000. Kinetics of pollutants sorption by biosorbents: review. Separation Purif. Methods 29, 189–232.
Hu, X., Liu, Y., Wang, H., Chen, A., Zeng, G., Liu, S., Guo, Y., Hu, X., Li, T., Wang, Y., Zhou, L., Liu, S., 2013. Removal of Cu(II) ions from aqueous solution using sulfonated magnetic graphene oxide composite. Sep. Purif. Technol. 108, 189–195.
Huang, Z.H., Zheng, X.Y., Lv, W., Wang, M., Yang, Q.H., Kang, F., 2011. Adsorption of Lead(II) ions from aqueous solution on low-temperature exfoliated graphene nanosheets. Langmuir 27, 7558–7562.
Karagoz, S., Tay, T., Ucar, S., Erdem, M., 2008. Activated carbons from waste biomass by sulfuric acid activation and their use on methylene blue adsorption. Bioresour. Technol. 99, 6214–6222.
Lagergren, S., 1898. About the theory of so-called adsorption of soluble substances. Kung. Svenska Vet. Hand. 24, 1–39.
Langmuir, I., 1918. The adsorption of gases on plane surfaces of glass, mica and platinum. J. Am. Chem. Soc. 40, 1361–1403.
Lee, Y.C., Yang, J.W., 2012. Self-assembled flower-like TiO_2 on exfoliated graphite oxide for heavy metal removal. J. Ind. Eng. Chem. 18, 1178–1185.
Lee, C.K., Low, K.S., Chew, S.L., 1999. Removal of anionic dyes by water hyacinth roots. Adv. Environ. Res. 3, 343–351.
Lei, Y., Chen, F., Luo, Y., Zhang, L., 2014. Three-dimensional magnetic graphene oxide foam/Fe3O4 nanocomposite as an efficient absorbent for Cr(VI) removal. J. Mater. Sci. 49, 4236–4245.
Li, S., Lu, X., Xue, Y., Lei, J., Zheng, T., Wang, C., 2012. Fabrication of polypyrrole/graphene oxide composite nanosheets and their applications for Cr(VI) in aqueous solution. PLoS One 7, e43328.
Li, L., Fan, L., Sun, M., Qiu, H., Li, X., Duan, H., Luo, C., 2013. Adsorbent for chromium removal based on graphene oxide functionalized with magnetic cyclodextrin–chitosan. Colloids Surf. B 107, 76–83.
Li, Y., Wang, C., Guo, Z., Liu, C., Wu, W., 2014. Sorption of thorium(IV) from aqueous solutions by graphene oxide. J. Radioanal. Nucl. Chem. 299, 1683–1691.
Machida, M., Mochimaru, T., Tatsumoto, H., 2006. Lead(II) adsorption onto the graphene layer of carbonaceous materials in aqueous solution. Carbon 44, 2681–2688.
Madadrang, C.J., Kim, H.Y., Gao, G., Wang, N., Zhu, J., Feng, H., Gorring, M., Kasner, M.L., Shifeng, Hou., 2012. Adsorption behavior of EDTA-graphene oxide for Pb (II) removal. ACS Appl. Mater. Interfaces 4, 1186–1193.
McKay, G., Allen, S.J., 1980. Surface mass transfer processes using peat as an adsorbent for dyestuffs. Canadian J. Chem. Eng. 58, 521–526.
Mi, X., Huang, G., Xie, W., Wang, W., Liu, Y., Gao, J., 2012. Preparation of graphene oxide aerogel and its adsorption for Cu^{2+} ions. Carbon 50, 4856–4864.
Mishra, A.K., Ramaprabhu, S., 2011. Functionalized graphene sheets for arsenic removal and desalination of sea water. Desalination 282, 39–45.
Mohan, D., Singh, K.P., 2002. Single- and multi-component adsorption of cadmium and zinc using activated carbon derived from bagasse—an agricultural waste. Water Res. 36, 2304–2318.
Nandi, D., Gupta, K., Ghosh, A.K., De, A., Banerjee, S., Ghosh, U.C., 2012. Manganese-incorporated iron(III) oxide–graphene magnetic nanocomposite: synthesis, characterization, and application for the arsenic(III)-sorption from aqueous solution. J. Nanopart. Res. 14, 1272.
Pan, N., Guan, D., He, T., Wang, R., Wyman, I., Jin, Y., Xia.F C., 2013. Removal of Th^{4+} ions from aqueous solutions by graphene oxide. J. Radioanal. Nucl. Chem. 298, 1999–2008.

Parham, H., Zargar, B., Shiralipour, R., 2012. Fast and efficient removal of mercury from water samples using magnetic iron oxide nanoparticles modified with 2-mercaptobenzothiazole. J. Hazard. Mater. 205–206, 94–100.
Pavasant, P., Apiratikul, R., Sungkhum, V., Suthiparinyanont, P., Wattanachira, S., Marhaba, T.F., 2006. Biosorption of Cu^{2+}, Cd^{2+}, Pb^{2+}, and Zn^{2+} using dried marine green macroalga *Caulerpa lentillifera*. Bioresour. Technol. 97, 2321–2329.
Redlich, O., Peterson, D.L., 1959. A useful adsorption isotherm. J. Phys. Chem. 63, 1024–1026.
Ren, Y., Yan, N., Wen, Q., Fan, Z., Wei, T., Zhang, M., Ma.F J., 2011. Graphene/δ-MnO_2 composite as adsorbent for the removal of nickel ions from wastewater. Chem. Eng. J. 175, 1–7.
Sahoo, A.K., Srivastava, S.K., Raul, P.K., Gupta, A.K., Shrivastava, R., 2014. Graphene nanocomposites of CdS and ZnS in effective water purification. J. Nanopart Res. 16, 2473.
Sarkar, D., Chattoraj, D.K., 1993. Activation parameters for kinetics of protein adsorption at silica water interface. J. Colloid Interface Sci. 157, 219–226.
Singh, D.B., Prasad, G., Rupainwar, D.C., 1996. Adsorption technique for the treatment of As(V)-rich effluents. Colloids Surf. A 111, 49–56.
Sips, R., 1948. On the structure of a catalyst surface. J. Chem. Phys. 16, 490–495.
Sun, Y., Wang, Q., Chen, C., Tan, X., Wang, X., 2012. Interaction between Eu(III) and graphene oxide nanosheets investigated by batch and extended X-ray absorption fine structure spectroscopy and by modeling techniques. Environ. Sci. Technol. 46, 6020–6027.
Temkin, M.J., Pyzhev, V., 1940. Recent modifications to Langmuir isotherms. Acta Physichim. USSR 12, 217–222.
Thomas, J.M., Thomas, W.J., 1997. Principle and practice of heterogeneous catalysis. VCH, Weinheim.
Tian, Y., Wu, M., Lin, X., Huang, P., Huang, Y., 2011. Synthesis of magnetic wheat straw for arsenic adsorption. J. Hazard. Mater. 193, 10–16.
Varshney, K.G., Khan, A.A., Gupta, U., Maheshwari, S.M., 1996. Kinetics of adsorption of phosphamidon on antimony(V) phosphate cation exchanger: evaluation of the order of reaction and some physical parameters. Colloid Surf. A 113, 19–23.
Wang, J., Chen, C., 2006. Biosorption of heavy metals by *Saccharomyces cerevisiae*: a review. Biotechnol. Adv. 24, 427–451.
Wang, X., Yu, J., 2014. Application of Fe_3O_4/graphene oxide composite for the separation of Cs(I) and Sr(II) from aqueous solution. J Radioanal. Nucl. Chem. 303, 807–813.
Wang, H., Yuan, X., Wu, Y., Huang, H., Zeng, G., Liu, Y., Wang, X., Lin, N., Qi, Y., 2013. Adsorption characteristics and behaviors of graphene oxide for Zn(II) removal from aqueous solution. Appl. Surf. Sci. 279, 432–440.
Weber, W.J., Morris, J.C., 1963. Kinetics of adsorption on carbon from solution. J. Sanit. Eng. Div. Am. Soc. Civ. Eng. 89, 31–60.
Wu, F.C., Tseng, R.L., Juang, R.S., 2000. Comparative adsorption of metal and dye on flake– and bead–types of chitosans prepared from fishery wastes. J. Hazard. Mater. 73, 63–75.
Wu, F.C., Tseng, R.L., Juang, R.S., 2001. Kinetic modeling of liquid-phase adsorption of reactive dyes and metal ions on chitosan. Water Res. 35, 613–618.
Wu, X.L., Wang, L., Chen, C.L., Xu, A.W., Wang, X.K., 2011. Water dispersible magnetite-graphene-LDH composites for efficient arsenate removal. J. Mater. Chem. 21, 17353–17359.
Wu, W., Yang, Y., Zhou, H., Ye, T., Huang, Z., Liu, R., Kuang, Y., 2013. Highly efficient removal of Cu(II) from aqueous solution by using graphene oxide. Water Air Soil Pollut. 224, 1372.
Yang, S.T., Chang, Y., Wang, H., Liu, G., Chen, S., Wang, Y., Liu, Y., Cao, A., 2010. Folding/aggregation of graphene oxide and its application in Cu^{2+} removal. J. Colloid Interface Sci. 351, 122–127.

Yuan, X., Wang, Y., Wang, J., Zhou, C., Tang, Q., Rao, X., 2013. Calcined graphene/MgAl-layered double hydroxides for enhanced Cr(VI) removal. Chem. Eng. J. 221, 204–213.
Zhang, N., Qiu, H., Si, Y., Wang, W., Gao, J., 2011. Fabrication of highly porous biodegradable monoliths strengthened by graphene oxide and their adsorption of metal ions. Carbon 49, 827–837.
Zhao, G., Ren, X., Gao, X., Tan, X., Li, J., Chen, C., Huang, Y., Wang, X., 2011a. Removal of Pb(II) ions from aqueous solutions on few-layered graphene oxide nanosheets. Dalton Trans. 40, 10945–10952.
Zhao, G., Li, J., Ren, X., Chen, C., Wang, X., 2011b. Few-layered graphene oxide nanosheets as superior sorbents for heavy metal ion pollution management. Environ. Sci. Technol. 45, 10454–10462.
Zhao, G., Wen, T., Yang, X., Yang, S., Liao, J., Hu, J., Shao, D., Wang, X., 2012. Preconcentration of U(VI) ions on few-layered graphene oxide nanosheets from aqueous solutions. Dalton Trans. 41, 6182–6188.
Zong, P., Wang, S., Zhao, Y., Wang, H., Pan, H., He, C., 2013. Synthesis and application of magnetic graphene/iron oxides composite for the removal of U(VI) from aqueous solutions. Chem. Eng. J. 220, 45–52.

CHAPTER 6

Sorption of Dyes on Graphene-Based Nanocomposites

6.1 ADSORPTION OF DYES

Dyes are a kind of organic compounds which can bring bright and firm color to other substances. Several industries such as textile, wool making, paper industries, paints and pigments use the dye on large amounts. On the way of processing the textile dyeing, it generates a large amount of wastewater effluents. The discharge of dye bearing wastewater into the aquatic media causes many detrimental effects on flora and fauna. This not only affects aesthetic merit but also inhibits sunlight penetration and reduces photosynthetic action of green plants. In addition, some dyes or their metabolites are either toxic or mutagenic, and carcinogenic. Therefore, there is an urgent need to remove dyes before effluent is discharged into receiving aquatic system. Currently, the most popular treatment methods for dye-laden wastewater are combinations of biological treatment (Sudarjanto et al., 2006), chemical coagulation (Shi et al., 2007; Lee et al., 2006), photocatalytic degradation (Sannino et al., 2013), electrochemical degradation (Aquino et al., 2013), photo-Fenton processes (Hernández-Rodríguez et al., 2014), and activated carbon adsorption (Gautam et al., 2013a, 2014; Liu et al., 2013b).

Efforts should be made to minimize the use and generation of hazardous chemicals, while increasing the treatment efficiencies of the wastewater generated. Therefore, investigation of alternative and appropriate technologies for the removal of toxic dyes is of utmost importance. Adsorption onto activated carbon is proven to be very effective in treating dye-laden industrial effluents. However, in view of high cost and associated problems of regeneration, there is a constant search for alternative low cost adsorbents. Such types of adsorbents includes mustard husk (Gautam et al., 2013b), chitosan (Sadeghi-Kiakhani et al., 2013; Demarchi et al., 2013), cactus (Barka et al., 2013), oil palm trunk fibers (Hameed and El-Khaiary, 2008), durian (*Durio zibethinus* Murray) peel (Hameed and Hakimi, 2008), almond shell

Nanomaterials for Wastewater Remediation
http://dx.doi.org/10.1016/B978-0-12-804609-8.00006-6

(Ardejani et al., 2008), sugarcane baggase (Gupta and Ali, 2004; Gupta et al., 2003), and sphagnum moss peat (Allen et al., 2004).

This chapter highlights the potential of graphene-based nanomaterials and nanocomposites for the adsorptive removal of dyes from aqueous solution for the economically viable and sustainable management of wastewater. Optimization parameters like pH, adsorbent dose, and initial concentration of dye, equilibrium time, rotation speed, and temperature on the removal of dyes have been described in this chapter.

Reduced graphene oxide was modified with SnO_2 and TiO_2 by a direct redox reaction between graphene oxide and $SnCl_2$ or $TiCl_3$ at 90°C (Zhang et al., 2011). The method offers several advantages, including (1) an extra reducing agent such as toxic hydrazine is not required for the reduction of graphene oxide to reduced graphene oxide, (2) the in situ growth of metal oxides leads to the formation of uniform nanoparticles on individual reduced graphene oxide sheets, and (3) the process can be carried out under mild conditions. Using the designed systems of reduced graphene oxide–SnO_2 and reduced graphene oxide–TiO_2, the authors (Zhang et al., 2011) have demonstrated that the photocatalytic degradation of Rhodium B under visible light irradiation. The reduced graphene oxide–SnO_2 nanocomposite exhibited a superior photocatalytic activity under visible light irradiation compared to reduced graphene oxide–TiO_2 and P25, resulting from the effective retardation of charge recombination in the presence of reduced graphene oxide.

Recently, a novel method for the preparation of β-cyclodextrin grafted graphene oxide (GO-β-CD) had been developed and used for the removal of fuchsin acid, methyl orange, and methylene blue from the solution (Wang et al., 2014). The GO-β-CD had an excellent adsorption capacity for fuchsin acid and could be recycled and reused. The adsorption capacities for the three dyes were in the order: fuchsin acid > methylene blue > methyl orange.

A simple one step solvothermal strategy using nontoxic and cost-effective precursors has been developed to prepare magnetite/reduced graphene oxide (MRGO) nanocomposites for removal of dye pollutants from aqueous solutions (Sun et al., 2011). The MRGO nanocomposites exhibited excellent removal efficiency for rhodamine B and malachite green, which was 91 and 94%, respectively. It showed rapid separation from aqueous solution by an external magnetic field. Interestingly, the performance of the MRGO composites is strongly dependent on both the loading of Fe_3O_4 and the pH value. It has been observed that the removal efficiency of malachite green increased in a pH range from 4 and 10, while the removal efficiency of rhodamine B was maximized at pH 7. As the pH increased, the number of negatively charged surface sites on the MRGO increased. Malachite green has an overall positive charge in aqueous solution (Hameed and El-Khaiary, 2008; Ngah et al., 2010), which increases the electrostatic attraction between the adsorbent and the positively charged dyes, and thus higher removal efficiency was observed with increasing pH. However, rhodamine B has a carboxylic group (Vijayakumar et al., 2010). Thus its higher removal was observed at low pH, and at high pH, the acidic group dissociated, and thus partly neutralized the positive charge of the dye. As a result, the removal efficiency of the dye decreased at high pH. The effect of adsorbent dose of MRGO nanocomposites was also reported. As indicated in Fig. 6.1a, the characteristic UV–vis absorption of rhodamine B at 554 nm and malachite green at 625 nm, Fig. 6.1c decreased on increasing the adsorbent dose of MRGO-4 nanocomposites. When the amount of MRGO-4 was increased to 0.7 g/L, over 91% of rhodamine B and over 94% of malachite green could be removed within 2 h at room temperature. The adsorption behavior of MRGO adsorbent fitted well with the Freundlich isotherm and the pseudo-second-order kinetic model. Table 6.1 shows the application of graphene-based materials as adsorbents for the removal of dyes from aqueous solutions.

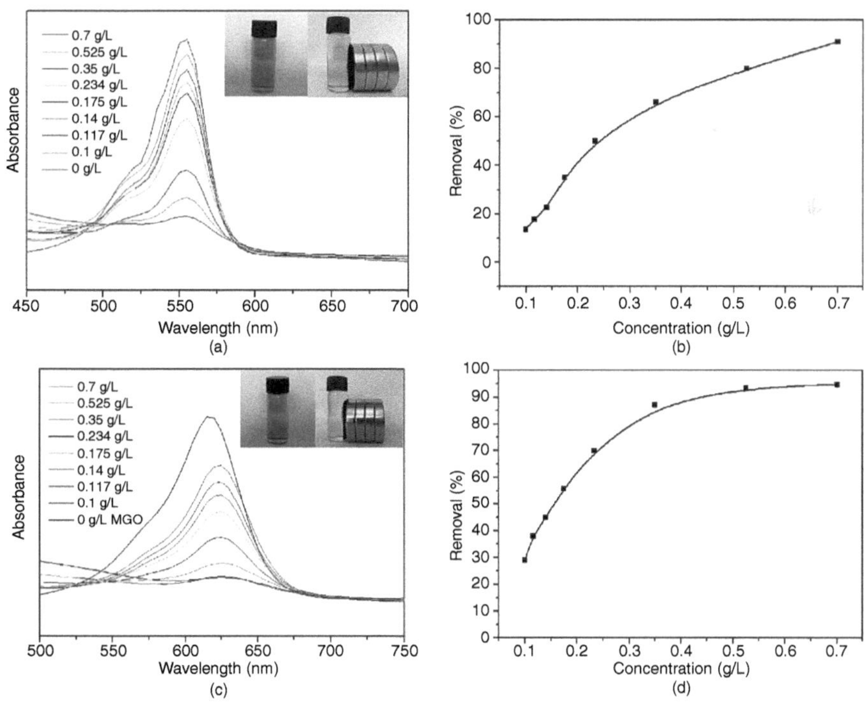

Figure 6.1 The removal of rhodamine B (a, b) and malachite green (c, d) at different concentrations of MRGO-4. Inserts in (a, c) show the photos of dye solutions before and after using MRGO-4 (0.7 g/L) *(Source: Sun et al., 2011, © Reproduced with kind permission of the Tsinghua University Press and Springer-Verlag Berlin Heidelberg).*

Table 6.1 Application of graphene-based materials as adsorbents for the removal of dyes from aqueous solutions

Materials	Dye	Adsorption capacity (mg/g)	References
Graphene	Methylene blue	153.85	Liu et al. (2012)
Graphene	Methylene blue	~1520	Wu et al. (2011)
Graphene oxide	Methylene blue	17.3	Ramesha et al. (2011)
Graphene oxide	Methyl violet	2.47	Ramesha et al. (2011)
Reduced graphene oxide	Orange G	5.98	Ramesha et al. (2011)
Graphene oxide	Methylene blue	714	Yang et al. (2011)
Modified graphene oxide	Acridine orange	3300	Sun et al. (2012)
Graphene oxide-Fe_3O_4	Methylene blue	190.14	He et al. (2010)
Graphene oxide-Fe_3O_4	Red cationic dyes	140.79	He et al. (2010)
Magnetite/reduced graphene oxide	Malachite green	22	Sun et al. (2011)
Magnetic graphene	Organic dye fuchsine	89.4	Wang et al. (2011)
Magnetite graphene	Methylene blue	43.82	Ai et al. (2011)
Magnetic chitosan–graphene oxide	Methyl blue	95.16	Fan et al. (2012)
Graphene oxide–Fe_3O_4–SiO_2	Methylene blue	111.1	Yao et al. (2012b)
Reduced graphene oxide–$MnFe_2O_4$	Methylene blue	35	Bai et al. (2012)
Reduced graphene oxide–$MnFe_2O_4$	Rhodamine B	23	Bai et al. (2012)
Reduced graphene oxide nanosheets	Methylene blue	54	Bai et al. (2012)
Reduced graphene oxide nanosheets	Rhodamine B	37	Bai et al. (2012)
CdS/graphene	Methylene blue	—	Sahoo et al. (2014)
ZnS/graphene	Methylene blue	—	Sahoo et al. (2014)
Magnetic graphene oxide	Methylene blue	~70	Zhang et al. (2015)
Polyacrylic acid functionalized magnetic Fe_3O_4 nanoparticle-graphene oxide nanocomposite	Methylene blue	~291	Zhang et al. (2015)
Impregnated Fe_3O_4/graphene oxide nanocomposites	Reactive Black 5	164	Kyzas et al. (2013)

Table 6.1 Application of graphene-based materials as adsorbents for the removal of dyes from aqueous solutions (*cont.*)

Materials	Dye	Adsorption capacity (mg/g)	References
Co-precipitated Fe_3O_4/ graphene oxide nanocomposite	Reactive Black 5	188	Kyzas et al. (2013)
Superparamagnetic graphene oxide–Fe_3O_4 hybrid composite	Methylene blue	167.2	Xie et al. (2012)
Superparamagnetic graphene oxide–Fe_3O_4 hybrid composite	Neutral red	171.3	Xie et al. (2012)
Graphene-Fe_3O_4	Pararosaniline	198.23	Wu et al. (2013)

6.2 GRAPHENE-BASED MAGNETIC NANOCOMPOSITES

The magnetic properties of the graphene-based magnetic nanocomposites were mostly investigated at room temperature (300 K) with an applied magnetic field—$15,000 \leq H \leq 15,000$ Oe. Table 6.2 shows the magnetic properties, including saturation magnetization (M_s = emu/g), coercivity (H_c = Oe), and remanence (M_r = emu/g) for the graphene-based magnetic nanocomposites.

A one-pot solvothermal synthesis method was developed to prepare reduced graphene oxide (RGO) supported ferrite (MFe_2O_4, M = Mn, Zn, Co, and Ni) hybrids using graphite oxide and metal ions (Fe^{3+} and M^{2+}) as starting materials (Bai et al., 2012). The prepared hybrids possess considerable saturation magnetization which was very helpful in the recycling of the adsorbent after the adsorption of dye in water. It was found that over 92% rhodamine B and 100% methylene blue with a concentration of 5 mg/L can be removed by the hybrids within 2 min when the adsorbent dose of the hybrids was 0.6 g/L.

Yao et al. (2012a) synthesized the magnetic Fe_3O_4@graphene composite and applied for the removal of dyes from aqueous media. Through a chemical deposition method, Fe_3O_4 nanoparticles in size of 30 nm were homogeneously dispersed onto graphene sheets. Adsorption isotherm and kinetics of methylene blue and Congo red onto magnetic Fe_3O_4@graphene composite were studied in a batch system. The maximum adsorption capacities of methylene blue and Congo red on magnetic Fe_3O_4@graphene composite were found to be 45.27 and 33.66 mg/g, respectively. The adsorption

Table 6.2 Magnetic properties of graphene based magnetic nanocomposites at room temperature

Samples	Saturation magnetization (M_s = emu/g)	Coercivity (H_c = Oe)	Remanence (M_r = emu/g)	References
Reduced graphene oxide–$MnFe_2O_4$	40.5	44.3	2.2	Bai et al. (2012)
Reduced graphene oxide–$ZnFe_2O_4$	34.5	39.4	1.5	Bai et al. (2012)
Reduced graphene oxide–$CoFe_2O_4$	48.8	117.4	3.1	Bai et al. (2012)
Reduced graphene oxide–$NiFe_2O_4$	32.3	48.4	1.5	Bai et al. (2012)
Reduced graphene oxide–$MnFe_2O_4$-1/2	28.9	76.2	3.6	Bai et al. (2012)
Reduced graphene oxide–$MnFe_2O_4$-2	64.9	82.9	5.1	Bai et al. (2012)
Superparamagnetic graphene oxide–Fe_3O_4 hybrid composite	15.0	—	—	Xie et al. (2012)
Graphene oxide-Fe_3O_4	48.5	—	—	Shen et al. (2010)
Fe	~90	—		Shen et al. (2010)
Manganese-incorporated iron(III) oxide (IMBO)–graphene nanocomposite at 5 Kelvin	3.9	230	0.42	Nandi et al. (2012)
IMBO at 5 Kelvin	7.3	232	0.77	Nandi et al. (2012)
IMBO–graphene nanocomposite at 100 Kelvin	3.2	92	0.21	Nandi et al. (2012)
IMBO at 100 Kelvin	5.8	96	0.36	Nandi et al. (2012)
IMBO–graphene nanocomposite at 300 Kelvin	2.4	32	0.09	Nandi et al. (2012)
IMBO at 300 Kelvin	4.5	38	0.17	Nandi et al. (2012)
Graphene-Fe_3O_4	61.2	—	—	Wu et al. (2013)

kinetics of removal processes fitted well with the pseudo-second order kinetic equation.

A superparamagnetic graphene–Fe_3O_4 nanocomposite (G/Fe_3O_4) was synthesized by a facile one-pot solvothermal method and used as an adsorbent to remove dye pararosaniline from water (Wu et al., 2013). In comparison with G/Fe_3O_4 prepared by the in situ chemical coprecipitation, the newly prepared G/Fe_3O_4 had higher adsorption efficiency for the dye. The adsorption kinetics, adsorption capacity of the adsorbent, and the effect of the adsorbent dosage and solution pH on the removal efficiency of pararosaniline were investigated in batch mode. The adsorption capacity of G/Fe_3O_4 for pararosaniline was evaluated using the Freundlich and Langmuir adsorption isotherm models. The Langmuir maximum adsorption capacity was found to 198.23 mg/g and the separation factor was 0.0523, which shows the favorable nature of adsorbate on adsorbent. The regeneration of G/Fe_3O_4 adsorbent was performed and the results showed that ethanol could be used as the desorption solvent for the regeneration of the G/Fe_3O_4 adsorbent; the desorption ratio for the dye was around 92% when ethanol at pH 3 was used, and the G/Fe_3O_4 adsorbent were regenerated and reused at least five times without a significant loss of the sorption capacity.

Magnetic β-cyclodextrin–chitosan/graphene oxide materials (MCCG) were fabricated through a facile chemical route and their application as excellent adsorbents for dye removal were demonstrated in batch mode by Fan et al. (2013). The results showed that, benefiting from the surface property of graphene oxide, hydrophobicity of β-cyclodextrin, the abundant amino and hydroxyl functional groups of chitosan, and from the magnetic property of Fe_3O_4, the adsorbent possesses quite a good and versatile adsorption capacity to the methylene blue. The saturation magnetization of the MCCG composite was 55.13 emu/g, indicated that the MCCG composite was highly magnetic and was able to recycle by the external magnet. The BET surface area and pore volume of MCCG estimated from Barret–Joyner–Halenda analysis of the isotherms were determined to be 402.1 m^2/g and 0.4152 cm^3/g, respectively. The average of the pore size distribution was 3.178 nm indicated that MCCG was a mesoporous material. The adsorption kinetics and equilibrium isotherm were well-described by pseudo-second-order kinetic and Langmuir isotherm model, respectively. The thermodynamic parameters suggested that the adsorption process was spontaneous and endothermic in nature.

A graphene-based magnetic nanocomposite was synthesized and used as an adsorbent for the removal of a dye fuchsine from aqueous solutions

(Wang et al., 2011). Effects of various process parameters such as adsorbent dosage, ionic strength, kinetics, and solution pH on the removal efficiency of fuchsine were investigated. The resultant kinetic data were well fitted by a pseudo second-order model. Freundlich model and Langmuir model were used to study the adsorption isotherms. The results indicated that the linear correlation coefficients for Langmuir and Freundlich models were 0.997 and 0.947, respectively. The higher correlation coefficient for Langmuir model indicated that the Langmuir model fitted the adsorption data better than the Freundlich. The Langmuir maximum adsorption capacity for fuchsine onto the graphene magnetic nanocomposite was 89.4 mg/g. The graphene magnetic nanocomposite proved to be a highly efficient adsorbent for separation purposes.

A functional hybrid of reduced graphene oxide (RGO)–Fe_3O_4 nanoparticles was chemically synthesized with exceptionally high yield and tunable RGO/Fe_3O_4 ratio (Geng et al., 2012). The adsorption behaviors of a series of rhodamine B using the hybrid as the adsorbent were systematically investigated in aqueous solutions. The results showed that, benefiting both from the surface property of RGO and from the magnetic property of Fe_3O_4, the hybrid possesses quite a good and versatile adsorption capacity and can be easily and rapidly extracted from water by external magnetic attraction. Since the molecular weight of RGO is much smaller than that of Fe_3O_4 nanoparticles and the former dominates the molecular adsorption, the adsorptive capacity of RGO/Fe_3O_4 adsorbent was greatly increased and meanwhile the RGO/Fe_3O_4 hybrid weight was greatly reduced by just simply increasing the atomic ratio of C/Fe in the hybrid. The maximum adsorption capacity was 44.4 mg/g for the rhodamine B adsorption onto the synthesized RGO/Fe_3O_4 hybrid. The RGO/Fe_3O_4 hybrid was also suitable for the adsorption of dyes from multidye solution without compromising the adsorption capacity, as compared with that measured separately. The inherent advantages of this nanostructured adsorbent, such as noncompromised adsorption capacity, low cost, easy, rapid extraction and regeneration, good tolerance, multiplex adsorbability, and handy operation, may pave a new, efficient, and sustainable way toward highly-efficient dye pollutant removal from aqueous solutions.

6.3 PHOTOCATALYTIC DEGRADATION

Photocatalyst has emerged as one of the most promising materials for environmental remediation. It represents an easy way to utilize the energy of either natural sunlight or indoor illumination light. The mechanism of these functions is based on the in situ generated highly reactive oxygen species ($OH^{•}$, $O_2^{•-}$) for mineralization of organic compounds. Photocatalysts, such as TiO_2, ZnO, CdS, WO_3, SnO_2, and ZnS, have demonstrated their efficiency in degrading a wide range of organic pollutants into readily biodegradable compounds, and eventually mineralizing them to carbon dioxide and water (Wang et al., 2015). Among the catalysts developed, TiO_2 has been extensively studied for the degradation of environmental pollutants from aqueous medium due to its low cost, high physicochemical stability, and super catalytic properties. However, the application of TiO_2 is limited by the fast recombination of the electron and photogenerated electron–hole pairs. The addition of graphene to TiO_2 prevents the recombination of the electron–hole pairs in the TiO_2@graphene composites. The TiO_2@graphene composite has a number of attributes, such as increased adsorptivity of pollutants, extended light absorption range, and facile charge transportation and separation, which were rarely reported for other composite carbon composites (Zhang et al., 2010a).

Recently, Gao et al. (2013) synthesized the graphene oxide (GO)–CdS composites and applied the as-prepared composites for the photocatalytic degradation of acid orange 7, methylene blue, and rhodamine B from aqueous solutions. The result showed that GO–CdS composites exhibit higher efficiency in photodegradation of dyes than pure CdS nanoparticles under visible light irradiation. More than 80% of acid orange 7 were degraded by GO–CdS composites, while only 50% of AO7 was decomposed by pure CdS nanoparticles within 60 min. In addition, the interactions between GO sheets and CdS nanoparticles inhibited the photocorrosion of CdS and leaching of Cd^{2+} was observed in the studied samples. However, only 3.5 wt.% Cd^{2+} of GO–CdS was leached out after photodegradation, while 38.6 wt.% Cd^{2+} of CdS was lost into aqueous solution. The excellent performances of GO–CdS composites was attributed to the effective charge transfer from CdS to GO reduces the recombination rate of photogenerated electron–hole pairs, uniform deposition of CdS on GO sheets eliminates aggregation of CdS nanoparticles, and the strong interactions between GO and CdS enhancing the durability of GO–CdS composites in aqueous medium.

The photocatalytic degradation of methylene blue from aqueous solution has been carried out in the absence/presence of CdS, ZnS, and corresponding graphene nanocomposites acting as catalysts, under dark/UV radiation (Sahoo et al., 2014). It has been reported that the degradation of methylene blue dye was very fast in the presence of catalysts and UV light.

These observations suggested that ~63 and 70% degradation of methylene blue in 40 min take place in the presence of CdS and ZnS nanoparticles, respectively. The degradation percentage was lower for CdS in comparison to ZnS, this might be due to the susceptibility of CdS toward the photocorrosion that could account for its lower photocatalytic activity in comparison to that of ZnS. Additionally, higher photocatalytic activity of ZnS could be attributed to the to its larger band gap of ZnS (3.68 eV) than CdS (2.42 eV) inhibiting the recombination of photogenerated electron–hole pair (Hu et al., 2011; Lee et al., 2005). When CdS/graphene and ZnS/graphene were used as a catalyst in the absence of UV radiation, only ~30% of methylene blue was degraded within 45 min on the simple stirring. Interestingly, methylene blue when subjected to UV radiation for 40 min in the presence of CdS/graphene (ZnS/graphene) composites corresponding to 0.25:1.00, 0.5:1.0, and 1:1 wt. ratio of Cd–MCDT: GO (ZnS–MCDT:GO) shows degradation of 81 (83), 84 (88), and 88 (93)%, respectively.

Recently, Samiee and Goharshadi (2014) synthesized graphene nanosheets as adsorbent for the azo dye reactive black 5 from aqueous solutions. The BET surface area and average pore diameter was 594.7 m^2/g and 6.28 nm, respectively, which shows that the synthesized material is a good adsorbent. The influence of different parameters including adsorbent dosage, pH, temperature, dye concentration, and ionic strength on the dye removal efficiency was investigated on the removal of reactive black 5 (Fig. 6.2). The experimental data were fitted well with the pseudo-second order kinetic model ($R^2 = 0.997$). The activation energy for adsorption of reactive black 5 on to the graphene nanosheets was 25.80 kJ/mol. It means the adsorption process was physically controlled. The adsorption equilibrium was fitted well to the Freundlich isotherm which indicated the multilayer adsorption of reactive black 5 on graphene nanosheets from aqueous solutions. The thermodynamic parameters confirmed that the reactive black 5 adsorption on graphene surface was spontaneous and endothermic in nature. Some significant features of using graphene as an adsorbent for reactive black 5 removal is the fast dye removal process, the low required amount of graphene, and the frequent usage of graphene with no change in its efficiency.

Reduced graphene oxide/potassium niobate composite nanoscrolls were prepared by incorporating reduced graphene oxide sheets into potassium niobate sheets at room temperature (Li et al., 2013b). Photocatalytic degradation of rhodamine B under UV light was carried out and the degradation efficiency of the composite nanoscrolls were reached to 68.3%

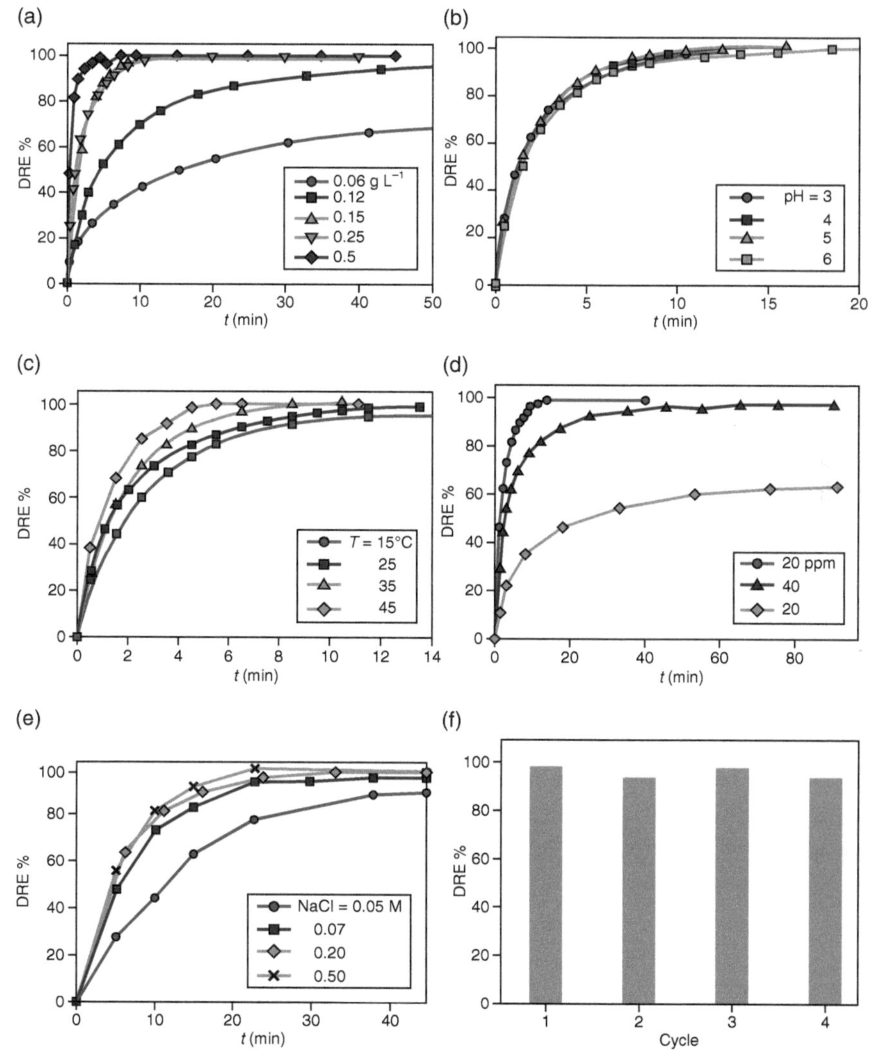

Figure 6.2 The effects of (a) graphene dosage (pH 3.0, $T = 25°C$, and dye concentration = 20 mg/L), (b) pH (0.25 g/L adsorbent, $T = 25°C$, and dye concentration = 20 mg/L), (c) temperature (0.25 g/L adsorbent, pH 3.0, and dye concentration = 20 mg/L), (d) initial dye concentration (0.25 g/L adsorbent, pH 3.0, and $T = 25°C$) on the reactive black 5 removal process, (e) ionic strength (0.25 g/L adsorbent, pH 3.0, $T = 25°C$, and initial dye concentration = 20 mg/L) and (f) reuse capacity of graphene (0.25 g/L adsorbent, pH 3.0, $T = 25°C$, and dye concentration = 20 mg/L) on the reactive black 5 dye removal efficiency percentage (DRE%) by graphene nanosheets. *(Source: Samiee and Goharshadi, 2014, © Reproduced with kind permission of the Springer Science + Business Media Dordrecht).*

after irradiation for 30 min, increased 34.3% relative to the pure potassium niobium nanotubes. The dramatically enhanced activity was attributed to the synergistic contribution of both a retarded charge recombination rate caused by a high electronic mobility of reduced graphene oxide and an increased surface adsorption of rhodamine B in water.

Well-dispersed iron oxide nanoparticles supported onto graphene oxide sheets (GO–Fe_3O_4 nanocomposite) were synthesized and used as heterogeneous Fenton-like catalyst for the degradation of acid orange 7 from aqueous solutions (Zubir et al., 2014). The degradation reaction was systematically investigated under various experimental conditions such as nanocomposite dosage, pH, temperature, oxidant, and dye concentrations. It has been found that the degradation was initially very fast and 80% degradation was achieved within 20 min, while ~98% of acid orange 7 was successfully removed after 180 min of reaction time. The degradation kinetics of acid orange 7 was most influenced by pH and temperature.

It is well known that pH has a significant impact in heterogeneous Fenton-like reaction, which directly affects the oxidative degradation of dyes by $HO^{\bullet}$ radicals. The optimal acid orange 7 degradation value of 98% was achieved at pH 3 within 180 min reaction. When the pH was lowered to 2.5, the acid orange 7 degradation decreased to nearly 80%. The result suggested that the protonation of H_2O_2 in forming an oxonium ion ($H_3O_2^+$) had occurred, which then enhanced H_2O_2 stability and reduced its reactivity toward $\equiv Fe^{2+}$ (Daud and Hameed, 2010). Moreover, the scavenging effect of HO^- radicals by the excessive H^+ was also enhanced at pH 2.5 (Navalon et al., 2011). The reduction in acid orange 7 degradation was also observed when the initial pH was higher than 3. The reaction kinetics was described by a pseudo-first-order reaction following the Langmuir–Hinshelwood mechanism. Analysis of the spent nanocomposite suggested that the phase of iron oxide nanoparticles remained unchanged while minor pore volume losses occurred via carbon deposition and/or restacking of graphene oxide sheets. The heterogeneous reaction was dominated by the rate of intrinsic chemical reactions on the surface of active sites rather than the rate of mass transfer. The graphene oxide-Fe_3O_4 nanocomposite delivered best acid orange 7 degradation of 98%. The spent catalysts was mainly stable, as the Fe_3O_4 phase remained similar to a pristine catalyst, though minor changes in surface area and pore volume were observed possibly attributed to the adsorption of intermediates or restacking of the GO–Fe_3O_4 sheets during the reaction.

Recently, catalyst P25–graphene was prepared and applied for degradation of dye from an aqueous solution (Wang et al., 2015). Three types of dyes were selected to determine the feasibility of the catalyst for the dye degradation, including sulfonic, azoic, and fluorescent dyes. P25–graphene catalyst showed better performance to degrade all selected dyes. They also investigated the influence of inorganic salts and surfactants on the photocatalytic degradation of rhodamine B using catalyst P25–graphene. The degradation of rhodamine B was suppressed by the presence of NaCl, but the effect of Na_2SO_4 was negligible. The degradation of rhodamine B was significantly suppressed by all three types of surfactant, namely anionic, cationic, and nonionic surfactants (Table 6.3).

Muthirulan et al. (2014) proposed a simple and proficient methodology for the preparation of hybrid photocatalyst based on TiO_2-graphene nanocomposite for acid orange 7 dye degradation under UV irradiation. HRTEM and SEM studies revealed that TiO_2 nanoparticles were uniformly dispersed on graphene surface. The photocatalytic degradation measurements demonstrated that the TiO_2-graphene composites exhibited an enhanced photocatalytic activity for acid orange 7 degradation under UV irradiation compared to pure TiO_2. This was due to the greater adsorptivity of dyes, extended light absorption and increased charge separation efficiency due to excellent electrical properties of graphene and the large surface contact between graphene and TiO_2 nanoparticles. As a result, the TiO_2-graphene composites can be widely used as a ternary composite photocatalyst for treating the organic contaminant in the field of environmental protection.

Reduced graphene oxide–titania (rGO–TiO_2) nanocomposites were prepared by hydrothermal method at different reaction temperatures and extensively characterized to decipher the chemistry (Chang et al., 2015). The photocatalytic properties of the nanocomposites were investigated toward the degradation of methylene blue under natural sunlight. The rGO–TiO_2 shows better photocatalytic activity due to the extended visible light absorption, excellent adsorptivity, and effective electron transfer process than the other controlled photocatalysts. In addition, rGO–TiO_2 showed good sustainability after subjecting it to five consecutive cycles of photodegradation. It showed the enhanced photocatalytic performance and good sustainability toward dye removal by rGO–TiO_2 nanocomposite as a potential candidate for wastewater treatment in textile and dyeing industries.

Table 6.3 Application of TiO_2@graphene composite for dye removal from aqueous solutions

Composites	Synthesis procedure	Dye	Light sources	References
P25@graphene	Reduced by hydrazine hydrate then annealed at 500°C	Methylene blue	UV light (220–280 nm) and visible light (400–1050 nm)	Liu et al. (2013a)
P25-reduced graphene oxide	Hydrothermal reaction	Rhodamine B, Methyl orange, Procion Red MX-5B, Terasil blue 3RL-02, Lanasol yellow CE4G, and Acid red 88	Simulated sunlight	Wang et al. (2015)
P25@reduced graphene oxide	Hydrothermal and annealed at 400°C under argon	Rhodamine B	UV light (365 nm) and LED lamp (>420 nm)	Wang et al. (2011)
P25@graphene	Hydrothermal in water	Methylene blue and reactive black	UV light (365 nm)	Li et al. (2013a)
P25@graphene oxide	Mixing and sonication	methyl orange	Mercury vapor lamp delivering near-UV/vis irradiation (>350 nm)	Morales-Torres et al. (2013)
P25@graphene (loading of YF3:Yb3+, Tm3+)	Hydrothermal and annealed at 500°C under argon	Methyl orange	High-pressure Xenon lamp to simulate sunlight	Ren et al. (2012)
P25/$CoFe_2O_4$/graphene	Hydrothermal in water and ethanol	Methylene blue, methyl orange, and neutral dark yellow	Xe lamp with UV cutoff filter (>420 nm)	Sun et al. (2013)
TiO_2-graphene	Mixing with ultrasonification	Acid orange 7	Photodegradation of acid orange 7 in an aqueous solution under UV light	Muthirulan et al. (2014)

Cai et al. (2016) developed a hydrothermal synthesis method to prepare graphene-supported bismuth vanadate ($BiVO_4$) particle ($BiVO_4$–graphene) using graphene oxide, $Bi(NO_3)_3 \cdot 5H_2O$, and NH_4VO_3 as raw materials. The photocatalytic activity of $BiVO_4$–graphene composite was investigated by the degradation of methylene blue in an aqueous solution under visible light irradiation. It was confirmed that the presence of graphene enhanced the photocatalytic activity of $BiVO_4$ particle, and the $BiVO_4$–1.0% graphene showed the best photocatalytic activity toward the dye degradation from aqueous solutions. It was shown that the improved visible light activity of $BiVO_4$–graphene was due to the enhancement of electron–hole separation by the electron trapping of graphene. The result showed that the graphene played an important role in $BiVO_4$–graphene composite as a photocatalyst for degradation of organic pollutant.

Titanium dioxide/graphene oxide composites were prepared using a simple colloidal blending method and used for adsorption and photocatalysis performance under both UV and visible radiation (Nguyen-Phan et al., 2011). It was observed that the gradually increasing content of graphene oxide up to 10 wt.% promoted the removal efficiency and correspondingly, facilitated the photodegradation rate of methylene blue in aqueous solutions. The good photocatalytic performance on the TiO_2–graphene oxide composite systems irrespective of light sources was attributed to a synergy effects including the increase in specific surface area with graphene oxide amount as well as the formation of both—conjugations between dye molecules and aromatic rings and the ionic interactions between methylene blue and oxygen-containing functional groups at the edges or on the surfaces of graphene nanosheets. Graphene oxide worked as the adsorbent, electron acceptor, and photosensitizer to efficiently enhance the methylene blue photodecomposition in water.

Yoo et al. (2011) synthesized photocatalytic TiO_2 films using RF magnetron sputtering and three kinds of graphene oxide (GO) solutions with different concentration of 0.05 mg GO, 0.03 mg GO, and 0.01 mg GO in 10 mL ethanol were coated on TiO_2 films, respectively. The efficiency of the photocatalytic degradation was enhanced when TiO_2 thin film was coated with GO, it was related to its giant π-conjugation system and two-dimensional planar structure and efficient charge separation by coating GO. The maximum photodegradation rate was observed for TiO_2 thin film coated with 0.03 mg GO. It was attributed to the increase of absorbance and scattering of photons through excess carbon in the photosystem during the photocatalytic degradation of methylene blue under UV and visible light.

6.4 GRAPHENE-BASED CARBON NANOTUBES COMPOSITES

The efficient removal of organic dye methylene blue from the aqueous solution with a self-assembled cylindrical graphene–carbon nanotube hybrid fabricated by a simple one step hydrothermal process was carried out by Ai and Jiang (2012). The hybrid showed good performance for the removal of methylene blue from aqueous solution with a maximum adsorption capacity of 81.97 mg/g. The removal efficiency was found to be 97% at the initial methylene blue concentration of 10 mg/L. The kinetics of adsorption followed the pseudo-second-order kinetic model, while the Freundlich adsorption isotherm was applicable to describe the equilibrium adsorption process.

Recently, Liu et al. (2015) presented a facile method for the preparation of MnO_2–graphene–carbon nanotube (MnO_2–G–CNT) hybrid material in a simple one-pot reaction process by a chemical method. The adsorption behaviors of methyl orange onto the MnO_2–G–CNT were first systematically investigated and experimental results indicated that the material with the highest loading amount of MnO_2 showed an excellent adsorption capacity toward methyl orange. The adsorption kinetics was well fitted to pseudo-second-order model and the Freundlich isotherm model showed a better fit with experimental data than the Langmuir model and the maximum adsorption capacity was determined to 476.19 mg/g. The overall adsorption rate process was apparently influenced by intraparticle diffusion and external mass transfer of dye onto the MnO_2–G–CNT. The thermodynamic parameters indicated that the adsorption was spontaneous and exothermic and that the physical adsorption mechanisms included electrostatic interaction, which played a dominant role in the adsorption mechanism between methyl orange and the hybrid material.

Elsagh et al. (2014) applied single-walled carbon nanotubes (SWCNTs), carboxylate group functionalized single-walled carbon nanotubes (SWCNT-COOH), graphene (G), and graphene oxide (GO) as adsorbents for the removal of Basic Red 46 from the aqueous solution. Various physicochemical parameters such as electrical conductivity behaviors, contact time, solution pH, and dye concentration were studied. The adsorption equilibrium behavior of Basic Red 46 was described reasonably well by the Langmuir isotherm and the maximum removal capacity was found to be 55.57 and 30.52 mg/g for graphene oxide and graphene, respectively. Adsorption kinetics data were modeled using the pseudo-first and pseudo-second order, and intraparticle diffusion models. It was found that the pseudo-first order kinetic model for SWCNTs, SWCNT-COOH, and the pseudo-second order for G and GO were fitted to the experimental data.

Graphene oxide (GO) and reduced graphene oxide (RGO) platelets were pillared with carbon nanotubes (CNTs) by using the chemical vapor deposition (CVD) method with acetonitrile as the carbon source and nickel nanoparticles as the catalyst (Zhang et al., 2010b). Characterization data showed that the composite materials with RGO layers pillared by CNTs formed a robust three-dimensional (3D) porous structure of specific surfaces as high as 352 m^2/g. The CNT-pillared RGO composite materials exhibited an excellent visible light photocatalytic performance in degrading dye rhodamine B because of the unique porous structure and the exceptional electron transfer property of graphene. Such CNT-RGO composites presented a new family of innovative carbon materials for visible-light-activated photocatalysis. The high adsorption ability toward organic dyes, strong π-π interaction with dye chromophores, efficient photosensitized electron injection, and retarded electron-radical recombination enable the composite material to be an excellent photocatalyst for photoassisted degradation of dye pollutants in water.

6.5 GRAPHENE-BASED SULFONIC MAGNETIC NANOCOMPOSITES

As graphene is an electron-rich, hydrophobic nanomaterial with large specific area, and p–p electrostatic stacking property (McAllister et al., 2007), it has been served as an extraordinarily wonderful adsorbent or extraction material. Sulfonation increases the surface of the graphene Zeta potential, thus has high dispersion property in aqueous solution, which ensures the high interaction of sulfonated graphene with aromatic pollutants and thereby assures the high adsorption of organic dyes on sulfonated graphene (Zhao et al., 2011). A magnetic-sulfonic graphene nanocomposite (G-SO_3H/Fe_3O_4) was synthesized and used as adsorbent for removal of three cationic dyes: safranine T, neutral red, victoria blue, and three anionic dyes: methyl orange, brilliant yellow, and alizarin red, from environmental water (Wang et al., 2013). Several optimization parameters such as solution pH, amount of adsorbent, adsorption kinetics, adsorption isotherms, and ionic strength. The results have established good potential of G-SO_3H/Fe_3O_4 for the adsorption of cationic dyes more efficiently and selectively than anionic dyes at pH 6.0. The adsorption rate was initially fast and more than 93% of cationic dyes were removed within 10 min of contact time. Adsorption kinetics followed the pseudo-second-order kinetic model. The adsorption isotherm fitted well with Langmuir and Freundlich adsorption models. The maximum adsorption capacities of G-SO_3H/Fe_3O_4 for safranine T, neutral red, and victoria blue dyes were 199.3, 216.8, and 200.6 mg/g. The adsorbed cationic dyes were eluted by ethanol as eluent using different pH values. The preparartion method was simple, sensitive, and rapid, and was suitable for the adsorption of cationic dyes in environmental water.

In a facile method the biopolymers mediated and synthesized 3D graphene oxide-based gels were reported by Cheng et al. (2013a). The obtained hybrid gels present well-defined and interconnected 3D porous network, which allows the pollutants pore diffusion in the adsorbent. The obtained porous graphene oxide-biopolymer gels were efficiently used to remove cationic dyes methylene blue and methyl violet from wastewater in batch mode and the influence of contact time, initial dye concentration, and solution pH value on the adsorption capacity of the prepared graphene oxide-biopolymer gels were studied. The graphene oxide-biopolymer gels displayed an adsorption capacity as high as 1100 mg/g for methylene blue and 1350 mg/g for methyl violet, respectively. Furthermore, the adsorption kinetics and isotherms of the methylene blue were studied in detail and the experimental data of methylene blue adsorption fitted well with the pseudo-second-order kinetic model and the Langmuir isotherm, and the results indicated that the adsorption process was controlled by the intraparticle diffusion.

6.6 GRAPHENE-BASED POLYMER NANOCOMPOSITES

Recently, it has been established that graphene oxide presented high adsorption capacities to various water contaminants (Kyzas et al., 2014; Guo et al., 2012; Wang and Zhang, 2011). However, the high water dispersibility of graphene oxide restricted its practical applications in wastewater treatment since ultrahigh centrifugation was necessary after the adsorption process to regenerate the used adsorbent and for the recovery of adsorbed adsorbate. Zhang et al. (2013) presented a facile method by enwrapping graphene oxide in porous particles to remove dyes and other pollutions from wastewater. Polyethersulfone was chosen as the matrix to enwrap graphene oxide through a liquid–liquid phase separation process, the polyethersulfone/graphene oxide porous particles showed a dense skin layer and porous structures. The BET analysis showed that the polyethersulfone/graphene oxide have surface area of 25.67 m^2/g, mesoporous volume of 0.0148 cm^3/g, BJH pore size was 5.59 and the total pore volume was 4.49 cm^3/g. Methylene blue was chosen as the adsorbate to investigate the adsorption capability, kinetics, and isotherms of the prepared particles, and the particles displayed an adsorption capacity up to 62.5 mg/g for methylene blue dye. Meanwhile, the effect of contacting time, initial concentration, temperature, and pH value on the adsorption capacity was studied. The experimental data of methylene blue adsorption fitted the pseudo-second-order kinetic model and the Langmuir isotherm very well, and the adsorption process was controlled by the intraparticle diffusion. The results indicated that the polyethersulfone/graphene oxide porous particles showed good selective adsorbability to cationic dyes, such as methylene blue and methyl violet.

Dopamine, a mussel adhesive protein inspired molecule have been widely used due to its remarkable performances in drug delivery, cellular imaging, bone tissue, biosensor, catalyst, and energy storage. It has been reported that dopamine could undergo self-polymerization in alkaline or oxidizing environment containing aqueous solutions and adhere onto almost any solid surface without surface pretreatment. In addition, the resultant polydopamine is rich in catechol groups like mussel adhesive proteins, which endows the polydopamine versatile chemical reactivity and can be used as a robust platform for diverse secondary reactions, such as biopolymer grafting, biomimetic mineralization, and in situ growth of metal nanoparticles.

Recently, Cheng et al. (2013b) employed the single layer graphene as an ultrathin substrate (<1 nm) to clarify the self-assembly process of polydopamine. The self-assembly of the polydopamine coated graphene nanosheets was conducted at different reaction concentrations and times, and monitored by atomic force microscopy images. Field emission scanning electron microscope results indicated that the polydopamine clusters exhibited surface dependent aggregation and roughness. The typical procedure for the synthesis of polydopamine coated reduced graphene oxide was divided into three steps: (1) dopamine converts graphene oxide into reduced graphene oxide; (2) the pH induced spontaneously polymerization of dopamine and deposition of polydopamine precursors on the reduced graphene oxide surface; (3) the further growth of polydopamine precursors and the self-assembly into large polydopamine clusters. Furthermore, the polydopamine layer assembled graphene sheets could function as versatile templates to construct 2D architectures with mineralized inorganic/metal nanoparticles; and the regarding catalytic ability and the adsorption properties were studied. Moreover, the dopamine was functional as reductant for low temperature assembly of porous 3D hydrogel via p–p interactions and polydopamine polymerization. The polydopamine assembled graphene aerogel was used as a promising adsorbent for the removal of oils, organic solvents, and dyes from contaminated water.

Debnath et al. (2014) evaluated the performance of graphene oxide in the removal of Congo red dye from aqueous solution in batch mode. The effect of process parameters such as solution pH, contact time, initial concentration of Congo red, and temperature on the adsorption of Congo red onto graphene oxide was investigated. The adsorption of Congo red onto graphene oxide was favored at neutral pH. The kinetics of the adsorption data was analyzed using four kinetic models namely pseudo-first-order, pseudo-second order, Elovich model, and intraparticle diffusion models. The results from the kinetic studies indicated that the rate of adsorption follows a pseudo-first-order with respect to the Congo red solution concentration and that in general the order of data fitted was pseudo-first-order > pseudo-second-order > Elovich equation. The kinetic parameters obtained from the kinetic studies suggested that the adsorption process was controlled by film diffusion. The equilibrium data were modeled with the Langmuir, Freundlich, and Redlich–Peterson models. The equilibrium data fitted well with the Redlich–Peterson model and characterized by a Langmuir type isotherm. The Langmuir maximum adsorption capacities were found to 546.1, 557.5, and 571.5 at 25°C, 35°C, and 45°C, respectively. The values of mean free energy of adsorption were in the range 8.0–16.0 kJ/mol which describes that the adsorption reaction was physically controlled. Hence, the adsorption of Congo red was due to the electrostatic interaction with the graphene and graphene oxide. Thermodynamic parameters revealed that the adsorption process was endothermic and spontaneous in nature. The values of $\Delta H°$ and $\Delta S°$ of the adsorption process were 8.19 kJ/mol and 0.10 kJ/mol, respectively. The low value of $\Delta H°$ (<40 kJ/mol) indicated that adsorption process occurred mainly through a physical interaction.

6.7 GRAPHENE-BASED SAND COMPOSITES

Sreeprasad et al. (2013) presented an in situ strategy for the preparation of graphene immobilized on sand using asphalt. The formation of graphene material was confirmed by the presence of G and D bands at 1578 cm^{-1} and 1345 cm^{-1} in Raman spectroscopy and the 2D sheet-like structure with wrinkles in transmission electron microscopy. The potential applicability of supported graphenic materials in the removal of organic dyes rhodamine-6G was carried out for purifying water. Adsorption studies were conducted in batch mode as a function of time, particle size, and adsorbent dose to optimize the process parameters. The adsorption capacity was found to be 75.4 mg/g. The continuous mode experiments were conducted in multiple cycles and they confirmed that the material can be used for water purification applications.

A suitable methodology was introduced to immobilize graphene on sand without the need of any binder, resulting in a composite, referred to as graphene sand composite (Gupta et al., 2012). Raman spectroscopy confirmed that the material is indeed graphenic in nature, having G and D bands at 1597 and 1338 cm^{-1}, respectively. The as-prepared graphene sand composite was used as adsorbent for the removal of rhodamine 6G from water. The spectroscopic and microscopic analyses coupled with adsorption experiments revealed that physical adsorption played a dominant role in the adsorption process. The maximum adsorption capacity for rhodamine 6G removal was 55 mg/g. The adsorbent was easily regenerated using acetone as eluent.

6.8 GRAPHENE-BASED CHITOSAN COMPOSITES

A hydrophilic and biocompatible three-dimensional chitosan–graphene mesostructures with large specific surface area (603.2 m^2/g) and unique mesoporosity were prepared by simply thermal treatment of graphene oxide nanosheets and chitosan (Cheng et al., 2012). Investigations showed that the three-dimensional mesostructures was able to remove reactive black 5 from aqueous solution. A removal efficiency of 97.5% was achieved readily at initial reactive black 5 concentrations of 1.0 mg/mL. These results indicated that the three-dimensional chitosan-graphene mesostructures can be used to treat the dye bearing wastewater, providing novel insights for the fabrication of 3D functionalized graphene with various building blocks for broad potential applications.

Magnetic graphene/chitosan nanocomposite was fabricated through a facile chemical route and its application as a new adsorbent for the removal of acid orange 7 was investigated in batch mode (Sheshmani et al., 2014). The results showed that, benefiting from the surface property of graphene oxide, the abundant amino and hydroxyl functional groups of chitosan, and from the magnetic property of Fe_3O_4, the adsorbent possesses quite a good and versatile adsorption capacity for the acid orange 7, and was easily and rapidly extracted from water by magnetic attraction. The maximum absorption capacity was highest in the acidic medium and reached at equilibrium within 120 min contact time. The batch adsorption experiments showed that the adsorption of the acid orange 7 was considerably dependent on pH of medium, amount of adsorbent, and contact time. The adsorption kinetic was well fitted to the pseudo-first order kinetic model. The equilibrium isotherm was fitted to the Langmuir isotherm model with maximum adsorption capacity of 42.7 mg/g. The study suggested that the magnetic graphene/chitosan is a promising nanoadsorbent for removal of anionic azo dyes from aqueous solution.

Du et al. (2014) prepared graphene oxide/chitosan/silica fibers by a wet spinning technique. Further silica nanoparticles were etched off from the graphene oxide/chitosan/silica fibers to produce graphene oxide/chitosan fibers with porous structure. Adsorption experiments with the porous graphene oxide/chitosan fibers were performed based on various parameters, such as initial Congo red concentration, solution pH, adsorbent dose, and contact time in batch mode. The equilibrium isotherm results indicated that the equilibrium data were perfectly represented by the Langmuir isotherm. The maximum adsorption capacity was found to 294.12 mg/g for the adsorption of Congo red by graphene oxide/chitosan fibers. The kinetics data were best represented by the pseudo-first-order model for the adsorption of Congo red onto the porous graphene oxide/chitosan fibers.

6.9 CONCLUSIONS

Graphene and graphene-based nanocomposites have been widely utilized for the adsorption of dyes and color from water/wastewater because of their high surface area to volume ratio, porosity, and physicochemical stability. It has been observed that the graphene and its oxides are generally aggregated in its bare form so the functionalization with inorganic or organic materials may enhance their dispersion in the aqueous media. Several graphene-based nanocomposites have been synthesized by the doping or functionalization with chelating agents. The functionalization of graphene by the chelating agents such as humic acids, fulvic acids, EDTA, or other synthetic polymeric compounds may increase the adsorption capacity and reduces the equilibrium adsorption time. It has been well established that the functional groups on the surface of graphene have a major role in the adsorption of anions or cations in the liquid phase due to the increased or decreased protonation and deprotonation. The solution pH played most important role in the surface groups protonation and deprotonation, according to the increase or decrease in solution pH. The contact time and solution pH also played a major role in the mass transfer of any pollutants from liquid phase to the boundary of adsorbent, which further facilitates the adsorption of pollutants. When the adsorption equilibrium is established, one can easily draw the conclusion about the adsorption kinetics and isotherms of the adsorbate and adsorbent. The kinetics of the adsorption processes of dyes have been modeled with the pseudo-first order, pseudo-second order, and intraparticle diffusion rate equations. However, in most of the studied adsorbate–adsorbent systems, the pseudo-second order model is most fitted. It is also found that the Langmuir isotherm model was most celebrated models for the study of equilibrium isotherm of adsorption processes. Several graphene-based nanocomposites such as magnetic nanocomposites, graphene-based chitosan composites, graphene-based carbon nanotubes, graphene-based sand composites, graphene-based catalysts, and graphene-based polymer composites have been extensively synthesized and utilized for their adsorption potential toward the removal of hazardous dyes from aqueous solutions. Among these, the graphene-based nanocomposites have their own superiority due to their magnetic properties and increased surface functionalization. The graphene-based magnetic nanocomposites can be easily synthesized via low cost precursors at room temperature and have increased surface functionalization and can be easily regenerated after their use from liquid phase with the help of external magnet. Recently, graphene-based photocatalysts have gain increased application in the photodegradation of hazardous dyes in water. However, the photodegradation processes involve two phases, the first phase involves the adsorption of dye on the surface of the catalysts, and the second is the ring cleavage of the dye. In some cases it has been pointed out that the photodegradation may generate secondary pollutants which may be more toxic than their precursor molecules. Therefore, the adsorptive separations on graphene-based magnetic nanocomposites are more suitable to remove the environmental contaminants from water. In addition, most of the reported work on the use of graphene-based nanocomposites for removal of dyes from aqueous media is in the batch mode at laboratory scale for single dye solutions. Hence, there is an urgent need to conduct the experiment in continuous column mode with multidye solutions. It will enhance our understanding of the suitability of the adsorbent at industrial scale to tackle the real wastewater pollution in water treatment plants.

REFERENCES

Ai, L., Jiang, J., 2012. Removal of methylene blue from aqueous solution with self-assembled cylindrical graphene–carbon nanotube hybrid. Chem. Eng. J. 192, 156–163.

Ai, L.H., Zhang, C.Y., Chen, Z.L., 2011. Removal of methylene blue from aqueous solution by a solvothermal-synthesized graphene/magnetite composite. J. Hazard. Mater. 192, 1515–1524.

Allen, S.J., Mckay, G., Porter, J.F., 2004. Adsorption isotherm models for basic dye adsorption by peat in single and binary component systems. J. Colloid Interface Sci. 280, 322–333.

Aquino, J.M., Rocha-Filho, R.C., Bocchi, N., Biaggio, S.R., 2013. Electrochemical degradation of the Disperse Orange 29 dye on a β-PbO_2 anode assessed by the response surface methodology. J. Environ. Chem. Eng. 1, 954–961.

Ardejani, F.D., Badii, K., Limaee, N.Y., Shafaei, S.Z., Mirhabibi, A.R., 2008. Adsorption of Direct Red 80 dye from aqueous solution onto almond shells: effect of pH, initial concentration and shell type. J. Hazard. Mater. 151, 730–737.

Bai, S., Shen, X., Zhong, X., Liu, Y., Zhu, G., Xu, X., Chen, K., 2012. One-pot solvothermal preparation of magnetic reduced graphene oxide-ferrite hybrids for organic dye removal. Carbon 50, 2337–2346.

Barka, N., Abdennouri, M., Makhfouk, M.E., Qourzal, S., 2013. Biosorption characteristics of cadmium and lead onto eco-friendly dried cactus (*Opuntia ficus indica*) cladodes. J. Environ. Chem. Eng. 1, 144–149.

Cai, X., Zhang, B., Shi, L., Liu, H., Zhang, J., Huang, L., Tan, S., 2016. Study the photocatalysis activity of hydrothermal-synthesized $BiVO_4$–graphene composite on methylene blue. Desalin. Water Treat. 57, 6365–6371, 2015.

Chang, B.Y.S., Mehmood, M.S., Pandikumar, A., Huang, N.M., Lim, H.N., Marlinda, A.R., Yusoff, N., Chiu, W.S., 2015. Hydrothermally prepared graphene-titania nanocomposite for the solar photocatalytic degradation of methylene blue. Desalin. Water Treat. 57, 238–245.

Cheng, J.S., Dub, J., Zhu, W., 2012. Facile synthesis of three-dimensional chitosan–graphene mesostructures for reactive black 5 removal. Carbohydr. Polym. 88, 61–67.

Cheng, C., Deng, J., Lei, B., He, A., Zhang, X., Ma, L., Li, S., Zhao, C., 2013a. Toward 3D graphene oxide gels based adsorbents for high-efficient water treatment via the promotion of biopolymers. J. Hazard. Mater. 263, 467–478.

Cheng, C., Li, S., Zhao, J., Li, X., Liu, Z., Ma, L., Zhang, X., Sun, S., Zhao, C., 2013b. Biomimetic assembly of polydopamine-layer on graphene: Mechanisms, versatile 2D and 3D architectures and pollutant disposal. Chem. Eng. J. 228, 468–481.

Daud, N.K., Hameed, B.H., 2010. Decolorization of acid Red 1 by Fenton-like process using rice husk ash-based catalyst. J. Hazard. Mater. 176, 938–944.

Debnath, S., Maity, A., Pillay, K., 2014. Impact of process parameters on removal of Congo red by graphene oxide from aqueous solution. J. Environ. Chem. Eng. 2, 260–272.

Demarchi, C.A., Campos, M., Rodrigues, C.A., 2013. Adsorption of textile dye Reactive Red 120 by the chitosan–Fe(III)-crosslinked: Batch and fixed-bed studies. J. Environ. Chem. Eng. 1, 1350–1358.

Du, Q., Sun, J., Li, Y., Yang, X., Wang, X., Wang, Z., Xia, L., 2014. Highly enhanced adsorption of congo red onto graphene oxide/chitosan fibers by wet-chemical etching off silica nanoparticles. Chem. Eng. J. 245, 99–106.

Elsagh, A., Moradi, O., Fakhri, A., Najafi, F., Alizadeh, R., Haddadi, V., 2014. Evaluation of the potential cationic dye removal using adsorption by graphene and carbon nanotubes as adsorbents surfaces. Arabian J. Chem. Available from: http://dx.doi.org/10.1016/j.arabjc.2013.11.013.

Fan, L.L., Luo, C.N., Li, X.J., Lu, F., Qiu, H., Sun, M., 2012. Fabrication of novel magnetic chitosan grafted with graphene oxide to enhance adsorption properties for methyl blue. J. Hazard. Mater. 215–216, 272–279.

Fan, L., Luo, C., Sun, M., Qiu, H., Li, X., 2013. Synthesis of magnetic β-cyclodextrin–chitosan/graphene oxide as nanoadsorbent and its application in dye adsorption and removal. Colloids Surf. B Biointerfaces 103, 601–607.

Gao, P., Liu, J., Sun, D.D., Ng, W., 2013. Graphene oxide–CdS composite with high photocatalytic degradation and disinfection activities under visible light irradiation. J. Hazard. Mater. 250–251, 412–420.

Gautam, R.K., Chattopadhyaya, M.C., Sharma, S.K., 2013a. Biosorption of heavy metals: recent trends and challenges. In: Sharma, S.K., Sanghi, R. (Eds.), Wastewater Reuse and Management. Springer, London, pp. 305–322.

Gautam, R.K., Mudhoo, A., Chattopadhyaya, M.C., 2013b. Kinetic, equilibrium, thermodynamic studies and spectroscopic analysis of Alizarin Red S removal by mustard husk. J. Environ. Chem. Eng. 1, 1283–1291.

Gautam, R.K., Mudhoo, A., Lofrano, G., Chattopadhyaya, M.C., 2014. Biomass-derived biosorbents for metal ions sequestration: adsorbent modification and activation methods and adsorbent regeneration. J. Environ. Chem. Eng. 2, 239–259.

Geng, Z., Lin, Y., Yu, X., Shen, Q., Ma, L., Li, Z., Pan, N., Wang, X., 2012. Highly efficient dye adsorption and removal: a functional hybrid of reduced graphene oxide–Fe_3O_4 nanoparticles as an easily regenerative adsorbent. J. Mater. Chem. 22, 3527–3535.

Guo, J., Wang, R., Tjiu, W.W., Pan, J., Liu, T., 2012. Synthesis of Fe nanoparticles@graphene composites for environmental applications. J. Hazard. Mater. 225–226, 63–73.

Gupta, V.K., Ali, I., 2004. Removal of lead and chromium from wastewater using bagasse fly ash—a sugar industry waste. J. Colloid Interface Sci. 271, 321–328.

Gupta, V.K., Jain, C.K., Ali, I., Sharma, M., Saini, V.K., 2003. Removal of cadmium and nickel from wastewater using bagasse fly ash—a sugar industry waste. Water Res. 37, 4038–4044.

Gupta, S.S., Sreeprasad, T.S., Maliyekkal, S.M., Das, S.K., Pradeep, T., 2012. Graphene from sugar and its application in water purification. ACS Appl. Mater. Interfaces 4, 4156–4163.

Hameed, B.H., El-Khaiary, M.I., 2008. Batch removal of malachite green from aqueous solutions by adsorption on oil palm trunk fibre: Equilibrium isotherms and kinetic studies. J. Hazard. Mater. 154, 237–244.

Hameed, B.H., Hakimi, H., 2008. Utilization of durian (*Durio zibethinus* Murray) peel as low cost sorbent for the removal of acid dye from aqueous solutions. Biochem. Eng. J. 39, 338–343.

He, F., Fan, J.T., Ma, D., Zhang, L., Leung, C., Chan, H.L., 2010. The attachment of Fe_3O_4 nanoparticles to graphene oxide by covalent bonding. Carbon 48, 3139–3144.

Hernández-Rodríguez, M.J., Fernández-Rodríguez, C., Doña-Rodríguez, J.M., González-Díaz, O.M., Zerbani, D., Pérez Peña, J., 2014. Treatment of effluents from wool dyeing process by photo-Fenton at solar pilot plant. J. Environ. Chem. Eng. 2, 163–171.

Hu, H., Wang, X., Liu, F., Wang, J., Xu, C., 2011. Rapid microwave assisted synthesis of graphene nanosheets-zinc sulfide nanocomposites: optical and photocatalytic properties. Synth. Met. 161, 404–410.

Kyzas, G.Z., Travlou, N.A., Kalogirou, O., Deliyanni, E.A., 2013. Magnetic graphene oxide: effect of preparation route on reactive black 5 adsorption. Materials 6, 1360–1376.

Kyzas, G.Z., Deliyanni, E.A., Matis, K.A., 2014. Graphene oxide and its application as an adsorbent for wastewater treatment. J. Chem. Technol. Biotechnol. 89, 196–205.

Lee, H.S., Woo, C.S., Youn, B.K., Kim, S.Y., Oh, S.T., Sung, Y.E., Lee, H.I., 2005. Bandgap modulation of TiO2 and its effect on the activity in photocatalytic oxidation of 2-isopropyl-6- methyl-4-pyrimidinol. Top. Catal. 35, 255–260.

Lee, J.W., Choi, S.P., Thiruvenkatachari, R., 2006. Evaluation of the performance of adsorption and coagulation processes for the maximum removal of reactive dyes. Dyes Pigm. 69, 196–203.

Li, J., l Zhou, S., Hong, G.B., Chang, C.T., 2013a. Hydrothermal preparation of P25–graphene composite with enhanced adsorption and photocatalytic degradation of dyes. Chem. Eng. J. 219, 486–491.

Li, X., Zhang, T., Gu, S., Kang, S.Z., Li, G., Mu, J., 2013b. Reduced graphene oxide/potassium niobate composite nanoscrolls with enhanced photocatalytic activity for dye degradation. Sep. Purif. Technol. 108, 139–142.
Liu, T., Li, Y., Du, Q., Sun, J., Jiao, Y., Yang, G., Wang, Z., Xia, Y., Zhang, W., Wang, K., Zhu, H., Wu, D., 2012. Adsorption of methylene blue from aqueous solution by graphene. Colloids Surf. B Biointerfaces 90, 197–203.
Liu, S., Sun, H., Liu, S., Wang, S., 2013a. Graphene facilitated visible light photo-degradation of methylene blue over titanium dioxide photocatalysts. Chem. Eng. J. 214, 298–303.
Liu, Z., Li, L., Tang, L., Shi, R., Gu, Q., Liang, X., Yao, X., 2013b. Investigation of adsorption performance on 1,2-dichloroethane by heat and acid modified activated carbon. J. Environ. Chem. Eng 1, 131–136.
Liu, Y., Tian, Y., Luo, C., Cui, G., Yan, S., 2015. One-pot preparation of a MnO_2–graphene–carbon nanotube hybrid material for the removal of methyl orange from aqueous solutions. New J. Chem. 39, 5484–5492.
McAllister, M.J., Li, J.L., Adamson, D.H., Schniepp, H.C., Abdala, A.A., Liu, J., Alonso, M.H., Milius, D.L., Car, R., Prud'Homme, R.K., Aksay, I.A., 2007. Single sheet functionalized graphene by oxidation and thermal expansion of graphite. Chem. Mater. 19, 4396–4404.
Morales-Torres, S., Pastrana-Martínez, L.M., Figueiredo, J.L., Faria, J.L., Silva, A.M.T., 2013. Graphene oxide–P25 photocatalysts for degradation of diphenhydramine pharmaceutical and methyl orange dye. Appl. Surf. Sci. 275, 361–368.
Muthirulan, P., Devi, C.K.N., Sundaram, M.M., 2014. Fabrication and characterization of efficient hybrid photocatalysts based on titania and graphene for acid orange seven dye degradation under UV irradiation. Adv. Mat. Lett. 5 (4), 163–171.
Nandi, D., Gupta, K., Ghosh, A.K., De, A., Banerjee, S., Ghosh, U.C., 2012. Manganese-incorporated iron(III) oxide–graphene magnetic nanocomposite: synthesis, characterization, and application for the arsenic(III)-sorption from aqueous solution. J. Nanopart. Res. 14, 1272.
Navalon, S., Dhakshinamoorthy, A., Alvaro, M., Garcia, H., 2011. Heterogeneous Fenton catalysts based on activated carbon and related materials. ChemSusChem. 4, 1712–1730.
Ngah, W.S.W., Ariff, N.F.M., Hashim, A., Hanaf, M.A.K.M., 2010. Malachite green adsorption onto chitosan coated bentonite beads: Isotherms, kinetics and mechanism. Clean-Soil Air Water 38, 394–400.
Nguyen-Phan, T.D., Pham, V.H., Shin, E.W., Pham, H.D., Kim, S., Chung, J.S., Kim, E.J., Hur, S.H., 2011. The role of graphene oxide content on the adsorption-enhanced photocatalysis of titanium dioxide/graphene oxide composites. Chem. Eng. J. 170, 226–232.
Ramesha, G.K., Kumara, A.V., Muralidhara, H.B., Sampath, S., 2011. Graphene and graphene oxide as effective adsorbents toward anionic and cationic dyes. J. Colloid Interface Sci. 361, 270–277.
Ren, L., Qi, X., Liu, Y., Huang, Z., Wei, X., Li, J., Yang, L., Zhong, J., 2012. Upconversion-P25– graphene composite as an advanced sunlight driven photocatalytic hybrid material. J. Mater. Chem. 22 (23), 11765–11771.
Sadeghi-Kiakhani, M., Arami, M., Gharanjig, K., 2013. Preparation of chitosan-ethyl acrylate as a biopolymer adsorbent for basic dyes removal from colored solutions. J. Environ. Chem. Eng. 1, 406–415.
Sahoo, A.K., Srivastava, S.K., Raul, P.K., Gupta, A.K., Shrivastava, R., 2014. Graphene nanocomposites of CdS and ZnS in effective water purification. J. Nanopart. Res. 16, 2473.
Samiee, S., Goharshadi, E.K., 2014. Graphene nanosheets as efficient adsorbent for an azo dye removal: kinetic and thermodynamic studies. J. Nanopart. Res. 16, 2542.
Sannino, D., Vaiano, V., Sacco, O., Ciambelli, P., 2013. Mathematical modelling of photocatalytic degradation of methylene blue under visible light irradiation. J. Environ. Chem. Eng. 1, 56–60.
Shen, X., Wu, J., Bai, S., Zhou, H., 2010. One-pot solvothermal syntheses and magnetic properties of graphene-based magnetic nanocomposites. J. Alloys Compd. 506, 136–140.

Sheshmani, S., Ashori, A., Hasanzadeh, S., 2014. Removal of Acid Orange 7 from aqueous solution using magnetic graphene/chitosan: A promising nano-adsorbent. Int. J. Biol. Macromol. 68, 218–224.

Shi, B., Li, G., Wang, D., Feng, C., Tang, H., 2007. Removal of direct dyes by coagulation: the performance of preformed polymeric aluminum species. J. Hazard. Mater. 143, 567–574.

Sreeprasad, T.S., Gupta, S.S., Maliyekkal, S.M., Pradeep, T., 2013. Immobilized graphene-based composite from asphalt: Facile synthesis and application in water purification. J. Hazard. Mater. 246–247, 213–220.

Sudarjanto, G., Keller-Lehmann, B., Keller, J., 2006. Optimization of integrated chemical–biological degradation of a reactive azo dye using response surface methodology. J. Hazard. Mater. 138, 160–168.

Sun, H., Cao, L., Lu, L., 2011. Magnetite/reduced graphene oxide nanocomposites: one step solvothermal synthesis and use as a novel platform for removal of dye pollutants. Nano. Res. 4 (6), 550–562.

Sun, L., Yu, H., Fugetsu, B., 2012. Graphene oxide adsorption enhanced by in situ reduction with sodium hydrosulfite to remove acridine orange from aqueous solution. J. Hazard. Mater. 203–204, 101–110.

Sun, J., Fu, Y., Xiong, P., Sun, X., Xu, B., Wang, X., 2013. A magnetically separable P25/CoFe2O4/graphene catalyst with enhanced adsorption capacity and visible-light-driven photocatalytic activity. RSC Adv. 3 (44), 22490–22497.

Vijayakumar, G., Yoo, C.K., Elango, K.G.P., Dharmendirakumar, M., 2010. Adsorption characteristics of rhodamine B from aqueous solution onto baryte. Clean-Soil Air Water 38, 202–209.

Wang, F., Zhang, K., 2011. Reduced graphene oxide–TiO2 nanocomposite with high photocatalystic activity for the degradation of rhodamine B. J. Mol. Catal. A Chem. 345 (1–2), 101–107.

Wang, C., Feng, C., Gao, Y., Ma, X., Wu, Q., Wang, Z., 2011. Preparation of a graphene-based magnetic nanocomposite for the removal of an organic dye from aqueous solution. Chem. Eng. J. 173, 92–97.

Wang, S., Wei, J., Lv, S., Guo, Z., Jiang, F., 2013. Removal of organic dyes in environmental water onto magnetic-sulfonic graphene nanocomposite. Clean-Soil Air Water 41 (10), 992–1001.

Wang, S., Li, Y., Fan, X., Zhang, F., Zhang, G., 2014. β-Cyclodextrin functionalized graphene oxide: an efficient and recyclable adsorbent for the removal of dye pollutants. Frontiers of Chemical Science and Engineering 9, 77–83.

Wang, J., Zhu, H., Hurren, C., Zhao, J., Pakdel, E., Li, Z., Wang, X., 2015. Degradation of organic dyes by P25-reduced graphene oxide: influence of inorganic salts and surfactants. J. Environ. Chem. Eng. 3, 1437–1443.

Wu, T., Cai, X., Tan, S., Li, H., Liu, J., Yang, W., 2011. Adsorption characteristics of acrylonitrile, p-toluenesulfonic acid, 1-naphthalenesulfonic acid and methyl blue on graphene in aqueous solutions. Chem. Eng. J. 173, 144–149.

Wu, Q., Feng, C., Wang, C., Wang, Z., 2013. A facile one-pot solvothermal method to produce superparamagnetic graphene–Fe_3O_4 nanocomposite and its application in the removal of dye from aqueous solution. Colloids Surf. B Biointerfaces 101, 210–214.

Xie, G., Xi, P., Liu, H., Chen, F., Huang, L., Shi, Y., Hou, F., Zeng, Z., Shao, C., Wang, J., 2012. A facile chemical method to produce superparamagnetic graphene oxide–Fe_3O_4 hybrid composite and its application in the removal of dyes from aqueous solution. J. Mater. Chem. 22, 1033–1039.

Yang, S.T., Chen, S., Chang, Y., Cao, A., Liu, Y., Wang, H., 2011. Removal of methylene blue from aqueous solution by graphene oxide. J. Colloid Interface Sci. 359, 24–29.

Yao, Y., Miao, S., Liu, S., Ma, L.P., Sun, H., Wang, S., 2012a. Synthesis, characterization, and adsorption properties of magnetic Fe_3O_4@graphene nanocomposites. Chem. Eng. J. 184, 326–332.

Yao, Y., Miao, S., Yu, S., Ma, L.P., Sun, H., Wang, S., 2012b. Fabrication of Fe_3O_4/SiO_2 core/shell nanoparticles attached to graphene oxide and its use as an adsorbent. J. Colloid Interface Sci. 379, 20–26.
Yoo, D.H., Cuong, T.V., Pham, V.H., Chung, J.S., Khoa, N.T., Kim, E.J., Hahn, S.H., 2011. Enhanced photocatalytic activity of graphene oxide decorated on TiO2 films under UV and visible irradiation. Curr. Appl. Phys. 11, 805–808.
Zhang, H., Lv, X., Li, Y., Wang, Y., Li, J., 2010a. P25–graphene composite as a high performance photocatalyst. ACS Nano. 4 (1), 380–386.
Zhang, L.L., Xiong, Z., Zhao, X.S., 2010b. Pillaring chemically exfoliated graphene oxide with carbon nanotubes for photocatalytic degradation of dyes under visible light irradiation. ACS Nano. 4, 7030–7036.
Zhang, J., Xiong, Z., Zhao, X.S., 2011. Graphene–metal–oxide composites for the degradation of dyes under visible light irradiation. J. Mater. Chem. 21, 3634–3640.
Zhang, X., Cheng, C., Zhao, J., Ma, L., Sun, S., Zhao, C., 2013. Polyethersulfone enwrapped graphene oxide porous particles for water treatment. Chem. Eng. J. 215–216, 72–81.
Zhang, J., Azam, M.S., Shi, C., Huang, J., Yan, B., Liu, Q., 2015. Poly(acrylic acid) functionalized magnetic graphene oxide nanocomposite for removal of methylene blue. RSC Adv. 5, 32272–32282.
Zhao, G.X., Li, J.X., Wang, X.K., 2011. Kinetic and thermodynamic study of 1-naphthol adsorption from aqueous solution to sulfonated graphene nanosheets. Chem. Eng. J. 173, 185–190.
Zubir, N.A., Yacou, C., Zhang, X., Diniz da Costa, J.C., 2014. Optimisation of graphene oxide–iron oxide nanocomposite in heterogeneous Fenton-like oxidation of Acid Orange 7. J. Environ. Chem. Eng. 2, 1881–1888.

CHAPTER 7

Functionalized Magnetic Nanoparticles: Adsorbents and Applications

7.1 MAGNETIC NANOPARTICLES

Hierarchically structured magnetic nanoparticles have become the very promising nanomaterials during the last 5 years, due to their magnetic properties, high physicochemical stability, low toxicity and ease in synthesis, and excellent recycling capability, and have been extensively studied to remove toxic metal ions and organic pollutants from water. Since these materials show higher removal capacities within low contact time than the bulk material. Their micrometer-sized overall structure provides the necessary mechanical robustness against wear and tear; whereas their nanostructure provides the high surface area as well as the high removal capacity for heavy metal ions, pharmaceuticals, volatile organic compounds, and dyes.

Therefore, magnetic nanoparticles and magnetic nanocomposites are of great interest for researchers from variety of fields, including magnetic fluids, catalysis, biomedicine, drug delivery, magnetic resonance imaging, data storage, and environmental remediation (Jung et al., 2011; Singh et al., 2011; Ambashta and Sillanpaa, 2010; Shen et al., 2009; Lu et al., 2007). Till now various appropriate methods have been developed for the synthesis of magnetic nanoparticles of different sizes and compositions, successful application of such magnetic nanoparticles in the areas listed previously is extremely dependent on the stability of the particles under a wide range of different conditions. In majority of the envisaged applications, the particles perform best when the size of the nanoparticles is below a critical value, which is dependent on the source material but is typically around 10–20 nm (Lu et al., 2007). The design and fabrication of low-cost and economically viable magnetic nanoparticle-based adsorbents has generated great interest in a variety of scientific communities ranging from chemical, biological, and environmental science to engineering. Magnetic nanoparticle-based adsorbents can be used in the separation and purification of biologically as well as environmentally relevant target species with high precision and accuracy

Nanomaterials for Wastewater Remediation
http://dx.doi.org/10.1016/B978-0-12-804609-8.00007-8

(Wang et al., 2012; Singh et al., 2011; Sun et al., 2011, Girginova et al., 2010; Koehler et al., 2009; Lee et al., 2009) due to their magnetic characteristics.

Functionalized magnetic nanoparticles are very promising for applications in sensing, catalysis, biolabeling, and separation and purification. Especially in liquid-phase extraction of heavy metal and dyes, such small and magnetically separable particles may be useful as they combine the advantages of high dispersion, high reactivity, high stability under acidic conditions, and easy separation. In the following, we focus mainly on recent developments in the synthesis of magnetic nanoparticles and their nanocomposites, and various strategies for the protection and stabilization of these particles against oxidation and acid erosion. Further, functionalization and application of such magnetic nanoparticles for the separation and purification of hazardous metal ions and dyes from the water and wastewater will be discussed in brief.

7.2 SYNTHESIS OF MAGNETIC NANOPARTICLES

During the past few years, a large number of the published articles about nanoparticles have described efficient routes to attain shape-controlled, highly stable, low cost, ecofriendly, and narrow size distribution magnetic nanoparticles. Currently, several popular methods including coprecipitation, microemulsion, thermal decomposition, sonochemical, assisted by microwave, chemical vapor deposition, combustion synthesis, carbon arc, laser pyrolysis synthesis, hydrothermal synthesis, sol–gel synthesis have been reported for synthesis of magnetic nanoparticles.

The synthesis of superparamagnetic nanoparticles is a complex process because of their colloidal nature. In the last decades, much research has been devoted to the synthesis of magnetic nanoparticles. Especially during the last few years, many publications have described efficient synthetic routes to shape-controlled, highly stable, functionalized, and monodisperse magnetic nanoparticles. These methods have been used to prepare particles with homogeneous composition and narrow size distribution. However, the most common method for the production of magnetite nanoparticles is the chemical coprecipitation technique of iron salts.

7.2.1 Coprecipitation

Coprecipitation is a very facile and convenient way to synthesize iron oxide nanoparticles (either Fe_3O_4 or γ-Fe_2O_3) from aqueous Fe^{2+}/Fe^{3+} salt solutions by the addition of a base under inert atmosphere at room temperature or at elevated temperature. The chemical reaction of Fe_3O_4 formation may be written as (Laurent et al., 2008):

$$Fe^{2+} + 2Fe^{3+} + 8OH^- \rightarrow Fe_3O_4 + 4H_2O \tag{7.1}$$

According to the thermodynamics of this reaction, complete precipitation of Fe_3O_4 should be expected at a pH between 8 and 14, with a stoichiometric ratio of 2:1 (Fe^{3+}/Fe^{2+}) in a nonoxidizing oxygen environment. The size, shape, and composition of the magnetic nanoparticles very much depends on the type of salts used (eg, chlorides, sulfates, nitrates), the Fe^{2+}/Fe^{3+} ratio, the reaction temperature, the pH value, and ionic strength of the media. With this synthesis, once the synthetic conditions are fixed, the quality of the magnetite nanoparticles is fully reproducible. However, magnetite nanoparticles are not very stable under ambient conditions, and are easily oxidized to maghemite or dissolved in an acidic medium.

$$Fe_3O_4 + 2H^+ \rightarrow \gamma - Fe_2O_4 + Fe^{2+} + H_2O \tag{7.2}$$

Since oxidation of maghemite is the lesser problem as it is a ferrimagnet. Therefore, magnetite particles can be subjected to deliberate oxidation to convert them into maghemite. This transformation is achieved by dispersing them in acidic medium, then addition of iron(III) nitrate. The maghemite particles obtained are then chemically stable in alkaline and acidic medium (Lu et al., 2007). The main advantage of the coprecipitation process is that a large amount of nanoparticles can be synthesized in a short time. However, the control of particle size distribution is limited, because only kinetic factors are controlling the growth of the crystal (Laurent et al., 2008). In the syntheses of magnetic nanoparticles by coprecipitation route, mainly two stages are involved: (1) a short burst of nucleation occurs when the concentration of the species reaches critical supersaturation, and then, (2) there is a slow growth of the nuclei by diffusion of the solutes to the surface of the crystal. To fabricate monodisperse iron-oxide magnetic nanoparticles, these two key stages should be separated; that is, nucleation should be avoided during the period of enlargement.

During the last few years, significant advances in preparing monodisperse magnetite nanoparticles, of different sizes, have been made by the use of organic additives as stabilization or functionalization agents. Recently, a novel magnetic nanoadsorbent for the removal of Cu^{2+} was firstly developed by the surface modification of chitosan-coated magnetic nanoparticles (CCMNPs) with α-ketoglutaric acid (α-KA), which is a natural, inexpensive, harmless, and environmental friendly biological reagent containing active functional groups such as carboxyl groups (Zhou et al., 2009). The TEM pattern of γ-Fe_2O_3 shows monodisperse rod-like and rhombohedral shape with the particle sizes of 110 and 10 nm, respectively, whereas the TEM picture of the CCMNPs shows discrete spherical shape. It was earlier reported that the shapes of the Fe_2O_3 products are rod like to spherical, and the particle size of the sample had been clearly changed after modification by the surfactants. TEM images also reveals that the CCMNPs synthesized in this study are multidispersed with a mean diameter of around 30 nm, and no marked aggregation occurs. It can be clearly found out that the chitosan coating process significantly altered the morphology of γ-Fe_2O_3 nanoparticles. It has been shown that the γ-Fe_2O_3 nanoparticles were coated by the chitosan, and the prepared CCMNPs exhibited good distribution.

Tai et al. (2011) fabricated magnetic porous microspheres of Fe_3O_4@poly(methylmethacrylate-co-divinylbenzene) by a simple strategy of co-precipitation reaction. The magnetic microspheres materials, consisting of polymer-coated iron oxide nanoparticles, were synthesized by the modified suspension polymerization of methacrylate and divinylbenzene in the presence of a magnetic fluid. The large surface area and stabilized magnetic porous polymeric microspheres of Fe_3O_4@poly(methylmethacrylate-co-divinylbenzene) used in the adsorptive removal of phenol from aqueous solution. The TEM and SEM images of Fe_3O_4@poly(methylmethacrylate-co-divinylbenzene) magnetic porous microspheres have been shown in the Fig. 7.1 and Fig. 7.2, respectively. It can be clearly seen from the Fig. 7.1f

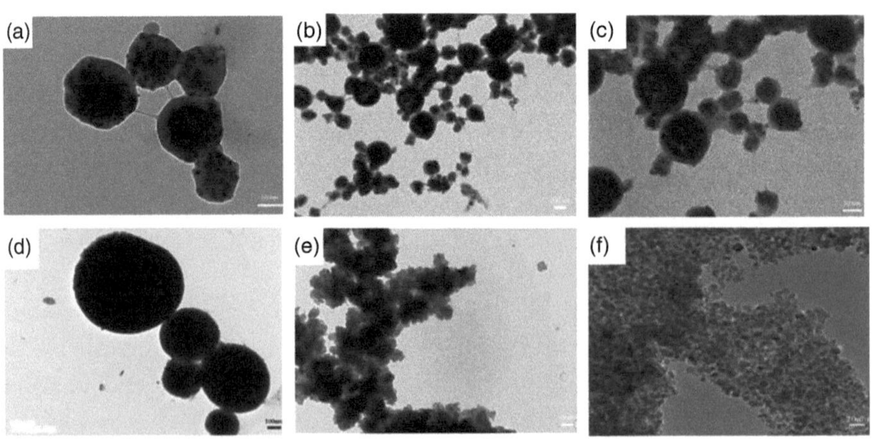

Figure 7.1 ***The TEM images of Fe_3O_4@poly(methylmethacrylate-co-divinylbenzene) magnetic porous microspheres.*** *(Source: Tai et al., 2011, © Elsevier).*

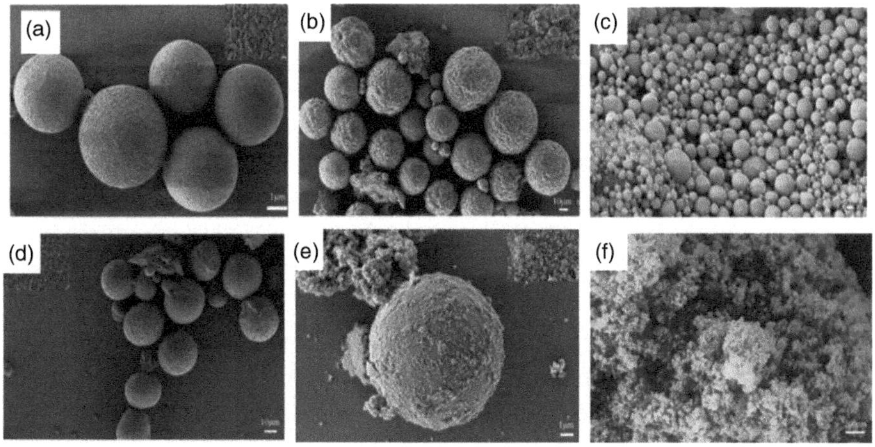

Figure 7.2 ***The SEM images of Fe_3O_4@poly(methylmethacrylate-co-divinylbenzene) magnetic porous microspheres.*** *(Source: Tai et al., 2011, © Elsevier).*

and Fig. 7.2f that the distribution of the oleic acid-coated Fe_3O_4 nanoparticles is comparatively uniform; in addition, the oleic acid-coated magnetite nanoparticles are easy to aggregate. Recent studies showed that oleic acid is an excellent candidate for the stabilization of magnetic Fe_3O_4 and γ-Fe_2O_3 nanoparticles (Tai et al., 2011; Willis et al., 2005).

Hong et al. (2008) demonstrated that precipitation of Fe_3O_4 nanoparticles with NH_4OH instead of NaOH gives a better crystallinity, higher saturation magnetization, and smaller size. Nitrogen passes during the precipitation not only protects oxidation of the Fe_3O_4 nanomaterials but also reduces the particle size when compared to methods without oxygen removal. Magnetic nanoparticles, Fe_3O_4 or γ-$Fe_2O_{3,}$ prepared under alkaline condition of NH_4OH, can be suitably stabilized with silica to form well-dispersed magnetic silica nanospheres. It gives several advantages over other stabilization materials. The change in the SiO_2/Fe_3O_4 ratio controls the particle size distribution.

7.2.2 Hydrothermal Syntheses

Several scientists have reported the hydrothermal synthesis of magnetic nanoparticles for a wide range of applications (Feng et al., 2012; Lee et al., 2011; Wan et al. 2007). Hydrothermal reactions are performed in aqueous solutions at high temperature and pressure. The reaction conditions, such as solvent, temperature, and time, usually have important effects on the products synthesized by hydrothermal method. There are mainly two routes for the synthesis of magnetic nanoparticles via hydrothermal conditions: hydrolysis and oxidation or neutralization of mixed metal hydroxides. Hydrothermal method offers high production yields with reduced production costs and is environment friendly as a result of the simple equipment, readily available source materials, does not have explosive or corrosive gases, and uses only deionized water as the reaction medium (Lee et al., 2011). Wan et al. (2007) have successfully synthesized Fe_3O_4 nanostructures with different morphologies such as Fe_3O_4 nanoparticles, nanorods, and nanowire bundles via a developed hydrothermal route. Fe_3O_4 nanoparticles with uniform size and relatively better monodispersity were synthesized at room temperature. They have reported that the synthesized product was mainly composed of uniform Fe_3O_4 nanoparticles with the average diameter of about 20 nm. The selected area electron diffraction (SAED) pattern indicated the good crystalline nature of the nanoparticles. It was worthwhile to note that the complexation between the surfactant molecules and ferrites leads to an improvement in the uniformity of crystallite size and morphology. As a result, the prepared Fe_3O_4 nanoparticles show uniform size and relatively better monodispersity.

Cao et al. (2006) reported a new reaction route to obtain rice- and cube-shaped single-crystalline α-Fe_2O_3 nanostructures via a reaction between $Fe(NO_3)_3 \cdot 9H_2O$ and $NH_3 \cdot H_2O$ in ethylene glycol at 200 and 280°C, respectively. The morphology of the products was studied using transmission electron microscopy and high-resolution TEM. The averaged size of a rice-shaped α-Fe_2O_3 nanocrystal was along its major axis, approximately 84.9 nm, and along its minor axis, approximately 41.2 nm; the length-to-diameter ratio was about two. The single-crystalline nature of synthesized α-Fe_2O_3 was confirmed by the strong spots of the selected area electron diffraction pattern. On the basis of the results, the authors have indicated that the reaction temperature plays a key role for varying the morphology of the synthesized α-Fe_2O_3. The sensitivity to the reaction temperature implies that the reaction route is under kinetic rather than thermodynamic control. The reaction between $Fe(NO_3)_3$ solution and $NH_3 \cdot H_2O$ is shown in equation:

$$Fe(NO_3)_3 + 3NH_3 \cdot H_2O \rightarrow Fe(OH)_3 + 3NH_4NO_3 \tag{7.3}$$

and α-Fe_2O_3 particles will form through a two-step phase transformation, according to the following equation:

$$Fe(OH)_3 \rightarrow \beta - FeOOH \rightarrow \alpha - Fe_2O_3 \tag{7.4}$$

Ethylene glycol will have an effect on preventing the $Fe(OH)_3$ nanoparticles from agglomeration (Cao et al., 2006). Therefore, uniform α-Fe_2O_3 nanocrystals were obtained after being heated at 200–280°C to complete the phase transformation.

7.2.3 Microemulsion

Microemulsion route is presented to obtain magnetic nanoparticles of different sizes, offering several advantages with respect to other similar methods (Lopez Perez et al., 1997a,b; Okoli et al., 2012; Vidal-Vidal et al., 2006). A microemulsion can be defined as a thermodynamically stable isotropic dispersion of two immiscible liquids consisting of nanosized domains of one liquid in the other, stabilized by an interfacial film of surface-active molecules. The surfactant molecules provide a confinement effect that limits particle nucleation, growth and agglomeration. In this method, coprecipitation occurs in tiny droplets of water (water pools) embedded with surfactant molecules and homogenously distributed in an oil phase. The size of these "water pools," so-called reverse micelles, which act as microreactors for the synthesis of the nanoparticles, is thermodynamically defined by the water-to-surfactant molar ratio (Makovec and Košak, 2005).

In this field, the main challenge is to create the suitable surface of magnetic nanoparticles in order to functionalize and allow strong interactions with specific biological components. In an effort to make ultrasmall, coated, and uniform size magnetic nanoparticles, Santra et al. (2001) performed water-in-oil microemulsion mediated sonochemical synthesis of superparamagnetic iron oxide nanoparticles in nonionic surfactants. The magnetic core was coated by a thin layer of silica for stabilization of magnetic nanoparticles. The synthesized magnetic nanoparticles have diameters as small as 1–2 nm, and the thickness of the silica coating is as thin as 1–2 nm.

Vestal and Zhang (2002) have used normal and reverse micelle microemulsion methods to synthesis single-phase $CoCrFeO_4$ nanoparticles with a controlled size range of 6–16 nm. Results obtained in this study showed that microemulsion methods provide a synthetic approach that allows for high-quality nanoparticles with a narrow size distribution. This method can be utilized for the synthesis of magnetic nanoparticles of various morphologies. However, this method needs large amounts of solvents and the yield is low.

7.2.4 Thermal Decomposition

Magnetic nanoparticles with smaller size and different morphology can be synthesized through the thermal decomposition of organometallic compounds in high-boiling organic solvents containing stabilizing surfactants (Fantechi et al., 2012; Lu et al., 2007). Iron oxide magnetic nanoparticles can also be synthesized by the thermal decomposition method. In thermal decomposition, two different approaches have been used for this purpose. First, thermal decomposition of metal carbonyl precursors followed by an oxidation step using air, or oxidation by using an oxidant at elevated temperatures. The second is the decomposition of precursors with a cationic metal center in the absence of reducing agents. The presence of reducing agents leads to metal nanoparticles even by the use of cationic precursors.

Thermal decomposition route appears the best synthesizing method for controlling size and morphology of magnetic nanoparticles. It gives scalable and high production yield. Although, one of the major disadvantages of this method is the production of organic soluble nanomaterials which limit the extent of application uses of them in biological and environmental fields besides surface treatment is needed after synthesis; also, thermal decomposition methods usually lead to complicated processes or require relatively high temperatures. However, success in precise particles size control of Fe_3O_4 nanoparticle has only been achieved through thermal decomposition (Chin et al., 2011) using large quantities of toxic and expensive precursors and surfactants in organic solvent.

7.3 MAGNETIC NANOPARTICLES IN WASTEWATER TREATMENT

Magnetic nanoparticles such as Fe_3O_4 and α-Fe_2O_3 have been investigated to resolve various environmental problems, such as removing toxic metal ions and radioactive elements, capturing of microbial pathogens and organic dyes, accelerating the coagulation of sewage, and remediation of contaminated soils (Lin et al., 2011; Yong-Gang et al., 2010; Yong-Mei et al., 2010). The magnetic nanoparticles possess high surface area, high chemical stability, and optimal magnetic properties, which lead to high adsorption efficiency, highly efficient removal of contaminants, and easy and rapid separation of adsorbent from solution via external magnetic field.

Magnetism of the nanomaterials is a unique property that autonomously assists in water/wastewater purification by influencing the physical properties of contaminants in aqueous solutions. Thus, magnetic separations of pollutants have been used extensively in wastewater treatment and environmental cleanup. Magnetic iron oxide nanomaterials are promising candidates for larger-scale wastewater treatment due to their easy functionalization, low cost, high adsorption capacity, strong physicochemical stability, and easy separations. Thus, these nanomaterials can be efficiently utilized for the treatment of wastewater containing hazardous dyes and toxic metal ions. In the following sections, we have described the application of magnetic nanoparticles for rapid removal of heavy metals from aqueous environments.

7.3.1 Magnetic Nanoparticles as Nanosorbents for Heavy Metals

Heavy metal contamination of aqueous environment is an increasing worldwide problem. Heavy metal is well known to cause serious environmental and health problems due to their bioaccumulation through the food chain, and causes serious damage to the nervous and endocrine systems of human beings. Therefore, the development of selective and efficient methodologies for detecting and removing heavy metals from aqueous media is in great demand.

Recently, Feng et al. (2012) synthesized monodisperse carboxyl functionalized superparamagnetic iron oxide nanoparticles, Fe_3O_4(np)—COOH. Representative TEM images of the Fe_3O_4(np)—COOH nanoparticles deposited from a water dispersion and dried under ambient conditions revealed the presence of clearly defined features of an average size of ~8–17 nm, with narrow-sized distribution. They examined their ability to adsorption of Cd^{2+} from aqueous solutions. The adsorption efficiency of the particles for Cd^{2+} was on the order of 0.17−0.20 μmol Cd/mg Fe_3O_4(np)—COOH, which was among the highest reported in the literature. As reported by the authors, the adsorption efficiency of the 8 nm particles was a bit higher than that for the 17 nm analogs. It had been clearly observed that an increase in the acidity of the media led to desorption of Cd^{2+}.

In order to provide long-term high-quality water or to enable water recycling, Girginova et al. (2010) recently synthesized silica coated magnetite particles for magnetic removal of Hg^{2+} from water. Silica coated magnetite particles derivatized with dithiocarbamate groups were evaluated as magnetic nanodrivers to remove trace levels of Hg^{2+} from aqueous solutions using a simple external NdFeB magnet. The authors reported that the uptake efficiency for Hg^{2+} was significantly higher for the particles bearing dithiocarbamate groups at the surface (74%), as compared to the nonderivatized silica coated magnetite (24%). On the basis of this study, one can suggest that this efficiency of nanomaterials is related to the high stability of the chelates formed between the dithiocarbamate group and Hg^{2+}.

To further facilitate the adsorption affinity of guest molecules on nanomaterials, surface modification, including inorganic and organic coating, and covalent binding, has often been explored to enable specific metal complexation. Notably, the amino-functionalized materials demonstrated outstanding ability to remove a wide variety of heavy metal ions such as Cu(II), Co(II), Ni(II), Zn(II), Pb(II), Cr(VI), and Cd(II) from aqueous solutions owing to the strong metal-complexing capability of amino groups (Wang et al., 2010). SiO_2 is stable under acidic conditions and inert to redox reactions, as compared with the organic coating materials, and hence functions as an ideal shell composite to protect the inner magnetite core. Silica-coated core–shell magnetite nanoparticles (Fe_3O_4@SiO_2) have recently been investigated for potential environmental and biomedical applications (Wang et al., 2010). The SiO_2 coating provides plenty of surface hydroxyl groups, which offers ease of succedent functionalization of magnetite nanoparticles.

Wang et al. (2010) have developed a novel amino-functionalized Fe_3O_4@SiO_2 magnetic nanomaterial with a core–shell structure for the adsorptive removal of heavy metals from aqueous solutions. The TEM images of the manufactured nanoparticles are shown in Fig. 7.3, displaying the core–shell structure of Fe_3O_4@SiO_2 and Fe_3O_4@SiO_2–NH_2 particles. The average particle sizes of Fe_3O_4, Fe_3O_4@SiO_2, and Fe_3O_4@SiO_2–NH_2 were observed to be 13.4, 16.6, and 18.4 nm, respectively. High adsorption affinity for aqueous Cu(II), Pb(II), and Cd(II) was achieved through the complexation of metal ions by amino groups grafted on the silica surface of the nanomaterial, and the adsorption was not much impacted by the presence of cosolute of humic acid or alkali/earth metal ions. Thus, the result observed shows that core–shell structure of the nanoadsorbent formed by a magnetite wrapped with an inert silica layer provides ease of magnetic separation and protection from acid leaching in regeneration. These unique characteristics of synthesized Fe_3O_4@SiO_2–NH_2 nanoparticles give a high potential for effective and rapid removal of toxic metal ions in water treatment.

Very recently, Song et al. (2011) prepared polyrhodanine-encapsulated magnetic nanoparticles via one-step chemical oxidation polymerization and investigated their potential for adsorptive removal of heavy metal ions from aqueous solutions. Fig. 7.4 shows the overall synthesis procedure for the polyrhodanine-encapsulated magnetic nanoparticles. TEM images showed that the polyrhodanine-encapsulated magnetic nanoparticles were synthesized with an average diameter of c. 10 nm. Typically, 5.0 mg of the

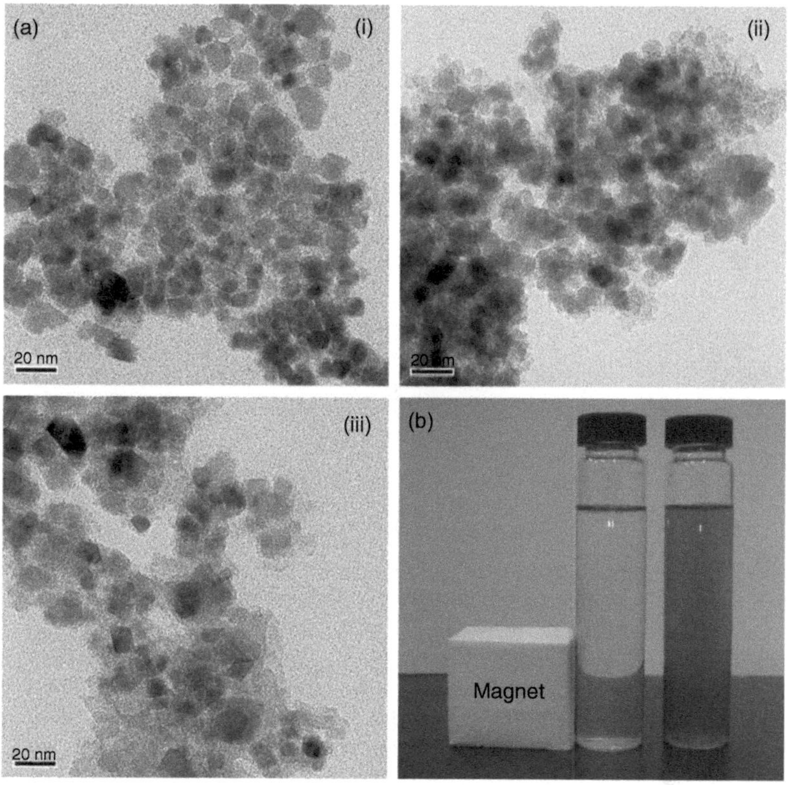

Figure 7.3 (a) Transmission electron microscopy (TEM) images of nanoparticles: (i) Fe_3O_4, (ii) Fe_3O_4@SiO_2, (iii) Fe_3O_4@SiO_2–NH_2. (b) The separation of Fe_3O_4@SiO_2–NH_2 nanoparticles from suspension under an external magnetic field and the same bottle of sample without the magnet. The average particle sizes of Fe_3O_4, Fe_3O_4@SiO_2, and Fe_3O_4@SiO_2–NH_2 were observed to be 13.4, 16.6, and 18.4 nm, respectively. *(Source: Wang et al., 2010, © Elsevier).*

PR-MNPs was added into 10 mL of Hg(II) ion solution (80 mg/L) at a pH value of 4.0. Then, the prepared samples were shaken at 350 rpm using a mechanical shaker at 25°C in a batch mode. After a desired contact time, the PR-MNPs were removed from the solution using an external magnetic field. The inductively coupled plasma (ICP) analysis were observed for the residual Hg(II) ion concentration in the aqueous media. Similarly, the adsorption capacities of Cd(II), Mn(II), and Cr(III) ions were also determined. It was found that the adsorption capacity increases sharply with increase in pH from 2.0 to 4.0 but increases slightly with a pH rise from 4.0 to 8.0. It means that the concentration of H^+ ion significantly affects the adsorption behavior of PR-MNPs. The negative surface charge enhances the affinity

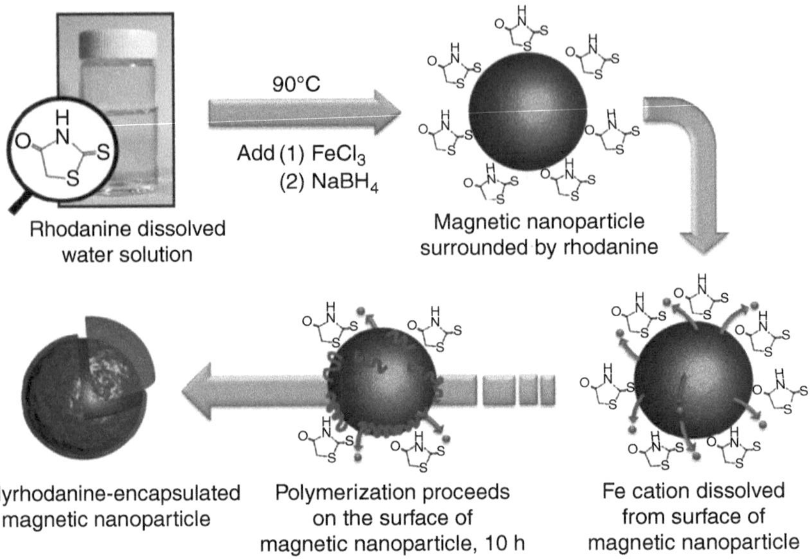

Figure 7.4 ***Schematic illustration of the fabrication process of polyrhodanine-encapsulated magnetic nanoparticles (PR-MNPs).*** When iron chloride was added into the rhodanine aqueous solution, the rhodanine molecules coordinated with Fe ions. After the injection of sodium borohydride ($NaBH_4$), magnetic nanoparticles were instantaneously formed with coordinated-rhodanine monomers as a stabilizer. As the reaction proceeded, the Fe ions were redissolved out from the surface of magnetic nanoparticles. Then, the Fe ions induced the oxidation of rhodanine monomers, leading to the chemical oxidation polymerization. In the chemical oxidation polymerization, Fe ion accepts an electron from rhodanine monomer by triggering the polymerization of rhodanine. At the end of the polymerization, the polyrhodanine-encapsulated magnetic nanoparticles (PR-MNPs) were obtained after washing with distilled water. *(Source: Song et al., 2011, © Elsevier).*

to positively charged metal ions. The increment rate of adsorption capacity diminished in the pH range of 4.0 to 8.0. The highest Hg(II) ion adsorptivity was 94.5% at the initial mercury ion concentration of around 1.3 mg/L. Two parameters isotherm model, Langmuir and Freundlich, were tested for the adsorption of Hg(II) ion onto the PR-MNPs. Comparing the R^2 value and F-test result of both models, it was concluded the experimental data were well fitted to the Freundlich model, which suggests the heterogeneous metal ion adsorption activity. It may result from the different adsorption sites (oxygen, nitrogen, and sulfur groups) of polyrhodanine-magnetic nanoparticles that have different metal-binding energies.

Recently, Qiu et al. (2012) studied the adsorption of Cu^{2+} from aqueous solution by polymer supported iron oxide nanoparticles and the effect of sulfate on the adsorption behavior was explained by X-ray photoelectron spectroscopy. In their study, the two hybrid sorbents (denoted as HFO–PS^- and HFO–PS^0, respectively) were prepared by loading hydrous ferric oxide (HFO) nanoparticles onto two polystyrene beads: PS^-, negatively charged with sulfonic acid groups, and PS^0, covalently bonded with neutral chloromethyl groups. X-ray photoelectron spectroscopy demonstrated that besides the electrostatic effects, the formation of Cu–SO_4 ternary complexes also accounted for the enhanced Cu^{2+} sorption on both bulky HFO and hybrid HFO sorbents in the presence of sulfate. The peak shift to higher binding energy was recognized in the sulfate system which indicated an increased adsorption of Cu^{2+} ions was due to the enhanced complexation (Fig. 7.5).

Chung et al. (2012) studied the sorption of Pb(II) and Cu(II) onto multiamine grafted mesoporous silica embedded with nanomagnetite and the steric factors affecting the mass transfer of Pb(II) and Cu(II) in mesocellular silica foams (MSU-F-S) were extensively investigated. The TEM images showed that the synthesized MSU-F-S has a three-dimensional interconnected structure with ~30 nm uniformly distributed pore size and ~20 nm window size (Fig. 7.6). The metal cations diffusion into the mesopores of MSU-FS/Fe_3O_4 was notably limited by the amine groups heterogeneously distributed throughout the pore channels during post-synthesis grafting. Due to the protonation on the adsorbent, the amine groups at edge sites were decreased during the adsorption process, which was the major steric factor. The maximum adsorption capacity for Pb(II) and Cu(II) onto monoamine functionalized MSU-FS/Fe_3O_4 were 222.4 and 57.9 mg/g, respectively.

Wang et al. (2011) synthesized the multishelled Co_3O_4—Fe_3O_4 hollow spheres with even magnetic phase distribution and their application in water treatment was investigated. The Langmuir maximum adsorption capacity of tripple shelled assemblies of Co_3O_4—Fe_3O_4 hollow spheres was found to be 125 mg/g for Congo red, which was 20 and 3.5 times higher than the case of commercial Co_3O_4 and Fe_2O_3 nanoparticles under the same conditions.

The chitosan and porous anodic alumina were used as an adsorbent for the removal of $Cr_2O_7^{2-}$ ions from aqueous solutions through electrostatic attraction among chitosan, porous anodic alumina, and $Cr_2O_7^{2-}$; lots of $Cr_2O_7^{2-}$ ions were adsorbed on the adsorbent and then reacted with iron. It has been concluded that most of the Cr^{6+} ions transformed into Cr^{3+} ions. The adsorption capacity was found to be 113.2 mg/g.

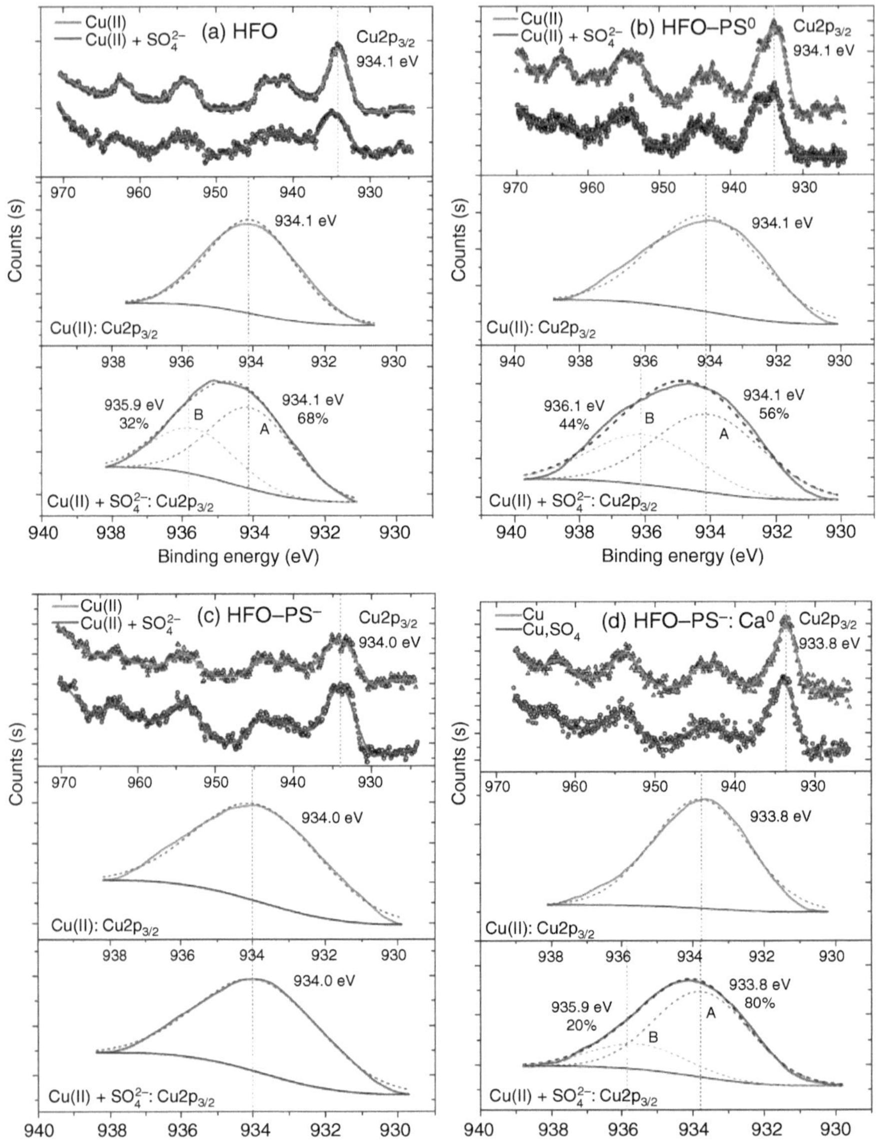

Figure 7.5 ***XPS spectra of Cu 2p region for Cu(II)-loaded sorbents in the absence or presence of sulfate.*** (a) HFO, 1 g/L; initial Cu(II), 1 mM; SO_4^{2-}, 5 mM. (b) HFO–PS^0, 1 g/L; initial Cu(II), 1 mM; SO_4^{2-}, 5 mM. (c) HFO–PS$^-$, 1 g/L; initial Cu(II), 1 mM; SO_4^{2-}, 5 mM. (d) HFO–PS$^-$, 1 g/L; initial Cu(II), 1 mM; SO_4^{2-}, 5 mM; background $Ca(NO_3)_2$, 10 mM. *(Source: Qiu et al., 2012, © Elsevier).*

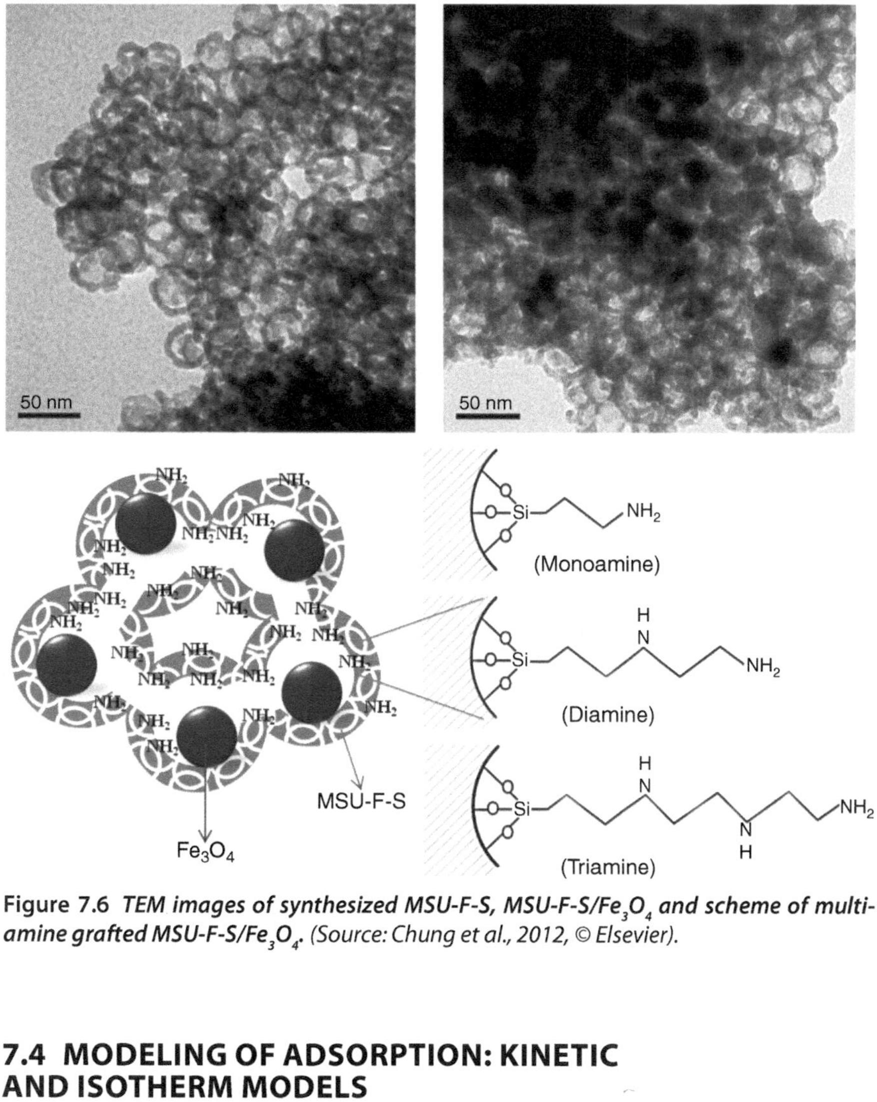

Figure 7.6 *TEM images of synthesized MSU-F-S, MSU-F-S/Fe_3O_4 and scheme of multi-amine grafted MSU-F-S/Fe_3O_4. (Source: Chung et al., 2012, © Elsevier).*

7.4 MODELING OF ADSORPTION: KINETIC AND ISOTHERM MODELS

Assessment of a solid–liquid sorption system is usually based on two types of investigations: equilibrium batch sorption tests and dynamic continuous-flow sorption studies. The kinetic study of adsorption processes provides useful data regarding the efficiency of the adsorption and the feasibility for scale-up operations. The kinetic data of adsorption can be evaluated using different types of mathematic models, of which one most widely used is Lagergren's rate equation (Lagergren, 1898). Equilibrium isotherm model equations such as, Langmuir, Freundlich, Temkin, and Redlich–Peterson are used to describe experimental adsorption data in batch mode (Langmuir, 1918; Freundlich, 1906; Temkin and Pyzhev, 1940; Redlich and Peterson, 1959). It is important to find best-fit isotherm to evaluate the efficacy of the prepared adsorbent to develop suitable industrial adsorption system designs.

Recently, a novel hybrid material zirconium polyacrylamide was synthesized by an environmental friendly sol–gel method and the removal of As(III) was carried out (Mandal et al., 2013). High removal percentage of As(III) was 98.22% was obtained within 120 min of contact time. The adsorption kinetics was best explained by second-order model and experimental data fitted well to the Freundlich isotherm model, which shows the multilayer adsorption of As(III) on zirconium polyacrylamide materials. The Langmuirs maximum adsorption capacity was found to be 41.48 mg/g.

Adsorption behavior of As on Fe–Al binary oxide was carried out by Hong et al. (2011) from aqueous solutions and the thermodynamic and kinetics were fully modeled to get highest removal percentage. Strong electrostatic interaction between the As(V) anion and the positive sites on Fe–Al binary oxide was found, As(V) was adsorbed rapidly and the adsorption rate was controlled by chemical adsorption. However, due to nonionic form the As(III) was not fully adsorbed on the Fe–Al binary oxide. The adsorption was exothermic and spontaneous in nature. Therefore, As(V) adsorption removal percentage decreased with an increase in temperature from 298 to 333 K, whereas As(III) adsorption capacity did not show any significant change with varying temperature. The kinetic studies showed that the adsorption process was well fitted to pseudo-second-order model and controlled by intraparticle diffusion model. Table 7.1 shows the functionalized iron oxide and other adsorbents for arsenic removal from aqueous solutions.

7.5 CONCLUSIONS AND FUTURE PERSPECTIVES

Magnetic nanoparticles and magnetic nanocomposites offer tremendous opportunity to treat wastewater-containing toxic metal ions, metalloids, pharmaceutical, and dyes because of their high surface area, excellent physicochemical stability, multifunctionality, ease in coating, and low cost of synthesis. Till now a variety of methods for the synthesis of magnetic nanoparticles and nanocomposites have been developed and practiced namely, coprecipitation, microemulsion, hydrothermal, and thermal decomposition. However, coprecipitation method using cheap iron oxide is one of

Table 7.1 Functionalized iron oxide and other adsorbents for arsenic removal

Adsorbents	Species	Adsorption capacity (mg/g)	References
Fe–Al binary oxide	As(V)	37.59	Hong et al. (2010)
	As(III)	40.65	Hong et al. (2010)
Iron oxide coated Fungal biomass	As(V)	1.01	Pokhrel and Viraraghavan (2008)
	As(III)	0.7	Pokhrel and Viraraghavan (2008)
Silica-containing iron oxide	As(V)	20	Zeng (2003)
	As(III)	20	Zeng (2003)
Iron coated pottery granules	As(V)	3.6	Dong et al. (2009)
Nanoscale Fe(0)	As(V)	1.8	Kanel et al. (2005)
Iron containing	As(V)	5.15	Gu and Deng (2007)
mesoporous carbon	As(III)	5.96	Gu and Deng (2007)
Zirconium polyacrylamide hybrid	As(III)	41.48	Mandal et al. (2013)
Fe–Al binary oxide	As(V)	—	Hong et al. (2011)

the most celebrated synthesis routes for the magnetic nanoparticles due to their low cost, easy operations, ease in design, and high yield.

The bare magnetic nanoparticles show high surface-to-volume ratio and have been used in a number of applications due to their mesoporous structures, which favorably facilitates the adsorption and diffusion of hazardous substances in the pores and on the active sites of magnetic nanoparticles. However, it has been found that the bare magnetic nanopaticles are very much prone to the air oxidation and acid corrosion. They may leach in highly acidic solution such as in case of acid mine drainage treatment and during the reduction of chromium present in tannery and leather wastewater. Arsenic contaminated industrial effluents may create problems during the treatment with the bare magnetic nanoparticles. Hence bare magnetic nanoparticle have been widely functionalized with various type of inorganic and organic materials to enhance the physicochemical stability, adsorption capacity and active surface site on adsorbents. The functionalized magnetic nanoparticles have shown better adsorption capacity toward the removal and recovery of As, Cr, Cd, Cu, Co, Zn, Hg, Pb, Ni, anionic and cationic dyes, and pharmaceuticals from wastewater.

The multifunctional magnetic nanoparticles have been successfully applied for the reduction of toxic metal ions up to ppb level in treated wastewater. Magnetic nanoparticle loaded with toxic metal ions in water can be effectively recovered and regenerated with the help of an external magnetic field. The regenerated nanoadsorbent can be successfully and effectively applied for recycling and treatment of wastewater. As reported in few literatures, the extreme acidic or basic conditions of solution can corrode the nanoparticles and loss of function groups can occur. Thus, the stability of the nanoparticles in the treatment of heavy metals strongly depends on pH of the solution.

Nanotechnology offers lots of advantageous potential for today's society but it also has some drawbacks like other techniques. Increasing use of nanoproducts and nanoadsorbents will increase the risk of nanopollutants in the environment. Hence, there is an essential need of research to solve the problems and overcome the challenges that have been discussed in this chapter. In addition, there is a further need for the study of nanoparticles behavior, fate, and effects in the different spheres of the ecosystem on a timely basis. Recent trends and widespread use of nanomaterials in the area of environment is extremely new and it needs further research before on site applications because of the nanotoxicity of materials. Although, the magnetic nanoparticles recommends tremendous potentials to treat wastewater containing toxic metal ions because of their high surface area, high physicochemical stability, multifunctionality, ease in coating, and low-cost of synthesis. Overall these nanomaterials can be successfully and effectively applied for the recycling, treatment, and sustainable management of water.

REFERENCES

Ambashta, R.D., Sillanpaa, M., 2010. Water purification using magnetic assistance: a review. J. Hazard. Mater. 180, 38–49.

Cao, H., Wang, G., Zhang, L., Liang, Y., Zhang, S., Zhang, X., 2006. Shape and magnetic properties of single-crystalline hematite (α-Fe_2O_3) nanocrystals. ChemPhysChem 7, 1897–1901.

Chin, S.F., Pang, S.C., Tan, C.H., 2011. Green synthesis of magnetite nanoparticles (via thermal decomposition method) with controllable size and shape. J. Mater. Environ. Sci. 2, 299–302.

Chung, J., Chun, J., Lee, J., Lee, S.H., Lee, Y.J., Hong, S.W., 2012. Sorption of Pb(II) and Cu(II) onto multi-amine grafted mesoporous silica embedded with nano magnetite: effects of steric factors. J. Hazard. Mater. 239–240, 183–191.

Dong, L., Zinin, P.V., Cowen, J.P., Ming, L.C., 2009. Iron coated pottery granules for arsenic removal from drinking water. J. Hazard. Mater. 168, 626–632.

Fantechi, E., Campo, G., Carta, D., Corrias, A., de Julián Fernández, C., Gatteschi, D., Innocenti, C., Pineider, F., Rugi, F., Sangregorio, C., 2012. Exploring the effect of Co doping in fine maghemite nanoparticles. J. Phys. Chem. C 116, 8261–8270.

Feng, Z., Zhu, S., Martins de Godoi, D.R., Samia, A.C.S., Scherson, D., 2012. Adsorption of Cd^{2+} on carboxyl-terminated superparamagnetic iron oxide nanoparticles. Anal. Chem. 84, 3764–3770.

Freundlich, H.M.F., 1906. Uber die adsorption in lasugen. J. Phys. Chem. 57, 385–470.

Girginova, P.I., Daniel-da-Silva, A.L., Lopes, C.B., Figueira, P., Otero, M., Amaral, V.S., Pereira, E., Trindade, T., 2010. Silica coated magnetite particles for magnetic removal of Hg^{2+} from water. J. Colloid Interface Sci. 345, 234–240.

Gu, Z., Deng, B., 2007. Use of iron-containing mesoporous carbon (IMC) for arsenic removal from drinking water. Environ. Eng. Sci. 24, 113–121.

Hong, R.Y., Li, J.H., Li, H.Z., Ding, J., Zheng, Y., Wei, D.G., 2008. Synthesis of Fe_3O_4 nanoparticles without inert gas protection used as precursors of magnetic fluids. J. Magn. Magn. Mater. 320, 1605–1614.

Hong, H.J., Farooq, W., Yang, J.S., Yang, J.W., 2010. Preparation and evaluation of Fe–Al binary oxide for arsenic removal: comparative study with single metal oxides. Separ. Sci. Technol. 45, 1975–1981.

Hong, H.J., Yang, J.S., Kim, B.K., Yang, J.W., 2011. Arsenic removal behavior by Fe–Al binary oxide: thermodynamic and kinetic study. Separ. Sci. Technol. 46, 2531–2538.

Jung, J.H., Lee, J.H., Shinkai, S., 2011. Functionalized magnetic nanoparticles as chemosensors and adsorbents for toxic metal ions in environmental and biological fields. Chem. Soc. Rev. 40, 4464–4474.

Kanel, S.R., Manning, B., Charlet, L., Choi, H., 2005. Removal of arsenic(III) from groundwater by nanoscale zero-valent iron. Environ. Sci. Technol. 39, 1291–1298.

Koehler, F.M., Rossier, M., Waelle, M., Athanassiou, E.K., Limbach, L.K., Grass, R.N., Gunther, D., Stark, W.J., 2009. Magnetic EDTA: coupling heavy metal chelators to metal nanomagnets for rapid removal of cadmium, lead and copper from contaminated water. Chem. Commun. 32, 4862–4864.

Lagergren, S., 1898. Zur theorie der sogenannten adsorption geloster stoffe. Kongliga Svenska Vetenskaps Academiens nya Handlingar 24, 1–39.

Langmuir, I., 1918. The adsorption of gases on plane surfaces of glass, mica and platinum. J. Am. Chem. Soc. 40, 1361–1403.

Laurent, S., Forge, D., Port, M., Roch, A., Robic, C., Elst, L.V., Muller, R.N., 2008. Magnetic iron oxide nanoparticles: synthesis, stabilization, vectorization, physicochemical characterizations, and biological applications. Chem. Rev. 108, 2064–2110.

Lee, P.L., Sun, Y.C., Ling, Y.C., 2009. Magnetic nano-adsorbent integrated with lab-on-valve system for trace analysis of multiple heavy metals. J. Anal. At. Spectrom. 24, 320–327.

Lee, S.J., Cho, J.H., Lee, C., Cho, J., Kim, Y.R., Park, J.K., 2011. Synthesis of highly magnetic graphite-encapsulated FeCo nanoparticles using a hydrothermal process. Nanotechnology 22, 375603.

Lin, Y.F., Chen, H.W., Chien, P.S., Chiou, C.S., Liu, C.C., 2011. Application of bifunctional magnetic adsorbent to adsorb metal cations and anionic dyes in aqueous solution. J. Hazard. Mater. 185, 1124–1130.

Lopez Perez, J.A., Lopez Quintela, M.A., Mira, J., Rivas, J., 1997a. Preparation of magnetic fluid with particles obtained in microemulsion. IEEE Trans. Magn. 33, 4359–4362.

Lopez Perez, J.A., Lopez Quintela, M.A., Mira, J., Rivas, J., Charles, S.W., 1997b. Advances in the preparation of magnetic nanoparticles by the microemulsion method. J. Phys. Chem. B 101, 8045–8047.

Lu, A.H., Salabas, E.L., Schuth, F., 2007. Magnetic nanoparticles: synthesis, protection, functionalization, and application. Angewandte Chemie Int. Ed. 46, 1222–1244.

Makovec, D., Košak, A., 2005. The synthesis of spinel–ferrite nanoparticles using precipitation in microemulsions for ferrofluid applications. J. Magn. Magn. Mater. 289, 32–35.
Mandal, S., Sahu, M.K., Patel, R.K., 2013. Adsorption studies of arsenic(III) removal from water by zirconium polyacrylamide hybrid material (ZrPACM-43). Water Res. Ind. 4, 51–67.
Okoli, C., Sanchez-Dominguez, M., Boutonnet, M., Jaärås, S., Civera, C., Solans, C., Kuttuva, G.R., 2012. Comparison and functionalization study of microemulsion-prepared magnetic iron oxide nanoparticles. Langmuir 28, 8479–8485.
Pokhrel, D., Viraraghavan, T., 2008. Arsenic removal in an iron oxide-coated fungal biomass column: analysis of breakthrough curves. Bioresource Technol. 99, 2067–2071.
Qiu, H., Zhang, S., Pan, B., Zhang, W., Lv, L., 2012. Effect of sulfate on Cu(II) sorption to polymer-supported nano-iron oxides: behavior and XPS study. J. Colloid Interface Sci. 366, 37–43.
Redlich, O., Peterson, D.L., 1959. A useful adsorption isotherm. J. Phys. Chem. 63, 1024–1026.
Santra, S., Tapec, R., Theodoropoulou, N., Dobson, J., Hebard, A., Tan, W., 2001. Synthesis and characterization of silica-coated iron oxide nanoparticles in microemulsion: the effect of nonionic surfactants. Langmuir 17, 2900–2906.
Shen, Y.F., Tang, J., Nie, Z.H., Wang, Y.D., Ren, Y., Zuo, L., 2009. Tailoring size and structural distortion of Fe_3O_4 nanoparticles for the purification of contaminated water. Bioresour. Technol. 100, 4139–4146.
Singh, K.P., Gupta, S., Singh, A.K., Sinha, S., 2011. Optimizing adsorption of crystal violet dye from water by magnetic nanocomposite using response surface modeling approach. J. Hazard. Mater. 186, 1462–1473.
Song, J., Kong, H., Jang, J., 2011. Adsorption of heavy metal ions from aqueous solution by polyrhodanine-encapsulated magnetic nanoparticles. J. Colloid Interface Sci. 359, 505–511.
Sun, L., Zhang, L., Liang, C., Yuan, Z., Zhang, Y., Xu, W., Zhang, J., Chen, Y., 2011. Chitosan modified Fe^0 nanowires in porous anodic alumina and their application for the removal of hexavalent chromium from water. J. Mater. Chem 21, 5877–5880.
Tai, Y., Wang, L., Gao, J., Amer, W.A., Ding, W., Yu, H., 2011. Synthesis of Fe_3O_4@poly(methylmethacrylate-co-divinylbenzene) magnetic porous microspheres and their application in the separation of phenol from aqueous solutions. J. Colloid Interface Sci. 360, 731–738.
Temkin, M.J., Pyzhev, V., 1940. Recent modifications to Langmuir isotherms. Acta Physico-Chimica USSR 12, 217–222.
Vestal, C.R., Zhang, Z.J., 2002. Synthesis of $CoCrFeO_4$ nanoparticles using microemulsion methods and size-dependent studies of their magnetic properties. Chem. Mater. 14, 3817–3822.
Vidal-Vidal, J., Rivas, J., Lopez-Quintela, M., 2006. Synthesis of monodisperse maghemite nanoparticles by microemulsion method. Colloids Surf. A 288, 44–51.
Wang, H., Yan, N., Li, Y., Zhou, X., Chen, J., Yu, B., Gong, M., Chen, Q., 2012. Fe nanoparticle-functionalized multi-walled carbon nanotubes: one-pot synthesis and their applications in magnetic removal of heavy metal ions. J. Mater. Chem. 22, 9230–9236.
Wan, J., Yao, Y., Tang, G., 2007. Controlled-synthesis, characterization, and magnetic properties of Fe_3O_4 nanostructures. Appl. Phys. A 89, 529–532.
Wang, J., Zheng, S., Shao, Y., Liu, J., Xu, Z., Zhu, D., 2010. Amino-functionalized Fe_3O_4@SiO_2 core–shell magnetic nanomaterial as a novel adsorbent for aqueous heavy metals removal. J. Colloid Interface Sci. 349, 293–299.
Wang, X., Zhong, Y., Zhai, T., Guo, Y., Chen, S., Ma, Y., Yao, J., Bando, Y., Golberg, D., 2011. Multishelled Co_3O_4-Fe_3O_4 hollow spheres with even magnetic phase distribution: synthesis, magnetic properties and their application in water treatment. J. Mater. Chem. 21, 17680–17687.

Willis, A.L., Turro, N.J., O'Brien, S., 2005. Spectroscopic characterization of the surface of iron oxide nanocrystals. Chem. Mater. 17, 5970–5975.

Yong-Gang, Z., Hao-Yu, S., Sheng-Dong, P., Mei-Qin, H., 2010. Synthesis, characterization and properties of ethylenediamine-functionalized Fe_3O_4 magnetic polymers for removal of Cr(VI) in wastewater. J. Hazard. Mater. 182, 295–302.

Yong-Mei, H., Man, C., Zhong-Bo, H., 2010. Effective removal of Cu (II) ions from aqueous solution by amino-functionalized magnetic nanoparticles. J. Hazard. Mater. 184, 392–399.

Zeng, L., 2003. A method for preparing silica-containing iron(III) oxide adsorbents for arsenic removal. Water Res. 37, 4351–4358.

Zhou, Y.T., Nie, H.L., Branford-White, C., He, Z.Y., Zhu, L.M., 2009. Removal of Cu^{2+} from aqueous solution by chitosan-coated magnetic nanoparticles modified with α-ketoglutaric acid. J. Colloid Interface Sci. 330, 29–37.

CHAPTER 8

Layered Double Hydroxides Nanomaterials for Water Remediation

Sushmita Banerjee, Pavan K. Gautam
Department of Chemistry, University of Allahabad, Allahabad, India

8.1 INTRODUCTION

The availability of adequate water in terms of both quantity and quality is essential for human existence. Water which was once regarded as a free gift of nature is becoming scarce, particularly in developing countries because of the variety of pollutants that are making their way to water resources (Farooqi et al., 2007). Recently, layered double hydroxides (LDHs) materials are gaining widespread recognition and importance from scientific and academic community due to its unique ability to remove a number of water pollutants simultaneously (El Gaini et al., 2009; Tuna et al., 2013). Due to the property of higher anion exchange capacity of LDH, it makes them promising materials for the elimination of ecologically undesirable inorganic and organic anions (Rocher et al., 2008; Benselka-Hadj Abdelkader et al., 2011). The removal of chromate (Goswamee et al., 1998), nitrate, phosphate (Mandel et al., 2013), arsenate, or vanadate anions (Rives and Ulibarri, 1999) as well as humic substances (Wang et al., 2014) and radioactive elements (Kulyukhin et al., 2010; Pshinko et al., 2011) from contaminated water or waste streams by LDH has been demonstrated. Also, numerous papers have reported the use of LDH or their calcined derivatives as organic contaminate adsorbents, such as in the adsorption of phenolic compounds (Subramanian et al., 2013; Ulibarri et al., 1995) and pesticides (Pavlovic et al., 2005; Inacio et al., 2001). In recent decades, LDHs have been widely examined for different environmental applications, like catalysts, ion-exchangers, and adsorbents to combat the problem of contamination of water (Li et al., 2009a; Chen et al., 2012a). Layered solids are those solids in which two dimensional sheets or lamellae are stacked together. While atoms of the same layer are held together by strong covalent bonds, atoms between two layers are held by weak van der Waals forces. LDHs consist of a divalent and a trivalent metal ion and adopt the layered brucite structure

Nanomaterials for Wastewater Remediation
http://dx.doi.org/10.1016/B978-0-12-804609-8.00008-X

with metal ions octahedrally coordinated to hydroxide ions, with exchangeable anions, usually carbonate, in the interlayer regions to balance the lattice positive charge associated with the trivalent ions (Zhang et al., 2004; Cavani et al., 1991). LDH belongs to a unique group of clays that have an anionic exchange capability. Additionally, they are anion-intercalated inorganic functional materials, also known as hydrotalcite-like compounds or anionic clays (Reichle, 1986). LDHs have a structure similar to the naturally occurring hydrotalcite clay, which has a chemical formula of $[Mg_6Al_2(OH)_{16}CO_3 \cdot 4H_2O]$ (Kloprogge and Frost, 2001). Layered solids exhibit a number of properties such as ion diffusion/mobility, surface acidity or basicity, intercalation and ion exchange, delamination and sorption because of which they can be used as catalysts, ion exchangers, and sorbents (Braterman et al., 2004). The first discovered LDH was the mineral hydrotalcite $Mg_6Al_2(CO_3)(OH)_{16} \cdot 4(H_2O)$. The importance of layered compound is based on their ability to retain chemical species with electrical charges compatible to those of the layers (Khan and O'Hare, 2002). LDHs possess anion exchange capabilities in their natural state. However, superior results have been obtained after thermal activation of LDHs. LDHs with different cations, such as Mg, Mn, Fe, Co, Ni, Cu, Zn for divalent cations and Al, Mn, Fe, Co, Ni, Cr, Ga for trivalent cations, and LDH derived materials, have been studied and possible wide applications have recommended them as anion exchangers, polymer stabilizers, antacids, anion scavengers, catalyst, catalyst supports, adsorbents, for filtration and controlled release of anions, electroactive and photoactive materials (Vaccari, 1999; Goh et al., 2008; Yu et al. 2008; Chen et al., 2011).

8.2 SYNTHESIS OF LAYERED DOUBLE HYDROXIDES

8.2.1 Coprecipitation Method

It is considered a simple and most widely used method for the synthesis of LDHs. LDHs can be synthesized easily in the laboratory by coprecipitation from solution of the corresponding metal nitrates (Zhang et al., 2007). The most common method of precipitation is by the use of sodium hydroxide and sodium carbonate, to form a carbonate interlayer (Dodds et al., 2007). Coprecipitation is often performed by the slow addition of metal nitrate solutions to a solution of hydroxide and carbonate. In general nitrate or chloride salts of the metal ion are used for the synthesis because of their low affinity to the positively charged layers, allowing easy modification of the interlayer anion fitting to the specific needs (Hamad et al., 2014). Usually two modifications of the method are applied. The first is precipitation at low supersaturation (Rojas et al., 2011) which is performed at a fix pH, thus it is also known as precipitation at constant pH. This process involves slow addition of the solution containing both the di- and trivalent metal ions to the aqueous solution of the anion, which will be intercalated in the precipitated LDHs (Delorme et al., 2006).

The extent of synthesis is stimulated by the simultaneous addition of an alkaline solution to the reactor at an appropriate rate. Precipitation at high supersaturation is somewhat simpler. The pH during the synthesis is variable until its end. Solution containing M(II) and M(III) ions is added to an alkaline solution, in which the LDH precipitates. Observation of the precipitates and X-ray diffraction characterization show that the coformation of the main layers and interlamellar domains takes place at a very early stage of the process without clear delaminated state of the layers (Millange et al., 2000). Some experimental parameters, such as temperature in the reactor, pH, concentration of metallic salts solution and alkaline solution, flow rate of reactants, ageing of precipitate, are important parameters in the preparation of LDH by coprecipitation. Other parameters, such as addition rate, stirring method, geometry of the reactor, including reactants injection pipes, are less important parameters (Gordon et al., 1959).

8.2.2 Urea Method

The technique is a modified version of precipitation method in which aqueous solution of urea and an aqueous solution of the desired metal salts are mixed together and gently heated which gradually hydrolyses to release hydroxide ions and gives very finely controlled precipitation (Willard and Tang, 1937). This process starts at low pH and increases gradually to pH 7.6. The only possible interlayer anion using the urea method is carbonate. Coprecipitation using urea as the base was developed to prepare monodisperse particles. Urea is a very weak Bronsted base ($pK_b = 13.8$), and is highly soluble in water. Hydrolysis of urea proceeds in two steps: (1) formation of ammonium cyanate (NH_4CNO) as the rate-determining step; and (2) fast hydrolysis of the cyanate into ammonium carbonate (Adachi-Pagano et al., 2003):

$$CO(NH_2)_2 + 2H_2O \rightarrow 2NH_4^+ + CO_3^{2-}$$

Large platelets of well-crystallized hydrotalcite with hexagonal shape are obtained by this method. Monodisperse particles between 1 and 5 mm, and up to 20 mm are obtained.

$$M_3^{2+}Al(OH)_8(CO_3)0.5nH_2O(M^{2+} = Mg^{2+}, Ni^{2+}, Zn^{2+})$$

The particles have generally a hexagonal shape, and the particle size of LDHs formed this way is usually larger than other synthesis methods (Hibino and Ohya, 2009). The size of the particles can be controlled by the concentration of the urea solution, and the temperature of the reaction.

8.2.3 Microwave Synthesis

The microwave synthesis of LDHs is when the aging step of the reaction takes place in a microwave. The microwave irradiation allows for a rapid aging process, which is usually around 15–60 min (Lee et al., 2001; Benito et al., 2009). Both urea and coprecipitation methods can be used in a microwave synthesis, with the microwave aging instead of a reflux aging (Benito et al., 2009). The microwave aging results in a homogeneous particle size, which is smaller than particles made from reflux aging and often has a higher specific surface area. Microwave irradiation time results in a well-crystallized material compared to conventional coprecipitation and extent of enhancement in crystallinity depends on the nature of the trivalent metal ion. In general, the surface area, porosity, and thermal stability of the synthetic materials are increased with the duration of microwave exposure (Fang et al., 2012). The shorter aging time can also help prevent impurities which may be formed during long aging methods.

8.2.4 Ion-Exchange Method

The ion-exchange method is used when the desired interlayer anions cannot be produced using the coprecipitation or urea methods (Meyn and Beneke, 1990). The desired anions replace the anions present in a previously prepared LDH. Chlorides and nitrate anions are often used in the prepared LDHs as they can be easily replaced with a wide range of inorganic and organic anions. The desired anion cannot be too large compared to the anion it is being exchanged with, otherwise it may not "fit" into the interlayer region (Ragavan et al., 2006). A variety of organo-LDHs, with different combinations of matrix cations as well as different interlayer organic anions, have been prepared using this direct anion-exchange route (Liu et al., 2006).

8.2.5 Sol–Gel Method

The process involves formation of a mobile colloidal suspension (sol) that then gels (gel) because of the internal crosslinking (Zhao et al., 2010). The preparation of LDH is the result of the hydrolysis and polymerization of a solution of metal alkoxides. The alkoxides are first dissolved in an organic solvent and refluxed. To this solution water is slowly added, causing occurrence of crosslinkage. Materials prepared by this process possess good homogeneity, comparatively good control of stoichiometry, and high surface area and high porosity characteristics. The sol–gel hydrotalcite exhibit excellent thermal stability up to 550°C (Zhang and Liu, 1999). However, LDH samples synthesized by coprecipitation are more crystalline than those prepared by the sol–gel method (Nilchi et al., 2010). The marked increase in specific surface area is attributed to the increase in mesopore volume. Sol–gel materials are also more basic due to different textural and morphological features (Wang et al., 1998). This technique limits the amount of alkaline required and thus is beneficial for industrial synthesis processes.

8.2.6 Memory Effect or Calcination–Rehydration Method

The conversion of the mixed metal oxides into LDHs has been referred to as regeneration, reconstruction, restoration, rehydration, or the calcination–rehydration process, structural memory effect or simply memory effect (Guo et al., 2013). The "memory effect" (also called the calcination–rehydration method) of LDHs involves calcination at mild temperatures (usually under 500°C) to form mixed metal oxides with high surface areas. The oxides are then placed into an aqueous solution containing the desired interlayer anion under an inert atmosphere (Marchi and Apesteguia, 1998). The LDH then reforms into the original layered structure, but with the new interlayer anion. If the sample is calcinated at a temperature that is too high, the "memory effect" no longer takes place (Palomares et al., 2004). It is also possible to reform some LDHs by flowing water vapor over the sample. Miyata (1983) was the first to describe the reconstruction of the original LDH structure by hydration of the calcined LDH. Marchi and Apesteguia (1998) have studied the reconstruction from Cu/Co/Zn/Al mixed oxides upon impregnation with aqueous alkaline solutions. This method is useful when the desired interlayer anion is large.

8.2.7 Postsynthesis Treatments

Postpreparative treatments are generally important to improve the crystallinity of the LDH materials developed by conventional techniques. Enhanced crystallinity is desirable in LDHs because it allows the greater stability of intercalated anions (Wang et al., 2010). Ageing the precipitation at or above room temperature is often used as a posttreatment method. In addition, hydrothermal, ultrasonic, and microwave treatment techniques have also been employed as posttreatments. Hydrothermal treatment is used to improve the crystallinity of the LDHs or to increase the anion exchange rate of low affinity anions such as alkyl carboxylates (Prevot et al., 2009). Hydrothermal treatment strongly influences the chemical composition (Mg^{2+}/Al^{3+}) of synthetic hydrotalcite (Cao et al., 2010). The crystallinity of the LDHs can also be improved by ultrasonic treatment (Luque de Castro and Priego-Capote, 2007). Larger LDH crystallites with enhanced adsorption capacities were observed using ultrasonic technique.

In view of environmental application LDHs attracted considerable attention due to their high-chemical versatility, compositional flexibility, and controllable functionalities. It can be easily tuned-up chemically in order to fulfill specific application through selection of different metal atoms and interlayered anions. Moreover, some of the unique properties, such as nontoxicity, biocompatibilty, large surface area, and high reactivity toward anionic species, envisioned its potential application as catalytic and adsorbent material that helps in solving water pollution–based problems. Therefore, with the help of LDHs, significant progress has been achieved toward removal of varieties of pollutants from contaminated water. In later section the overview of potential applications of LDHs that were reportedly investigated for scavenging of various water pollutants are briefly discussed.

8.3 POTENTIAL APPLICATIONS OF LDHs

8.3.1 Removal of Phenolic Compounds

LDHs were found to be extensively investigated for the removal of phenol from wastewater. Ulibarri et al. (1995) investigated sorption study on removal of two phenols namely trichlorophenol (TCP) and trinitrophenol (TNP) on Mg/Al-LDH and its calcined product HT-500. The adsorption efficiency was reportedly high for HT-500. Solution pH significantly influences the sorption performance of both LDHs; an interesting study was reported for adsorption of phenols by LDH and HT-500. In case of LDH the adsorption of TCP increases slightly as the pH increases from 4 to 13.0 while for TNP maximum adsorption of 55% was reported at pH 2.0. The reason for this was explained on the basis of pK_a value of TCP (6.9) and TNP (3.8). TCP exists in ionized form as phenolate ion at pH > pK_a value of TCP but in such basic condition the CO_3^{2-} displacement is not favored as the removal was due to adsorption that occurred only on external surface of LDH. For TNP that exists exclusively in the anionic form acts as Bronsted acid and easily displaces CO_3^{2-} at lower pH; thus observed removal was due to ion exchange. The anion exchange reaction that occurs between CO_3^{2-} of LDH and phenolate ions is represented as:

$$[Mg_3Al(OH)_8]_2\,CO_3 \cdot xH_2O + 2\,PhO^- \rightarrow 2[Mg_3Al(OH)_8]PhO \cdot xH_2O + CO_3^{2-}$$

where PhO^- represents phenolate ions. The extent of anion exchange was greater for TNP than TCP as the maximum adsorption amount of TNP corresponds to 25% of total anion exchange capacity and for TCP it was only 0.2%. The effect of the solution pH on the adsorption of phenols by HT-500 reveals that for both phenols, adsorption increased as the pH of the initial solution decreased. Eighty-four and seventy-seven percent of adsorption was observed for TCP and TNP at pH 4.0 and 2.0, respectively. As for calcined LDH, when it was released in aqueous condition, it rehydrated by consuming protons that resulted in formation of hydroxyl ions which favored adsorption of phenolate ions by anion exchange. The reaction mechanism representing exchange of phenolate ions taken place through reconstruction process is as follows:

$$Mg_3AlO_4(OH) + xH_2O + PhO^- \rightarrow [Mg_3Al(OH)_8](PhO) \cdot xH_2O + OH^-$$

It has been investigated that for removal of nonionic compounds simple LDHs were quite ineffective, therefore in order to enhance their adsorption efficiency these synthetic clay compounds were intercalated with long chain cationic surfactants that results into transformation of LDH's hydrophilic character into hydrophobic character (Backhaus et al., 2001). Zaghouane-Boudiaf et al. (2011) investigated removal of 2,4,5-trichlorophenol (TCP) using Mg/Al-LDH intercalated with sodium dodecyl benzene sulfonate (SDBS). The adsorption process was highly dependent on solution pH, initial TCP concentration and temperature. The adsorption of TCP as a function of pH was investigated at pH range 3–8 and maximum uptake amount of 79 mg/g of TCP (initial concentration 100 mg/L) was observed at pH 4.0. The most interesting part of this study is that the adsorption of TCP occurs in its molecular form rather than ionized form. The observed removal was as a result of hydrophobic–hydrophobic partition interaction and adsorption of TCP molecules mainly occurs in the free intermolecular space of Mg/Al-SDS LDH. It was also observed that initial TCP concentration also influences the adsorption process. Fig. 8.1 clearly represents that for all concentrations, rapid adsorption was observed during initial time period but as the time proceeded adsorption process slowed down and

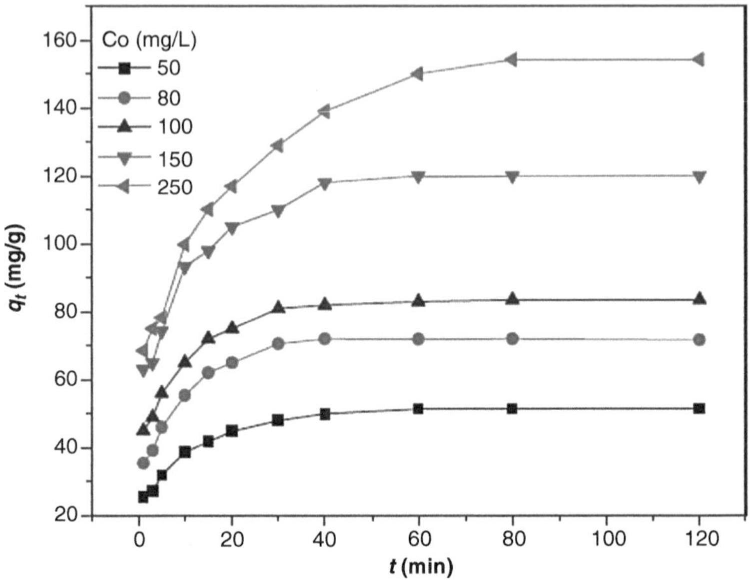

Figure 8.1 ***Effect of initial TCP concentration and contact time on the adsorption of TCP on MgAl–SDBS (V_{sol} = 50 mL; adsorbent mass = 20 mg; agitation speed = 100 rpm; temperature = 298 K; pH 4).*** *(Source: Zaghouane-Boudiaf et al., 2011, © Elsevier).*

attained state of equilibrium. For low TCP concentration saturation phase was observed within 30–40 min; however, for high concentration it was somewhat between 40 and 60 min. Kinetic investigation suggested that the sorption results agreed well with pseudosecond order model.

The endothermic nature of the TCP sorption process confirmed from the positive value of $\Delta H°$ evaluated from the thermodynamic investigation. The adsorption was monolayer which was further substantiated from the isotherm equilibrium investigation which suggests Langmuir as the best fitted model and maximum adsorption capacity reported as 240.5 mg/g at 298 K.

The removal of phenolic compounds from aqueous solution through photocatalytic degradation has also been investigated extensively in recent decades. Principally the process involves generation of holes and electron that allows various redox reactions and the targeted toxic reactants are transformed into nontoxic products through series of reactions. Paredes et al. (2011) reported mineralization of phenol through photocatalytic action using photocatalysts TiO_2/MgAl-LDH. The mechanism of phenol degradation is illustrated in Fig. 8.2. The reaction was initiated when phenol solution was irradiated with fluorescent black light for 8 h that results in

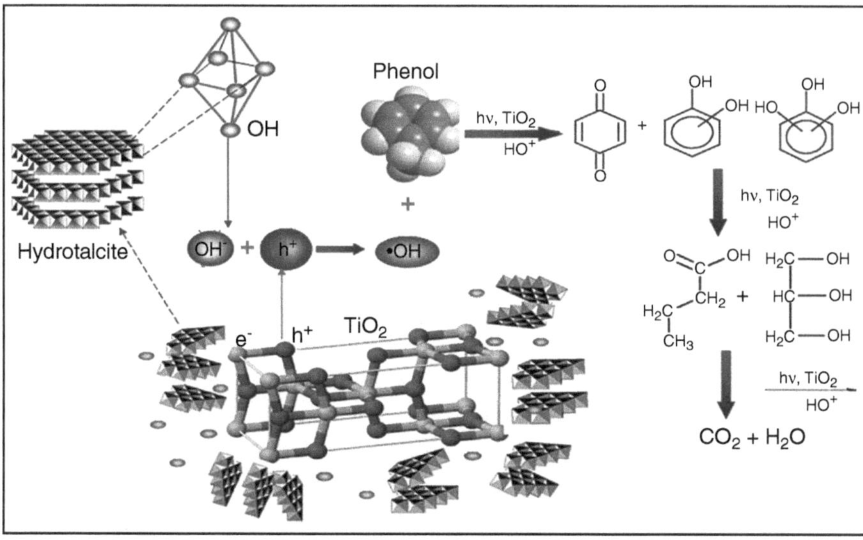

Figure 8.2 ***Mechanistic pathway for elimination of phenol by photocatalytic activity using TiO_2/MgAl-LDH as photocataylsts.*** *(Source: Paredes et al., 2011, © Elsevier).*

generation of holes from TiO_2. These holes then react with the hydroxyl ions of Mg/Al-LDH that result in formation of free hydroxyl radicals which promotes mineralization of phenol through series of reactions as given in Fig. 8.2 and finally converts into CO_2 and H_2O. It was also investigated through experiments that materials rich in OH groups such as LDH enhances titania activity. The phenol degradation of nearly 99% has been observed when calcined LDH was used along with titania in 1:1 ratio. The photocatalytic elimination of phenol was also attempted by Prince et al. (2015) using Zn(Ga)Al-LDH. The impregnation of Ga^{3+} into ZnAl-LDH may increase the electron–hole recombination rate through lowering of band gap to 3.18 eV below than ZnO whose band gap is 3.18 eV, thus facilitating electron mobility with ease. Phenol degradation of nearly 80% of initial 40 ppm and 60% of initial 80 ppm of phenol were observed in 6 h for ZnGaAl-5% photocatalysts.

8.3.2 Removal of Pesticides

Pesticides and herbicides that were widely used during agricultural related activities often get leached into subsurface water reservoirs and subsequently contaminated nearby surface water bodies. Several studies were reported regarding successful utilization of LDHs for capturing of these organic pollutants from waste streams. Some of the important investigation were presented as: You et al. (2002) investigated removal of dicamba (3,6-dichloro-2-methoxy benzoic acid), an ionizable organic pesticide using calcined Mg/Al-LDH. The adsorption kinetics of dicamba suggested that the adsorption rate was divided into fast and very slow reaction phases. More than 80% of adsorption was reported during first 10 min and reached equilibrium condition within 30 min. Adsorption isotherm for dicamba uptake reveals typical S-type curve indicating involvement of hydrophobic interaction between dicamba molecules. Moreover, presence of interfering ions affects the adsorption of dicamba molecules negatively, among which sulfate, hypophosphate and carbonate ions exhibit prominent affect and the same ions were responsible for complete desorption of dicamba from calcined LDH. This suggests that adsorption of dicamba on calcined LDH occurs through ion-exchange process. For better adsorption of pesticides, LDHs were interlayered with organic anions, such as surfactants, in order to infuse the hydrophobic character in LDHs. Bruna et al. (2006) employed organolayered LDH for the removal of two broadly used pesticides namely Carbetamida and Metamitron. The LDH with more dodecyl sulfate (DDS) content shows greater uptake capacity. Kinetic study for Carbetamide adsorption suggested rapid adsorption and equilibrium was reached in 2 h, however for Metamitron, rate of adsorption was slower and equilibrium was attained in 8 h. The slow adsorption rate for Metamitrom attributed to the steric hindrance caused due to its bulky size which results in slow diffusion of the pesticide molecules into the interlayer region of LDH. pH condition also influences sorption process for Carbetamide, with maximum adsorption of 17% observed at pH value 3 while for Metamitron highest adsorption of 14% is observed at pH 8. Removal of herbicides using LDHs was also investigated by several researchers. Adsorption of 2,4-dichlorophenoxyacetic acid (2,4-D) by Cu/Fe-LDH was investigated by Nejati et al. (2013). pH based adsorption study indicated that maximum removal of 78% was observed in acidic condition at pH 4. The adsorption system reached equilibrium condition within 50 min and data were represented accurately by pseudosecond order kinetic model. Equilibrium isotherm investigation indicated Langmuir model as the best-fitted model and maximum sorption capacity was evaluated as 1667 mg/g at 298 K. Investigation of thermodynamic parameters suggested spontaneous and exothermic nature of the sorption process.

8.3.3 Removal of Dyes

The adsorption behavior of LDHs has been massively investigated for the treatment of colored contaminants from wastewater. Ni et al. (2007)

successfully decolorized methyl orange (MO) from aqueous solutions using calcined Zn/Al-LDH. The removal of MO increases with increasing contact time and finally attains equilibrium within 120 min, further the percentage of MO removal decreased from 96.15 to 80.02% as the initial MO concentration increases from 50 to 200 mg/L. This was due to the fact that at higher initial concentration, the mole numbers available for MO adsorption was relatively higher than that of sites available for dye adsorption. Temperature also plays significant role in the sorption process since MO uptake increases with temperature hence time taken to reach equilibrium decreases as temperature increases from 298 to 338 K. Equilibrium isotherm data are consistent with both Langmuir and Freundlich isotherm model. Kinetic data for MO adsorption are fitted well the pseudosecond-order model indicating chemisorption mechanism being the rate-determining step. The mechanism of MO adsorption was also ascertained by XRD analysis of LDH after MO adsorption. The analysis of the XRD peaks indicated about the increase of interlayer distance to 0.856 nm which include thickness of LDH as 0.48 nm and gallery height after MO adsorption as 0.376 nm. Upon analysis the long axis (1.315 nm), short axis (0.497 nm) and molecular thickness (0.356 nm) of MO^- was consistent with that of gallery height of LDH. Fig. 8.3 elucidated the mechanism of accommodation of MO^- in between the interlayer region of LDH. Solution pH and competitive anions also affect the dye sorption process; maximum removal of 90.95% observed at pH 6, besides presence of CO_3^{2-} and PO_4^{3-} reduces MO removal to 37.45 and 23.75%, respectively. Likewise removal of dyes such as Orange G, Acid

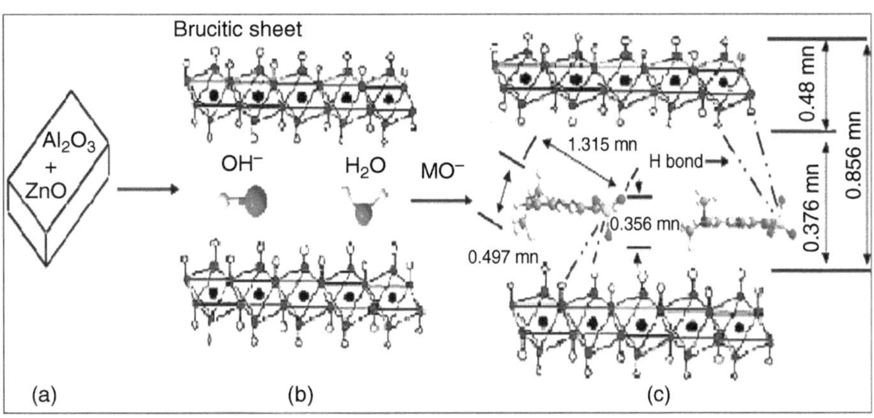

Figure 8.3 ***The pictorial representation of MO adsorption over Zn/Al-LDH.*** (a) The calcined Zn/Al-LDH. (b) The layered structure gets restructured during adsorption. (c) MO – intercalated between the brucite sheets. *(Source: Ni et al., 2007, © Elsevier).*

Orange 10, Remazol Blue 19, Sunset Yellow FCF, Acid Red, Direct Red 16 was successfully attempted by Benselka-Hadj Abdelkader et al. (2011), Extremera et al. (2012), Elkhattabi et al. (2013), Pereira de Sá et al. (2013), Xue et al. (2014), Abdolmohammad-Zadeh et al. (2015), using Mg/Fe-CO_3, Mg/Al-CO_3, Zn/Al-Cl, Ca/Al-NO_3, Zn/Al-NO_3 LDHs, respectively. In recent years the applicability of the magnetic LDH were also explored for dye removal studies. Chen et al. (2011) synthesized Fe_3O_4-Mg/Al LDH colloidal nanohybrids for sequestration of Congo red from aqueous phase. The dye removal rate was very fast and more than 96% of removal achieved in less than 5 min of contact time while complete equilibrium was established within 15 min. The removal of 100% was observed at initial dye concentration of 350 mg/L. The efficacy of LDH nanohybrid was also tested for other dyes such as Orange II, Rose bengal, Methyl orange, and Eosin B at their initial concentration of 100 mg/L at fixed contact time of 15 min. Hundred percent removal of Orange II and Rose bengal was attained at adsorbent dosages of 0.67 and 0.89 g/L, whereas for Methyl orange and Eosin B maximum removal of 95% was achieved at the dosage of 0.89 g/L. For recycling of the Congo red loaded spent adsorbent, sulfate radical-based advanced oxidation process was employed. The recycled sample exhibits 80% of dye removal efficacy even after 6 runs. In another study adsorption of methyl orange (MO) was accomplished using novel magnetic Fe_3O_4/ZnCr LDH (Chen et al., 2012). For synthesis of LDH electroplating, wastewater and pickling waste liquor were used. The maximum adsorption capacity of MO was found to be 240.16 mg/g at initial dye concentration of 50 g/L. The adsorbent can be easily separated by external magnetic field.

It has been found from the literature survey that most of the studies were based on the removal of anionic dyes, however, very less literature is available on the adsorption of cationic dyes which therefore indicated that ion-exchange was possibly favored when LDHs were used for removal studies. Photocatalytic assisted degradation of dyes was also successfully tested by number of researchers. Parida et al. (2007) reportedly used Copperphthalocyanine immobilized Zn/Al LDH as photocatalyst under solar radiation for removal of cationic dye methylene blue (MB). The photocatalytic degradation process of MB was highly influenced by the process parameters such as pH, adsorbate concentration, catalyst dosage, H_2O_2, $K_2S_2O_8$, and NaCl amount. Maximum MB removal of 75 and 33% was achieved at 4 h time and catalyst dose of 1 g/L under solar radiation and dark control, respectively. The presence of copperphthalocyanine on LDH released singlet oxygen which further promotes decomposition of MB. Likewise H_2O_2 and $K_2S_2O_8$

also promoted formation of large amount of singlet oxygen that enhances the photodecolorization of MB, however, NaCl exhibited negative trend. The photodegradation study was also conducted for removal of methyl orange and fast green using Zn/Al and Mg/Al based LDHs (Morimoto et al., 2011). The experimental results suggested that maximum adsorption capacity and photocatalytic activity were obtained for Mg/Al-LDH which has lowest particle size due to the presence of large LDH reactive surface area which facilitates photoinduced degradations of dyes.

8.3.4 Removal of Surfactants

Lots of studies have been devoted regarding removals of surfactants by using LDHs. Pavan et al. (1998) investigated efficiency of Mg/Al-CO_3 LDH for the removal of one of the widely used anionic surfactant, sodium dodecyl sulfate (SDS). The adsorption isotherm was studied and it was found that adsorption follows two step models. The first step reflects electrostatic interaction between the positively charged LDH particles with the negatively charged sulfate groups of SDS and second step represents the attraction between the surfactant's hydrophobic tails, as a result of van der Waals forces. The adsorption amount of SDS increases with elevated temperature which indicated that enthalpy rather than entropy as the driving force. The diffraction pattern of the SDS adsorbed LDH revealed that there was no change in the basal spacing which recommended that SDS anions were not exchanged with carbonate ions. In another study Pavan et al. (2000) investigated the adsorption of SDS, sodium octyl sulfate (SOS), SDBS, and sodium octyl benzene sulfonate (SOBS), on Mg/Al-CO_3 LDH. For all initial concentration of the surfactants the amount adsorbed increased with increasing surfactant concentration. The efficiency of LDH in terms of adsorption of SDS, SOS, SDBS, and SOBS reported as 20, 9.5, 25, and 25%, respectively. Inverse relationship has been investigated between adsorption isotherms and electrokinetic potential. Adsorption equilibrium has been attained after critical micelles concentration (CMC). dos Reis et al. (2004) investigated the effect of temperature, pH, and ionic strength on adsorption of SDBS on Mg/Al-CO_3 LDH. The adsorption of SDBS as a function of temperature was compared at two different temperatures (298 and 313 K). A small decrease in the adsorption amount was reported with the increase of the temperature at equilibrium concentrations below the CMC value. However an increase in the uptake amount of SDBS was observed for higher ionic strength until equilibrium concentrations equivalent to CMC value of 4.0×10^{-4} mol/dm while at concentrations higher than CMC, the adsorption was more efficient in smaller ionic strength. The effect of pH on SDBS adsorption was investigated at two different pH values (pH 7 and 9). It was examined that adsorption was almost independent of pH at equilibrium concentration lower than CMC value while for concentrations above CMC value the amount of SDBS uptake increases at higher pH value. Anbarasan et al. (2005) tested adsorption and intercalation potential of LDH using various anionic surfactants containing sulfonate, carboxyl, phosphonate, and sulfate end group. Through XRD interpretation it was ascertained that the sulfonate group containing surfactants were adsorbed on LDH whereas the sulfate, carboxyl, and phosphonate group containing surfactant were intercalated in between the brucite sheets of LDH.

8.3.5 Removal of Oxyanions

The adsorbate species belonging to oxyanion groups were extensively researched for removal studies using LDHs as an adsorbent material. The removal of trace levels of As (V) and Se (IV) from aqueous phase using calcined and uncalcined LDHs were studied at a length by Yang et al. (2005). It was investigated that the uptake amount of As(V) and Se(IV) on the calcined LDH were higher than the uncalcined LDH. The removal percent for As(V) and Se (VI) reached nearly 100 and 96% at adsorbent dosage of 1.0 and 1.4 g/L for initial adsorbate concentration of 20 ppb, respectively. Se(IV) adsorption increased with increasing temperature when calcined LDHs were used, however, uncalcined LDHs show adsorption independent of temperature whereas for As(V) reverse trend has been investigated. pH of the solution also influences the sorption study; maximum adsorption of Se(VI) occurred at pH 4.0 and As (V) at pH 6.0 for uncalcined LDH whereas for calcined LDH most of adsorption reported below pH 4.0. The presence of competitive anions impedes the sorption performance of calcined LDHs for both adsorbate species. Lv et al. (2007) treated water contaminated with high-fluoride concentration using Mg/Al-CO_3 LDH. Considerable removal of fluoride was observed in pH range 6–10. The fluoride removal in the presence of other anions was found decreased in the order $HCO_3^- > Cl^- > H_2PO_4^- > SO_4^{2-}$. Adsorption isotherm data found to be in accordance with Langmuir–Freundlich model and maximum adsorption capacity was reported as 319.8 mg/g. Kinetic models such as pseudo-first order, pseudosecond order, modified multiplex and double exponential models were used for the interpretation of the interaction mechanism between adsorbate and adsorbent. The experimental data were found more consistent with modified multiplex model which described that interaction between fluoride and LDH involves a rapid first order step which was supposed to be controlled by diffusion processes and a slow second order step responsible for reaction of fluoride with LDHs. Li et al. (2009) reported efficient removal of chromate anions by adsorption on Mg/Al LDH. The adsorption of Cr (VI) was highly dependent on Mg/Al molar ratio, pH, contact time, and initial Cr (VI) concentration, and temperature. Highest chromium removal of more than 98% was observed at highest Mg/Al molar ratio due to the smallest particle size. The removal was most effectual at pH range 2.5 and 5.0. The adsorption of Cr(VI) reached equilibrium within 150 min of contact time. The highest adsorption capacity was evaluated as 112.0 mg/g at 303 K. Ion exchange and surface complexation process interpreted as predominant mechanism for LDH and chromate ion interaction. In order to enhance the adsorption performance of LDHs, Triantafyllidis et al. (2010) modified Mg/Al-LDH with Fe and the same being used for the removal of phosphate ion from aqueous solutions. Removal of phosphate ions mainly occurred through ion exchange process as chlorides or nitrates were easily exchanged by phosphates as compared to carbonates. It was further examined that increase of the Fe loading in LDH structure led to the decrease of the sorption efficiency. The adsorption data fitted in Langmuir and Freundlich isotherm model. The maximum adsorption capacity from the Langmuir model evaluated for Fe modified LDH was 350 mg P/g. The regeneration possibility of the modified LDH was also tested. The phosphate loaded samples were treated with mixed aqueous solution of NaOH and NaCl. The desorption efficiency turned out to be 75%. The regenerated material was further characterized with XRD. The XRD peaks revealed about reduction in crystallinity due to partial deformation of LDH structure as a result of chemical treatment. The removal of molybdate ions from aqueous solution through sulfate-molybdate exchange process within the LDH interlayer was investigated by Ardau et al. (2012). Maximum Mo removal of 54% was observed for initial Mo concentration of 4 mmol/L. It was further analyzed that adsorption of Mo was highly influenced by the ratio of Zn^{2+}/Al^{3+} ion of the brucite sheets, higher the ratio more easily the sulfate gets released and thus facilitates movement of molybdate ions into the interlayers. Exchange of antimonate ions with the intercalated anions such as NO_3^-, Cl^-, SO_4^{2-}, and CO_3^{2-} of Mg/Al-LDH was investigated by Kameda et al. (2012). The removal percent of Sb increases with increasing Al/Sb molar ratio (ie, ratio between amounts of Al in Mg/Al-NO_3 LDH to that of Sb in the solution). More than 90% of removal was observed within 120 min of contact time. The Sb exchange was prominent when intercalated ion was NO_3^- and Cl^-, whereas for SO_4^{2-} exchange was somewhat reasonable but for CO_3^{2-} exchange was very meager. The uptake of trace level of bromate ions from aqueous solution at initial BrO_3^- concentration of 0.78 μmol/dm^3 was attempted by Chitrakar et al. (2011), using calcined and uncalcined LDH. The uncalcined LDH shows negligible BrO_3^- uptake, however, for calcined LDH the removal rate was slow and achieved equilibrium in 48 h. The bromate removal of 78% observed within 4 h beyond this removal occurs at much slower rate and over 48 h of contact time 94% removal was observed. The pH assisted uptake of bromate ion was investigated at initial pH 4.0–10.5 and uptake was almost same at the pH range 4.0–9.0.

8.3.6 Removal of Heavy Metal and Rare Earth Cations

The removal of heavy metal cations was first investigated by Fujii et al. (1992) by employing Mg/Al-LDHs with different combinations of interlayered anions for the sorption of Pb^{2+}, Cu^{2+}, and Zn^{2+}. However, on the basis of literature survey it was examined that LDHs-based removal of heavy metal cation was not investigated so comprehensively. Moreover, it was reported that for metal adsorption isomorphic substitution, chelation, and surface complexation are the main mechanistic process that are responsible for adsorbate–adsorbent interaction. Pérez et al. (2006) reported uptake of Cu^{2+}, Cd^{2+}, and Pb^{2+} on Zn/Al-EDTA LDH. The adsorption reaction is very fast for Cu^{2+} and Pb^{2+} and reaches equilibrium within 2 h of contact time while for Cd^{2+} adsorption occurs slowly and attains saturation phase within 24 h. At pH value 5.5 maximum adsorption was investigated. The mechanism of removal involved chelation between edta and metal cations. Interaction of Mn^{2+}, Fe^{2+}, and Cu^{2+} with Zn/Al-LDH was investigated by Jaiswal and Chattopadhyaya (2011). Adsorption was most effective at pH range 5.2–5.5. The removal of 16.55, 85, and 95.2% was observed for Mn^{2+}, Fe^{2+}, Cu^{2+}. Ionic radii plays an important role in the sorption process; the metal cation with lowest radii exhibits maximum removal. In another investigation Jaiswal and Chattopadhyaya (2013) utilized Co/Bi-LDH for the removal of Pb^{2+} from aqueous solutions. Maximum removal of 89.86% was reported within 120 min of contact time. The LDHs with magnetic character was reportedly employed by various researchers for the sequestration of metal cations. For instance, Shan et al. (2015) fabricated Fe_3O_4/MgAl-LDH magnetic composites for the decontamination of Co^{2+} from synthetic wastewater. The adsorption of Co^{2+} was found to be strongly influenced by contact time, humic acid (HA), fulvic acid (FA), pH, and ionic strength.

The adsorption of Co^{2+} increases with increasing pH and highest removal of more than 95% is observed at pH 10.3. The effect of ionic strength was investigated using $NaNO_3$ as background electrolyte. The adsorption was affected at low pH value due to background electrolyte, however, adsorption remains unaffected at high pH value. Moreover, adsorption percent increased in presence of HA and FA at pH range 2.4–6.8 whereas as the solution pH increases the adsorptive removal of Co^{2+} gets decreased in presence of HA and FA. This was due to strong complexation ability of Co^{2+} when HA and FA got adsorbed on the magnetic LDH composite at low pH while at high pH due to electrostatic repulsion HA and FA did not get adsorbed on the magnetic LDH composite thus the complexation of Co^{2+} was not favored. Similarly, Shou et al. (2015) also reported magnetic Fe_3O_4/MgAl-CO_3 LDH for adsorption of Cd from aqueous solutions. The adsorption of Cd^{2+} was maximum at pH value between 4 to 9, however adsorption process occurred slowly and reached equilibrium at 300 min.

The mechanistic processes (Fig. 8.4) that are responsible for adsorption of Cd^{2+}, interpreted by Shan et al. (2015), are given as: first, formation of $CdCO_3$ precipitate due to the reaction between Cd^{2+} and interlayered carbonate ion present on the surface and edge; second, intrusion of Cd^{2+} onto the brucite structure of LDH by substitution of the Mg^{2+} ions of the layers; third, formation of outer sphere surface complexes between Cd^{2+} with some deprotonated OH groups; and fourth, adsorption of Cd^{2+} on the surface of LDH electrostatically with OH group. The adsorption isotherm data fitted well with Langmuir model with maximum adsorption capacity

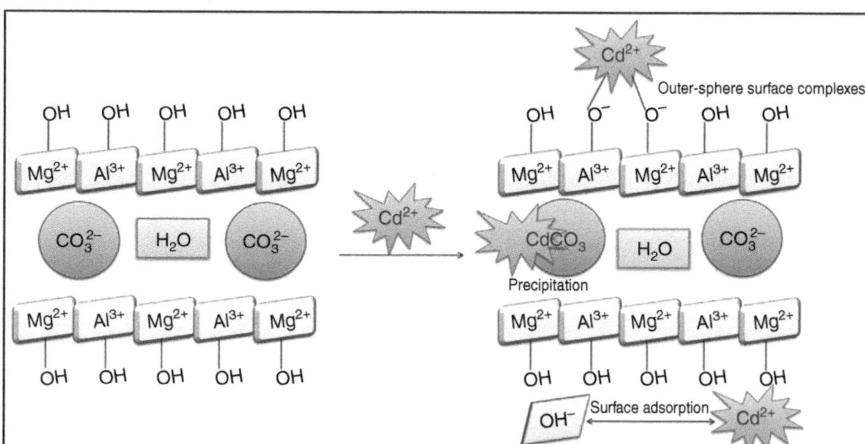

Figure 8.4 ***Schematic representation of the plausible adsorption mechanism of Cd^{2+} with Mg/Al-CO_3 LDH.*** *(Source: Shan et al., 2015, © Elsevier).*

reported as 54.7 mg/g at 323 K. The adsorption was feasible and endothermic in nature which was further confirmed by the evaluation of thermodynamics parameters. The magnetic LDH was quickly and easily separated using a magnet before and after the adsorption process.

In addition to heavy metal cation, removal of rare earth ions also found to be efficiently removed by LDH. Some of the important studies in this area reported as: Li et al. (2004) incorporated rare earth Eu $(EDTA)^{4-}$ through intercalation using Mg/Al-NO_3 LDH. The intercalation of Eu $(EDTA)^{4-}$ confirmed from the X-ray diffraction patterns which indicated that gallery height gets increased after intercalation. Kameda et al. (2011) reported uptake of Sc^{3+} and La^{3+} by EDTA intercalated Cu/Al LDH. The adsorptive uptake of Sc^{3+} and La^{3+} was promoted at pH around 6–6.5. The removal of rare earths (>95% for Sc^{3+} and >40% for La^{3+}) principally occurred through complex formation as $Sc(edta)^-$ and $La(edta)^-$. Moreover Cu/Al LDH form strong chelate complex with Sc^{3+} than La^{3+} due to formation of stable complex with $Sc(edta)^-$. In other study Kameda et al. (2013) investigated uptake efficiency of Cu/Al-EDTA LDH for rare earth ions, including Sc^{3+}, Y^{3+}, and La^{3+}. The uptake of cations is mainly attributed to chelating functions of the edta ion in the interlayer of Cu/Al-LDH. The removal was observed in the order $Sc^{3+} > Y^{3+} > La^{3+}$ due to their relative stabilities as $Sc(edta)^- > Y(edta)^- > La(edta)^-$ during formation of chelate complexes.

8.3.7 Removal of Nuclear Wastes

LDH materials facilitating large-scale removal and immobilization of radionuclides thus provide efficient way for the disposal and management of nuclear wastes. Lots of works related to removal of radionuclides using LDHs have been investigated by several workers. Removal of radionuclide anion, pertechnetate (TcO^{-4}) using Ni/Al-LDH has been systematically investigated by Wang and Gao (2006). Solution pH affects adsorption of pertechnetate anions as maximum uptake of 307 mg/g was observed at pH 8.0. An interesting study related to the sorption behavior of radionuclide anions has been reported. It was investigated that sorption of TcO^{-4} over LDH was a typical process and was strongly governed by the selection of both divalent and trivalent cations. The sorption performance of LDH materials follows the orders Al > Cr > Ga > Bi for trivalent cations and Ni > Co > Mg > Zn > Fe > Cu for divalent cations. Basal spacing of LDH materials was found to regulate the distribution coefficient of the radionuclides. A small increase in basal spacing enhances the distribution

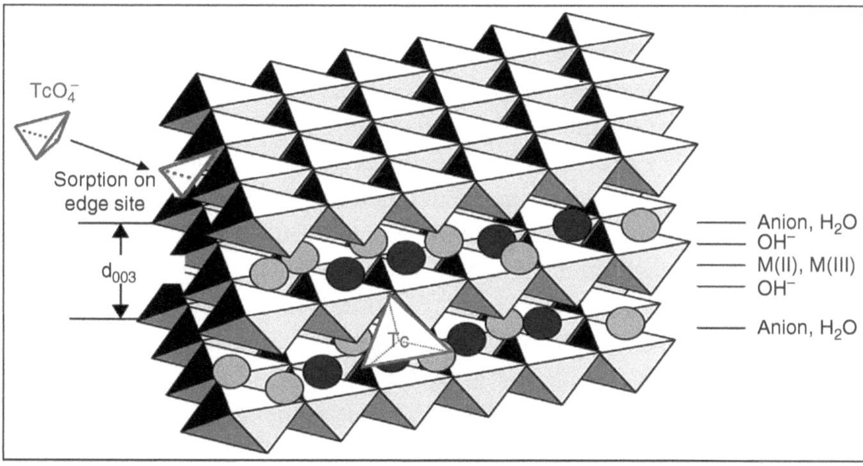

Figure 8.5 ***Schematic representation of sorption of TcO^{-4} along the edge sites of the LDH layer.*** The radionuclide anions easily displaced interlayered anions and get fitted into the caged space. *(Source: Wang and Gao, 2006, © Elsevier).*

coefficient by more than two orders of magnitude. Thus it was presumed that sorption of TcO^{-4} takes place at the edge sites or cage spaces of the LDH layers as represented in Fig. 8.5.

Sorption of ^{137}Cs, $^{85,90}Sr$, and ^{90}Y onto Mg/Al and Mg/Nd LDHs was investigated by Kulyukhin et al. (2008). The sorption of Cs and Sr radionuclides were reportedly insignificant when Mg/Al LDHs was used. However when EDTA was incorporated into the LDH the distribution coefficient of Sr increases by a factor of more than 40 indicating complexation of Sr with EDTA. Mg/Nd LDH also exhibits efficient sorption capacity for Sr and Y radionuclides. Efficiently capturing of $^{36}Cl^-$ with Ca/Al/Fe-LDH was investigated by Phillips and Vandeperre (2011). The removal of $^{36}Cl^-$ using LDH was based on the memory effect or rehydration mechanism. For which the LDHs was calcined to layered double oxides (LDO) and these LDO were employed for the capturing of the radioactive contaminants from the aqueous solution. The potential of LDO for removal of Cl^- ions in presence of other anions such as NO_3^- and CO_3^{2-} were also investigated. According to the results the relative affinity for three anions was $Cl^- \approx CO_3^{2-} > NO_3^-$. The result further suggested that the utmost affinity reported for Cl^- and CO_3^{2-} could be easily used for holdup and storage of radioactive $^{36}Cl^-$ and ^{14} during leaching treatment of graphite.

One of the most dangerous radioactive material U(VI) was efficiently removed from aqueous phase using magnetic Mg/Al- LDH intercalated with citrate by Zhang et al. (2012). Maximum adsorption amount of 180 mg/g of U(IV) was reported at pH 6.0, as at this pH value the functional groups of magnetic Mg/Al-citrate LDH becomes deprotonated and forms negative charge. Thus due to electrostatic forces the negatively charged sites and cationic U(VI) becomes attracted to each other that results in formation of metal-ligand magnetic composite complexes. Rate of adsorption was very fast at initial time period, later adsorption process proceeded slowly and equilibrium condition was reached after 4 h. Adsorption amount increased from 180 to 250 mg/g as temperature was increased from 298 to 328 K thus endothermic nature of the adsorption process was recommended. Adsorption data found to be consistent with Freundlich isotherm model and kinetic study advocated that pseudosecond model represents the adsorption of U(VI) in more appropriate manner.

8.3.8 Removal of Toxic Vapor From Water

Ma et al. (2014b) investigated an efficient method regarding capturing of toxic vapor using LDHs. For this very purpose the Mg/Al-LDH was intercalated with polysulfides (S_x^{2-} where $x = 2,4,5$) which exhibit enormous potential to capture large quantities of mercury (Hg^0) vapor. The capturing of Hg vapor mainly occurs through ion exchange as the interlayered S_x^{2-} was reacted readily with the Hg^0 through their S—S bond that leads to formation of HgS. The affinity of various anions for intercalation with LDH was tested and reported in the decreasing order of $CO_3^{2-} > SO_4^{2-} > OH^- > F^-Cl^- > Br^- > NO_3^- > I^-$. The successful intercalation of S_x^{2-} and Hg captured by weight through S—S bonding was ascertained through various analytical techniques such as FT-IR, Raman, and XPS spectra. The S_2-LDH, S_4-LDH, and S_5-LDH exhibits high adsorption capacity of 4.9×10^5, 7.4×10^5, and 1.0×10^6 μg/g, respectively, which corresponds to 50 − 100% adsorption rates. Overall adsorption of Hg^0 on S_x-LDH occurred via chemisorption mechanism. Similarly in another study Ma et al. (2014a) reported removal of I_2 vapor using same adsorbent, that is, Mg/Al layered double hydroxide (MgAl-LDH) nanocomposites intercalated with polysulfide (S_x^{2-}) groups (S_x-LDH, $x = 2, 4, 6$). The I_2 molecules were captured through exchange process with S_x^{2-} and were adsorbed in reduced form as I_3^- meanwhile S_x^{2-} group also oxidized to S_8. FT-IR, Raman, and XPS spectra confirm adsorption of I_2. The I_2 adsorption capacities of S_2-LDH, S_4-LDH, and S_6-LDH were evaluated as 1.32, 1.52, and 1.43, respectively.

8.3.9 Miscellaneous Application of LDHs as Adsorbent

8.3.9.1 *Air Pollution Control*

The combustion of fossil fuels such as coal or natural gas releases large volumes of CO_2, which is one of the major global warming gases, resulting in pollution problem. Therefore, the removal and recovery of these gases from gas mixture is very important for environment point of view. LDHs have property for the adsorption of gas molecules like SO_X (Cantu et al., 2005), NO_X (Constantino and Pinnavaia, 1994), and for CO_2 (Di Cosimo et al., 1998). The LDHs which comprises NO_x storage and reduction properties comprises three main components: a support material containing larger surface area (eg, Al_2O_3), an alkali compound, or alkaline earth metals (eg, Ca, Sr, Ba, K, or Na) which acts as a NO_x storage component, and a noble metal (eg, Pt, Rh, or Pd) as a catalytic redox component. LDH has been reported as efficient adsorbent for NO_x. In a mixed oxide derived from an Mg-Al LDH compound, Al_2O_3 acts as the support; MgO acts as a NO_x storage component. Cu-containing calcined Mg-Al LDHs have been found to be most effective and selective catalysts for NO_x removal (Di Cosimo et al., 1998; Climent et al., 1995). The studies showed that these LDHs derived catalysts significantly performed for NO_x storage/reduction, particularly at lower temperatures (Occelli et al., 2003). LDHs have also been proved effective for the reduction of SO_x, however the removal sequestration mechanism is slightly different from the case of NO_x. The sorbent LDH must have the property to oxidize SO_2 to SO_3 to produce a metal sulfate. Finally, the metal sulfide can be hydrolyzed by steam in the stripper to form the original metal oxide.Various materials like MgO, Al_2O_3, and Mg-Al spinel have showed their potential to scavenge SO_x but performance is limited.

8.3.9.2 *CO_2 Sequestration*

The removal and recovery of CO_2 from industrial gas stream is becoming increasingly significant in the field of ecofriendly energy production as well as to control the hazards of global warming. Numerous LDHs adsorbents, namely zeolite based (zeolite 5A, zeolite 13X), carbon molecular sieve, alumina, calcium oxide, etc., show low adsorption capacity at lower temperature (Yong and Rodrigues, 2002). However, the performance can be improved by optimizing their structural, textural, and thermal behavior, which is further determined by the synthetic parameters (Yong et al., 2001). Addition or coating of alkali salts such as K_2CO_3 increases both the adsorption capacity and the stability of the LDHs. Modification by rare earth elements has also enhanced the adsorption capacity (Ram Reddy et al., 2006). The presence of water vapor also facilitates the adsorption of CO_2 onto LDHs at high temperature while water vapor is known to compete with CO_2 in adsorption behavior onto zeolites (Hutson et al., 2004). Hydrotalcite has been also used as coating material on the commercial zeolites to improve its adsorption.

8.4 CONCLUSIONS

LDHs bear enormous potential for the removal of varieties of contaminants from water or wastewater. These materials can be synthesized easily in laboratory condition and its structure can be easily tailored for desired properties and applications in various fields. This chapter offers brief piece of information related to the recent trends and opportunities of LDHs materials for application in treatment of wastewater. In case of anionic contaminants adsorption process was mainly controlled by ion-exchange mechanism and exchange mostly depends on the charge of anions, greater the charge of the anions more easily it will replace the interlayered anions of LDH. For cationic pollutants, removal process was principally driven by precipitation, surface complexation, isomorphic substitution, and chelation. Adsorption of contaminants on LDHs was greatly influenced by various process parameters such as pH, adsorbent dosage, contact time, ionic strength, and temperature. However, some of the limitations restrict the applicability of LDHs in wide scale such as structural unsteadiness when the material is exposed to the environment of low pH, it may render treated water alkaline, particularly when calcined LDHs were employed and lastly regeneration of LDHs incurred high cost. If these challenges would meet up then LDHs materials would certainly add a new dimension to water treatment.

REFERENCES

Abdolmohammad-Zadeh, H., Nejati, K., Ghorbani, E., 2015. Synthesis, characterization and application of Zn-Al layered double hydroxide as a nano-sorbent for the removal of Direct Red 16 from industrial wastewater effluents. Chem. Eng. Commun. 202, 1349–1359.

Adachi-Pagano, M., Forano, C., Besse, J., 2003. Synthesis of Al-rich hydrotalcite-like compounds by using the urea hydrolysis reaction-control of size and morphology. J. Mater. Chem. 13, 1988–1993.

Anbarasan, R., Lee, W.D., Im, S.S., 2005. Adsorption and intercalation of anionic surfactants onto layered double hydroxides-XRD study. Bull. Mater. Sci. 28, 145–149.

Ardau, C., Frau, F., Dore, E., Lattanzi, P., 2012. Molybdate sorption by Zn–Al sulphate layered double hydroxides. Appl. Clay Sci. 65–66, 128–133.

Backhaus, W.K., Klumpp, E., Narres, H.D., Schwuger, M.J., 2001. Adsorption of 2,4-dichlorophenol on montmorillonite and silica: influence of non-ionic surfactants. J. Colloid Interf. Sci. 242, 6–13.

Benito, P., Labajos, F.M., Mafra, L., Rocha, J., Rives, V., 2009. Carboxylate-intercalated layered double hydroxides aged under microwave-hydrothermal treatment. J. Solid State Chem. 182, 18–26.

Benselka-Hadj Abdelkader, N., Bentouami, A., Derriche, Z., Bettahar, N., De Ménorval, L.C., 2011. Synthesis and characterization of Mg–Fe layer double hydroxides and its application on adsorption of Orange G from aqueous solution. Chem. Eng. J. 169, 231–238.

Braterman, P.S., Xu, Z.P., Yarberry, F., 2004. Handbook of Layered Materials. Marcel Dekker, New York, pp. 373–474.

Bruna, F., Pavlovic, I., Barriga, C., Cornejo, J., Ulibarri, M.A., 2006. Adsorption of pesticides carbetamide and metamitron on organohydrotalcite. Appl. Clay Sci. 33, 116–124.

Cantu, M., Lopez-Salinas, E., Valente, J.S., Montiel, R., 2005. SO_x removal by calcined MgAlFe hydrotalcite-like materials: effect of the chemical composition and the cerium incorporation method. Environ. Sci. Technol. 39, 9715–9720.

Cao, C.Y., Cui, Z.M., Chen, C.Q., Song, W.G., Cai, W., 2010. Ceria hollow nanospheres produced by a template-free microwave-assisted hydrothermal method for heavy metal ion removal and catalysis. J. Phys. Chem. C 114, 9865–9870.

Cavani, F., Trifiro, F., Vaccari, A., 1991. Hydrotalcite-type anionic clays: preparation, properties and applications. Catal. Today 11, 173–301.

Chen, C., Gunawan, P., Xu, R., 2011. Self-assembled Fe_3O_4-layered double hydroxide colloidal nanohybrids with excellent performance for treatment of organic dyes in water. J. Mater. Chem. 21, 1218–1225.

Chen, D., Li, Y., Zhang, J., Li, W., Zhou, J., Shao, L., Qian, G., 2012a. Efficient removal of dyes by a novel magnetic Fe_3O_4/ZnCr-layered double hydroxide adsorbent from heavy metal wastewater. J. Hazard. Mater. 243, 152–160.

Chen, D., Li, Y., Zhang, J., Zhou, J., Guo, Y., Liu, H., 2012b. Magnetic Fe_3O_4/ZnCr-layered double hydroxide composite with enhanced adsorption and photocatalytic activity. Chem. Eng. J. 185, 120–126.

Chitrakar, R., Sonoda, A., Makita, Y., Hirotsu, T., 2011. Calcined Mg-Al layered double hydroxides for uptake of trace levels of bromate from aqueous solution. Ind. Eng. Chem. Res. 50, 9280–9285.

Climent, M.J., Corma, A., Iborra, S., Primo, J., 1995. Base catalysis for fine chemicals production: claisen-schmidt condensation on zeolites and hydrotalcites for the production of chalcones and flavanones of pharmaceutical interest. J. Catal. 151, 60–66.

Constantino, V.R.L., Pinnavaia, T.J., 1994. Structure reactivity relationships for basic catalysts derived from a $Mg^{2+}/A1^{3+}/CO^{3-}$ layered double hydroxide. Catal. Lett. 23, 361–367.

Delorme, F., Seron, A., Bizi, M., Jean-Prost, V., Martineau, D., 2006. Effect of time on the reconstruction of the $Mg_4Al_2(OH)_{12}CO_3{\cdot}3H_2O$ layered double hydroxide in $NaCO_3$ solution. J. Mater. Sci. 41, 4875–4882.

Di Cosimo, J.I., Diez, V.K., Xu, M., Iglesia, E., Apesteguia, C.R., 1998. Structure and surface and catalytic properties of Mg-Al basic oxides. J. Catal. 178, 499–510.

Dodds, J., Espitalier, F., Louisnard, O., Grossier, R., David, R., Hassoun, M., Baillon, F., Gatumel, C., Lyczko, N., 2007. The effect of ultrasound on crystallisation-precipitation processes: some examples and a new segregation model. Part. Part. Syst. Char. 24, 18–28.

dos Reis, M.J., Silverio, F., Tronto, J., Valim, J.B., 2004. Effects of pH, temperature, and ionic strength on adsorption of sodium dodecylbenzenesulfonate into Mg–Al–CO_3 layered double hydroxides. J. Phys. Chem. Solids 65, 487–492.

El Gaini, L., Lakraimi, M., Sebbar, E., Meghea, A., Bakasse, M., 2009. Removal of indigo carmine dye from water to Mg-Al-CO(3)-calcined layered double hydroxides. J. Hazard. Mater. 161, 627–632.

Elkhattabi, E.H., Lakraimi, M., Badreddine, M., Legrouri, A., Cherkaoui, O., Berraho, M., 2013. Removal of Remazol Blue 19 from wastewater by zinc–aluminium–chloride-layered double hydroxides. Appl. Water Sci. 3, 431–438.

Extremera, R., Pavlovic, I., Pérez, M.R., Barriga, C., 2012. Removal of acid orange 10 by calcined Mg/Al layered double hydroxides from water and recovery of the adsorbed dye. Chem. Eng. J. 213, 392–400.

Fang, J., Li, M., Li, Q., Zhang, W., Shou, Q., Liu, F., Zhang, X., Cheng, J., 2012. Microwave assisted synthesis of CoAl-layered double hydroxide/graphene oxide composite and its application in supercapacitors. Electrochim. Acta 85, 248–255.

Farooqi, A., Masuda, H., Kusakabe, M., Naseem, M., Firdous, N., 2007. Distribution of highly arsenic and fluoride contaminated groundwater from East Punjab, Pakistan, and the controlling role of anthropogenic pollutants in the natural hydrological cycle. Geochem. J. 41, 213–234.

Fujii, S., Sugie, Y., Kobune, M., Touno, A., 1992. Uptakes of Cu^{2+}, Pb^{2+} and Zn^{2+} on synthetic hydrotalcite in aqueous solution. J. Jpn. Chem. Soc. 1992, 1504–1507.

Goh, K.H., Lim, T.T., Dong, Z., 2008. Application of layered double hydroxides for removal of oxyanions: a review. Water Res. 42, 1343–1368.

Gordon, L., Salutsky, M.L., Willard, H.H., 1959. Precipitation from Homogeneous Solution. John Wiley & Sons Inc., Chapman & Hall Limited, New York, London.

Goswamee, R.L., Sengupta, P., Bhattacharyya, K.G., Dutta, D.K., 1998. Adsorption of Cr(VI) in layered double hydroxides. Appl. Clay Sci. 13, 21–34.

Guo, Y., Zhu, Z., Qiu, Y., Zhao, J., 2013. Enhanced adsorption of acid brown 14 dye on calcined Mg/Fe layered double hydroxide with memory effect. Chem. Eng. J. 219, 69–77.

Hamad, H.A., Abd El-latif, M.M., Kashyout, A.B., Sadik, W.A., Feteha, M.Y., 2014. Study on synthesis of superparamagnetic spinel cobalt ferrite nanoparticles as layered double hydroxides by co-precipitation method. Russ. J. Gen. Chem. 84, 2205–2210.

Hibino, T., Ohya, H., 2009. Synthesis of crystalline layered double hydroxides: precipitation by using urea hydrolysis and subsequent hydrothermal reactions in aqueous solutions. Appl. Clay Sci. 45, 123–132.

Hutson, N.D., Speakman, S.A., Payzant, E.A., 2004. Structural effects on the high-temperature adsorption of CO_2 on a synthetic hydrotalcite. Chem. Mater. 16, 4135–4143.

Inacio, J., Taviot-Gueho, C., Forano, C., Besse, J.P., 2001. Adsorption of MCPA pesticide by Mg-Al-layered double hydroxides. Appl. Clay Sci. 18, 255–264.

Jaiswal, A., Chattopadhyaya, M.C., 2011. Interaction of Mn^{2+}, Fe^{2+} and Cu^{2+} heavy metal ions from aqueous solution by zaccagnaite, a hydrotalcite-like compound. Desalin. Water Treat. 29, 252–257.

Jaiswal, A., Chattopadhyaya, M.C., 2013. Synthesis and characterization of novel Co/Bi-layered double hydroxides and their adsorption performance for lead in aqueous solution. Arab. J. Chem. (in press). doi:10.1016/j.arabjc.2013.09.010.

Kameda, T., Hoshi, K., Yoshioka, T., 2011. Uptake of Sc^{3+} and La^{3+} from aqueous solution using ethylenediaminetetraacetate-intercalated Cu-Al layered double hydroxide reconstructed from Cu-Al oxide. Solid State Sci. 13, 366–371.

Kameda, T., Hoshi, K., Yoshioka, T., 2013. Preparation of Cu-Al layered double hydroxide intercalated with ethylenediaminetetraacetate by coprecipitation and its uptake of rare earth ions from aqueous solution. Solid State Sci. 17, 28–34.

Kameda, T., Nakamura, M., Yoshioka, T., 2012. Removal of antimonate ions from an aqueous solution by anion exchange with magnesium–aluminum layered double hydroxide and the formation of a brandholzite-like structure. J. Environ. Sci. Health Part A 47, 1146–1151.

Khan, A.I., O'Hare, D., 2002. Intercalation chemistry of layered double hydroxides: recent developments and applications. J. Mater. Chem. 12, 3191–3198.

Kloprogge, J.T., Frost, R.L., 2001. Infrared and Raman spectroscopic studies of layered double hydroxides (LDHs). In: Rives, V. (Ed.), Layered Double Hydroxides: Present and Future. Nova Sci. Pub. Inc, New York, pp. 139–192, (Chapter 5).

Kulyukhin, S.A., Krasavina, E.P., Gredina, I.V., Rumer, I.A., Mizina, L.V., 2008. Sorption of cesium, strontium, and yttrium radionuclides from the aqueous phase on layered double hydroxides. Radiochemistry 50, 493–501.

Kulyukhin, S.A., Krasavina, E.P., Gredina, I.V., Rumer, I.A., Mizina, L.V., 2010. Sorption of U(VI) from aqueous solutions on layered double hydroxides of Mg, Al, and Nd. Radiochemistry 52, 653–661.

Lee, J.H., Kim, C.K., Katoh, S., Murakami, R., 2001. Microwave-hydrothermal versus conventional hydrothermal preparation of Ni- and Zn-ferrite powders. J. Alloys Compd. 325, 276–280.

Li, L., Feng, Y., Li, Y., Zhao, W., Shi, J., 2009a. Fe_3O_4 core/layered double hydroxide shell nanocomposite: versatile magnetic matrix for anionic functional materials. Angew. Chem. Int. Ed. 48, 5888–5892.

Li, Y., Gao, B., Wu, T., Sun, D., Li, X., Wang, B., Lu, F., 2009b. Hexavalent chromium removal from aqueous solution by adsorption on aluminum magnesium mixed hydroxide. Water Res. 43, 3067–3075.

Li, C., Wang, G., Evans, D.G., Duan, X., 2004. Incorporation of rare-earth ions in Mg–Al layered double hydroxides: intercalation with an [Eu(EDTA)] chelate. J. Solid State Chem. 177, 4569–4575.

Liu, Z., Ma, R., Osada, M., Iyi, N., Ebina, Y., Takada, K., Sasaki, T., 2006. Synthesis, anion exchange, and delamination of Co–Al layered double hydroxide: assembly of the exfoliated nanosheet/polyanion composite films and magneto-optical studies. J. Am. Chem. Soc. 128, 4872–4880.

Luque de Castro, M.D., Priego-Capote, F., 2007. Ultrasound-assisted crystallization (sonocrystallization). Ultrason. Sonochem. 14, 717–724.

Lv, L., He, J., Wei, M., Evans, D.G., Zhou, Z., 2007. Treatment of high fluoride concentration water by MgAl-CO_3 layered double hydroxides: kinetic and equilibrium studies. Water Res. 41, 1534–1542.

Ma, S., Islam, S.M., Shim, Y., Gu, Q., Wang, P., Li, H., Sun, G., Yang, X., Kanatzidis, M.G., 2014a. Highly efficient iodine capture by layered double hydroxides intercalated with polysulfides. Chem. Mater. 26, 7114–7123.

Ma, S., Shim, Y., Islam, S.M., Subrahmanyam, K.S., Wang, P., Li, H., Wang, S., Yang, X., Kanatzidis, M.G., 2014b. Efficient Hg vapor capture with polysulfide intercalated layered double hydroxides. Chem. Mater. 26, 5004–5011.

Mandel, K., Drenkova-Tuhtan, A., Hutter, F., Gellermann, C., Steinmetz, H., Sextl, G., 2013. Layered double hydroxide ion exchangers on superparamagnetic microparticles for recovery of phosphate from waste water. J. Mater. Chem. A 1, 1840–1848.

Marchi, A.J., Apesteguia, C.R., 1998. Impregnation-induced memory effect of thermally activated layered double hydroxides. Appl. Clay Sci. 13, 35–48.

Meyn, M., Beneke, K., 1990. Anion-exchange reactions of layered double hydroxides. Inorg. Chem. 29, 5201–5207.

Millange, F., Walton, R.I., O'Hare, D., 2000. Time-resolved in situ X-ray diffraction study of the liquid-phase reconstruction of Mg–Al–carbonate hydrotalcite-like compounds. J. Mater. Chem. 10, 1713–1720.

Miyata, S., 1983. Anion-exchange properties of hydrotalcite-like compounds. Clays Clay Miner. 31, 305–311.

Morimoto, K., Tamura, K., Iyi, N., Ye, J., Yamada, H., 2011. Adsorption and photodegradation properties of anionic dyes by layered double hydroxides. J. Phys. Chem. Solids 72, 1037–1045.

Nejati, K., Davary, S., Saati, M., 2013. Study of 2,4-dichlorophenoxyacetic acid (2,4-D) removal by Cu-Fe-layered double hydroxide from aqueous solution. Appl. Surf. Sci. 280, 67–73.

Ni, Z.M., Xia, S.J., Wang, L.G., Xing, F.F., Pan, G.X., 2007. Treatment of methyl orange by calcined layered double hydroxides in aqueous solution: adsorption property and kinetic studies. J. Colloid Interf. Sci. 316, 284–291.

Nilchi, A., Garmarodi, S.R., Darzi, S.J., 2010. Adsorption behavior of nano sized sol–gel derived TiO_2–SiO_2 binary oxide in removing Pb^{2+} metal ions. Separ. Sci. Technol. 45, 801–808.
Occelli, M.L., Olivier, J.P., Auroux, A., Kalwei, M., Eckert, H., 2003. Basicity and porosity of a calcined hydrotalcite-type material from nitrogen porosimetry and adsorption microcalorimetry methods. Chem. Mater. 15, 4231–4238.
Palomares, A.E., Prato, J.G., Rey, F., Corma, A., 2004. Using the "memory effect" of hydrotalcites for improving the catalytic reduction of nitrates in water. J. Catal. 221, 62–66.
Paredes, S.P., Valenzuela, M.A., Fetter, G., Flores, S.O., 2011. TiO_2/MgAl layered double hydroxides mechanical mixtures as efficient photocatalysts in phenol degradation. J. Phys. Chem. Solids 72, 914–919.
Parida, K.M., Baliarsingh, N., Patra, B.S., Das, J., 2007. Copperphthalocyanine immobilized Zn/Al LDH as photocatalyst under solar radiation for decolorization of methylene blue. J. Mol. Catal. A 267, 202–208.
Pavan, P.C., Crepaldi, E.L., Valim, J.B., 2000. Sorption of anionic surfactants on layered double hydroxides. J. Colloid Interf. Sci. 229, 346–352.
Pavan, P.C., de, G., Gomes, A., Valim, J.B., 1998. Adsorption of sodium dodecyl sulfate on layered double hydroxides. Micropor. Mesopor. Mater. 21, 659–665.
Pavlovic, I., Barriga, C., Hermosin, M.C., Cornejo, J., Ulibarri, M.A., 2005. Adsorption of acidic pesticides 2,4-D, clopyralid and picloram on calcined hydrotalcite. Appl. Clay Sci. 30, 125–133.
Pereira de Sá, F., Cunha, B.N., Nunes, L.M., 2013. Effect of pH on the adsorption of Sunset Yellow FCF food dye into a layered double hydroxide (CaAl-LDH-NO_3). Chem. Eng. J. 215–216, 122–127.
Pérez, M.R., Pavlovic, I., Barriga, C., Cornejo, J., Hermosín, M.C., Ulibarri, M.A., 2006. Uptake of Cu^{2+}, Cd^{2+} and Pb^{2+} on Zn–Al layered double hydroxide intercalated with EDTA. Appl. Clay Sci. 32, 245–251.
Phillips, J.D., Vandeperre, L.J., 2011. Anion capture with calcium, aluminium and iron containing layered double hydroxides. J. Nucl. Mater. 416, 225–229.
Prevot, V., Caperaa, N., Taviot-Guého, C., Forano, C., 2009. Glycine-assisted hydrothermal synthesis of NiAl-layered double hydroxide nanostructures. Cryst. Growth Des. 9, 3646–3654.
Prince, J., Tzompantzi, F., Mendoza-Damian, G., Hernandez-Beltran, F., Valente, J.S., 2015. Photocatalytic degradation of phenol by semiconducting mixed oxides derived from Zn(Ga)Al layered double hydroxides. Appl. Catal. B 163, 352–360.
Pshinko, G.N., Kosorukov, A.A., Puzyrnaya, L.N., Goncharuk, V.V., 2011. Layered double hydroxides intercalated with EDTA as effective sorbents for U(VI) recovery from wastewater. Radiochemistry 53, 303–307.
Ragavan, A., Khan, A.I., O'Hare, D., 2006. Isomer selective ion-exchange intercalation of nitrophenolates into the layered double hydroxide $[LiAl_2(OH)_6]Cl{\cdot}xH_2O$. J. Mater. Chem. 16, 602–608.
Ram Reddy, M.K., Xu, Z.P., (Max) Lu, G.Q., Diniz da Costa, J.C., 2006. Layered double hydroxides for CO_2 capture: structure evolution and regeneration. Ind. Eng. Chem. Res. 45, 7504–7509.
Reichle, W.T., 1986. Synthesis of anionic clay minerals (mixed metal hydroxides hydrotalcite). Solid State Ionics 22, 135–141.
Rives, V., Ulibarri, M.A., 1999. Layered double hydroxides (LDH) intercalated with metal coordination compounds and oxometalates. Coord. Chem. Rev. 181, 61–120.
Rocher, V., Siaugue, J.M., Cabuil, V., Bee, A., 2008. Removal of organic dyes by magnetic alginate beads. Water Res. 42, 1290–1298.
Rojas, R., Bruna, F., De Pauli, C., Ulibarri, M., Giacomelli, C., 2011. The effect of interlayer anion on the reactivity of Mg–Al layered double hydroxides: improving and extending the customization capacity of anionic clays. J. Colloid Interf. Sci. 359, 136–141.

Shan, R., Yan, L., Yang, K., Hao, Y., Du, B., 2015. Adsorption of Cd(II) by Mg–Al–CO_3- and magnetic Fe_3O_4/Mg–Al–CO_3-layered double hydroxides: kinetic, isothermal, thermodynamic and mechanistic studies. J. Hazard. Mater. 299, 42–49.

Shou, J., Jiang, C., Wang, F., Qiu, M., Xu, Q., 2015. Fabrication of Fe_3O_4/MgAl-layered double hydroxide magnetic composites for the effective decontamination of Co(II) from synthetic wastewater. J. Mol. Liquids 207, 216–223.

Subramanian, T., Dhakshinamoorthy, A., Pitchumani, K., 2013. Amino acid intercalated layered double hydroxide catalyzed chemoselective methylation of phenols and thiophenols with dimethyl carbonate. Tertahedron Lett. 54, 7167–7170.

Triantafyllidis, K.S., Peleka, E.N., Komvokis, V.G., Mavros, P.P., 2010. Iron-modified hydrotalcite-like materials as highly efficient phosphate sorbents. J. Colloid Interf. Sci. 342, 427–436.

Tuna, A.O.A., Ozdemir, E., Simsek, E.B., Beker, U., 2013. Removal of As(V) from aqueous solution by activated carbon-based hybrid adsorbents: impact of experimental conditions. Chem. Eng. J. 223, 116–128.

Ulibarri, M.A., Pavlovic, I., Hermosin, M.C., Cornejo, J., 1995. Hydrotalcite-like compounds as potential sorbents of phenols from water. Appl. Clay Sci. 10, 131–145.

Vaccari, A., 1999. Clays and catalysis: a promising future. Appl. Clay Sci. 14, 161–198.

Wang, Y., Gao, H., 2006. Compositional and structural control on anion sorption capability of layered double hydroxides (LDHs). J Colloid Interf. Sci. 301, 19–26.

Wang, J.A., Novaro, O., Bokhimi, X., López, T., Gómez, R., Navarrete, J., Llanos, M.E., López-Salinas, E., 1998. Characterizations of the thermal decomposition of brucite prepared by sol-gel technique for synthesis of nanocrystalline MgO. Mater. Lett. 35, 317–323.

Wang, R.X., Wen, T., Wu, X.L., Xu, A.W., 2014. Highly efficient removal of humic acid from aqueous solutions by Mg/Al layered double hydroxides-Fe_3O_4 nanocomposites. RSC Adv. 4, 21802–21809.

Wang, J., Zhou, J.D., Li, Z.S., Song, Y.C., Liu, Q., Jiang, Z.H., Zhang, M.L., 2010. Magnetic, luminescent Eu-doped Mg–Al layered double hydroxide and its intercalation for Ibuprofen. Chemistry 16, 14404–14411.

Willard, H.H., Tang, N.K., 1937. A study of the precipitation of aluminum basic sulfate by urea. J. Am. Chem. Soc. 59, 1190–1196.

Xue, T., Gao, Y., Zhang, Z., Umar, A., Yan, X., Zhang, X., Guo, Z., Wang, Q., 2014. Adsorption of acid red from dye wastewater by Zn_2Al-NO_3 LDHs and the resource of adsorbent sludge as nanofiller for polypropylene. J. Alloys Compd. 587, 99–104.

Yang, L., Shahrivari, Z., Liu, P.K.T., Sahimi, M., Tsotsis, T.T., 2005. Removal of trace levels of arsenic and selenium from aqueous solutions by calcined and uncalcined layered double hydroxides (LDH). Ind. Eng. Chem. Res. 44, 6804–6815.

Yong, Z., Mata, V., Rodrigues, A.E., 2001. Adsorption of carbon dioxide onto hydrotalcite-like compounds (HTLCs) at high temperatures. Ind. Eng. Chem. Res. 40, 204–209.

Yong, Z., Rodrigues, A.E., 2002. Hydrotalcite-like compounds as adsorbents for carbon dioxide. Energ. Conserv. Manage. 43, 1865–1876.

You, Y., Zhao, H., Vance, G.F., 2002. Adsorption of dicamba (3,6-dichloro-2-methoxy benzoic acid) in aqueous solution by calcined–layered double hydroxide. Appl. Clay Sci. 21, 217–226.

Yu, Z., Chen, D., Ronning, M., Torbjorn, V., Ochoa-Fernandez, E., Holmen, A., 2008. Large-scale synthesis of carbon nanofibers on Ni–Fe–Al hydrotalcite derived catalysts. I. Preparation and characterization of the Ni–Fe–Al hydrotalcites and their derived catalysts. Appl. Catal. A 338, 136–146.

Zaghouane-Boudiaf, H., Boutahala, M., Tiar, C., Arab, L., Garin, F., 2011. Treatment of 2,4,5-trichlorophenol by MgAl–SDBS organo-layered double hydroxides: kinetic and equilibrium studies. Chem. Eng. J. 173, 36–41.

Zhang, X., Ji, L., Wang, J., Li, R., Liu, Q., Zhang, M., Liu, L., 2012. Removal of uranium(VI) from aqueous solutions by magnetic Mg–Al layereddouble hydroxide intercalated with citrate: kinetic and thermodynamic investigation. Colloid. Surf. A 414, 220–227.

Zhang, G., Liu, M., 1999. Preparation of nanostructured tin oxide using a sol–gel process based on tin tatrachloride and ethylene glycol. J. Mater. Sci. 34, 3213–3321.

Zhang, H., Qi, R., Evans, D.G., Due, X., 2004. Synthesis and characterization of a novel-scale magnetic solid base catalyst involving a layered double hydroxide supported on a ferrite core. J. Solid State Chem. 177, 772–780.

Zhang, H., Wen, X., Wang, Y., 2007. Synthesis and characterization of sulfate and dodecylbenzenesulfonate intercalated zinc-iron layered double hydroxides by one-step coprecipitation route. J. Solid State Chem. 180, 1636–1647.

Zhao, Y., He, S., Wei, M., Evans, D.G., Duan, X., 2010. Hierarchical films of layered double hydroxides by using a sol–gel process and their high adaptability in water treatment. Chem. Commun. 46, 3031–3033.

CHAPTER 9

Magnetic Nanophotocatalysts for Wastewater Remediation

9.1 INTRODUCTION

During the last decade, extensive efforts have been made in the development of photocatalytic materials for their use in physical and chemical applications on a wide scale. Catalysts such as TiO_2, CdS, Cu_2O, and Fe_3O_4 have been synthesized, and their catalytic properties have been harnessed in material science and pharmaceutical industries, petrochemical operations, and environmental cleanup. These materials often fall in the range of semiconductors of transition elements. In a variety of semiconductors employed, the anatase TiO_2 is the most preferable material for the photocatalytic applications in the degradation and reduction of organic contaminants and toxic metal ions due to its high photosensitivity, nontoxic nature, large bandgap, and high physicochemical stability. Photocatalytic reactions take place on the submicron surface site semiconductors of these catalysts, which enhance the degradation or remediation of environmental contaminants. However, when a catalyst is applied to the remediation of pollutants, it must be separated or recollected by filtration or sedimentation and reused in the separation and purification processes. However, in most cases, it is very difficult to collect the catalysts, which also causes a burden on manpower, reagents, and lapse of time.

However, the most significant factor in the use of TiO_2 is its crystal form. TiO_2 is typically used as a photocatalyst in two crystal structures: rutile and anatase (Fu et al., 2005). It has been reported that the anatase TiO_2 crystals generally have much higher activity than the rutile phase (Tanaka et al., 1991). Recently, magnetic nanoparticles of iron oxides and their spinel-structured ferrite particles have been coated with TiO_2 to synthesize magnetic photocatalytic particles to recover the photocatalyst particles from the treated water stream by applying an external magnetic field (Fu et al., 2005; Beydoun et al., 2002). The most popular magnetic core material is made up of iron oxide such as magnetite, maghemite, hematite, or ferrite because of their strong ferromagnetism, superparamagnetism, and ease of preparation of the colloidal particles in nanoscale.

Nanomaterials for Wastewater Remediation
http://dx.doi.org/10.1016/B978-0-12-804609-8.00009-1

However, it has been revealed that the photocatalyst TiO_2 particles did not form a uniform coating on the surface of the magnetic cores; rather they formed patches on the surface when crystalline TiO_2 was directly deposited, which decreases the recycle ratio of the magnetic photocatalyst. On the other hand, it has been found that the smaller the core particle size in the nanostructure, the higher the efficiency of magnetic photocatalyst as it enhances the surface area of nanomaterials.

9.2 SYNTHESIS AND CHARACTERIZATION

Recently, a photocatalyst of Ag–TiO_2/SiO_2/Fe_3O_4 that can photodegrade orange II was synthesized and applied by Xu et al. (2005). The effect of Ag contents on the photocatalytic activity for the dye was investigated. The synthesized product was characterized by XRD and SEM-EDX. The XRD patterns of Fe_3O_4, SiO_2/Fe_3O_4, TiO_2/SiO_2/Fe_3O_4, and 0.5% Ag–TiO_2/SiO_2/Fe_3O_4 showed the main phases of magnetic core were pure magnetite (Fig. 9.1). Comparison of XRD spectra of Fe_3O_4, with an XRD pattern of the SiO_2/Fe_3O_4 shows a very broad peak of SiO_2 and lower intensity of Fe_3O_4 than that of Fig. 9.1a. It was revealed from Fig. 9.1c that the crystalline anatase titanium dioxide coated on the magnetic core was present, and the intensity of Fe_3O_4 peak dramatically diminished as compared with the Fe_3O_4 in Fig. 9.1a. However, in the case of 0.5% Ag–TiO_2/SiO_2/Fe_3O_4,

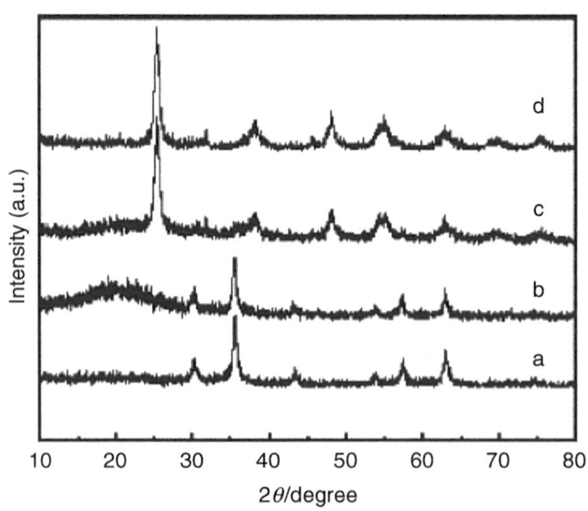

Figure 9.1 XRD patterns of (a) Fe_3O_4 (b) SiO_2/Fe_3O_4 (c) TiO_2/SiO_2/Fe_3O_4, and (d) 0.5% Ag–TiO_2/SiO_2/Fe_3O_4. *(Source: Xu et al., 2005; © Elsevier).*

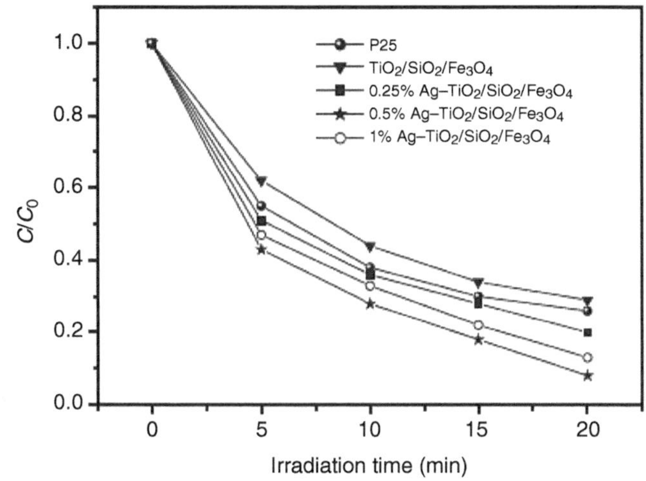

Figure 9.2 ***Effects of commercial TiO_2 (Degussa P25) and as-prepared catalysts on degradation of orange II solution (the initial concentration of dye was 5×10^{-5} mol/L).*** *(Source: Xu et al., 2005; © Elsevier).*

there was only anatase TiO_2 without any peak for Ag. It was stated that the Ag coating was small and dispersed well on the TiO_2 surface. The effects of loading Ag to $TiO_2/SiO_2/Fe_3O_4$ on the photodegradation of orange II were investigated thoroughly (Fig. 9.2). It was clear that the photocatalytic efficiency of Ag–$TiO_2/SiO_2/Fe_3O_4$ catalysts was higher than that of $TiO_2/SiO_2/Fe_3O_4$ catalyst. Their results showed that the 0.5% Ag–$TiO_2/SiO_2/Fe_3O_4$ has the highest photocatalytic efficiency, and its photocatalytic activity remained well even after three cycles.

In a study, Fu et al. (2005) reported $TiO_2/CoFe_2O_4$ composite nanoparticles with a core-shell structure synthesized via coprecipitation, and the shell TiO_2 nanocrystals were derived by sol–gel technology followed by heat treatment at 450°C. The morphology and the crystalline structure of composite nanoparticles were characterized by TEM and XRD, respectively. In a simple synthesis method, the shell TiO_2 nanocrystals were derived via sol–gel hydrolysis precipitation of titanium isopropoxide [$Ti(OC_3H_7)_4$] followed by calcination treatment at different temperatures. After that, a certain amount of modified-cobalt ferrite nanoparticles was dispersed in the mixture solution of water–ethanol with a molar ratio of 1:10, and the pH of the mixture was maintained at 5–6. In the aforementioned mixture different concentrations of $Ti(OC_3H_7)_4$ ethanol solutions were slowly added, and the mixture was stirred for 8 h to ensure the complete coating. Fig. 9.3 shows the structure of composite precursors of $TiO_2/CoFe_2O_4$

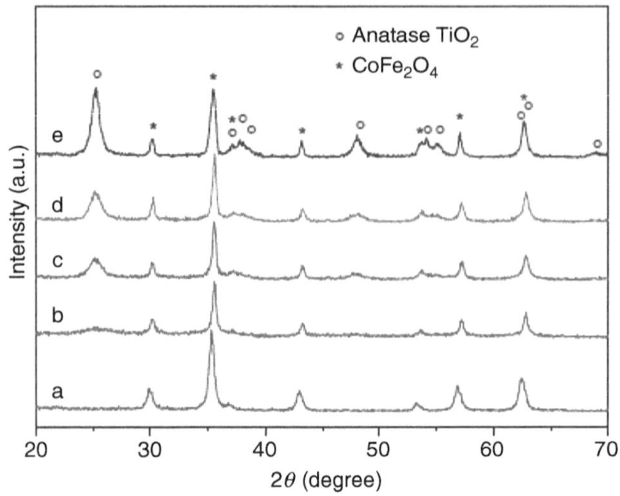

Figure 9.3 ***The XRD patterns of composite precursors of $TiO_2/CoFe_2O_4$ nanoparticles.*** (a) Before and (b–e) after being calcined at various temperatures for 1 h: (b) 200°C, (c) 300°C, (d) 400°C, and (e) 450°C. *(Source: Fu et al., 2005; © Elsevier).*

nanoparticles before and after being calcined at various temperatures. The XRD pattern of the composite precursor sample peaks appeared at 30.1, 35.4, 37.1, 43.1, 53.4, 56.9, and 62.6° attributed to {220}, {311}, {222}, {400}, {422}, {511}, and {440} reflections respectively of the spinel phase of $CoFe_2O_4$, indicating the titanium hydroxide as coating on $CoFe_2O_4$ was amorphous (Fig. 9.3a). The magnetic study shows that the nanocomposites were of good order and able to be recycled in an external magnetic field (Table 9.1). The as-prepared $TiO_2/CoFe_2O_4$ nanocomposite was used for the degradation of dye Procion Red MX-5B as a function of TiO_2 content under UV irradiation. Fig. 9.4 shows the catalytic degradation of dye by the $TiO_2/CoFe_2O_4$ nanocomposite, and the result was compared by commercial photocatalyst Degussa P25 and pure $CoFe_2O_4$ nanoparticles. Highest degradation was observed for the $CoFe_2O_4$ with 95% TiO_2 content.

9.3 APPLICATIONS OF MAGNETICALLY RECYCLABLE NANOPHOTOCATALYSTS

9.3.1 Degradation of Dyes

The photocatalytic activity of the $TiO_2/BaFe_{12}O_{19}$ nanoparticles for the decomposition of Procion Red MX-5B as a function of TiO_2 content under UV irradiation was investigated by Fu et al. (2006b), and the results are

Table 9.1 Magnetic properties and weight ratios of a few magnetic nanoparticles coated with TiO_2

Magnetic properties	Hc (Oe)	Ms (emu/g)
Weight ratio of $TiO_2/CoFe_2O_4$ (wt.%)		
0	550	63
0.10	580	56
0.50	540	53
0.70	520	51
0.90	535	52
Content of TiO_2 in $TiO_2/BaFe_{12}O_{19}$ (%)		
0	4141	75
10	4283	59
30	4280	41
50	4282	17

Source: Fu et al., 2005, 2006b.

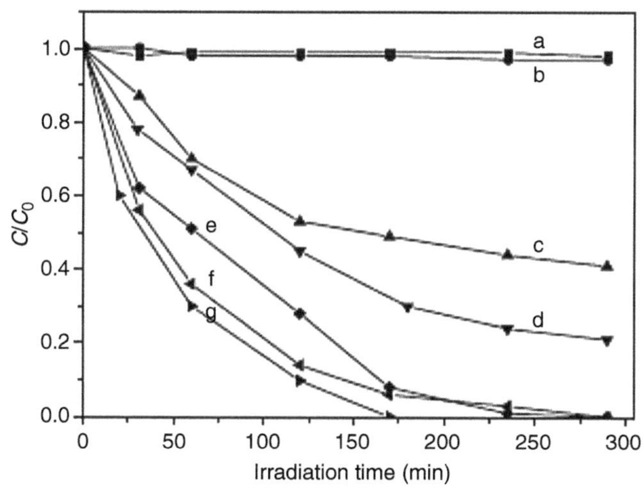

Figure 9.4 ***Concentration changes of Procion Red MX-5B vs UV irradiation time with different photocatalysts.*** (a) Without photocatalysts; and (b–f) are $CoFe_2O_4/TiO_2$ nanoparticles with different contents of TiO_2: (b) 0%, (c) 10%, (d) 50%, (e) 70%, and (f) 95%; and (g) Degussa P25. *(Source: Fu et al., 2005; © Elsevier).*

shown in Fig. 9.5. The obtained result was compared with photocatalytic performance of commercial photocatalyst Degussa P25 and pure $BaFe_{12}O_{19}$ nanoparticles. Curve (a) shows the degradation of Procion Red MX-5B that indicates that the dye is partially degraded in the absence of the photocatalyst under UV irradiation. In Fig. 9.5, curves (b–e) are of the same TiO_2

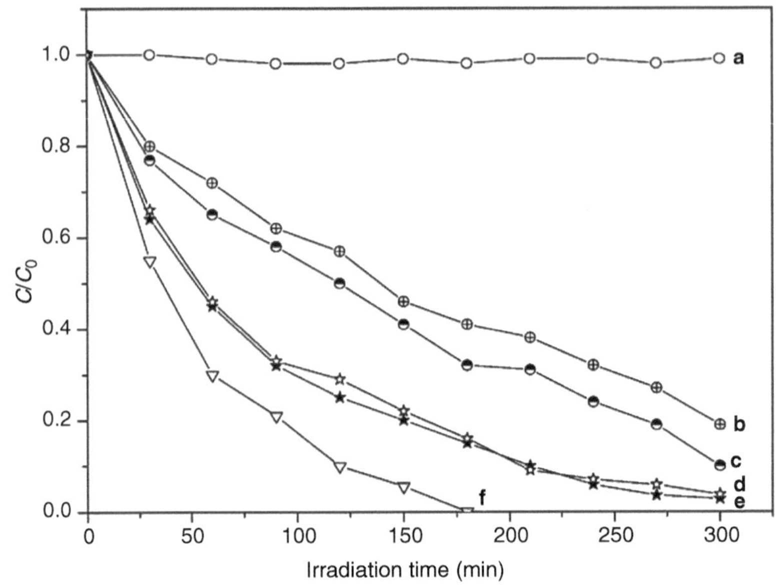

Figure 9.5 ***Concentration changes of Procion Red MX-5B vs UV irradiation time with different photocatalysts.*** (a–e) $TiO_2/BaFe_{12}O_{19}$ nanoparticles with different content of TiO_2: 0, 10, 20, 30, and 100%, respectively; and (f) is Degussa P25. *(Source: Fu et al., 2006b; © Elsevier).*

amount in the composite of $TiO_2/BaFe_{12}O_{19}$ nanoparticles with weight ratios of $TiO_2/BaFe_{12}O_{19}$: 10, 20, 30, and 100%, respectively. The photocatalytic activity of Degussa P25 under UV irradiation is shown in curve (f). While the content of TiO_2 was 30% under UV light irradiation the photodegradation ratio of Procion Red MX-5B by the composite of $TiO_2/BaFe_{12}O_{19}$ was 96% similar to that of pure TiO_2. However, the degradation activity of the synthesized composites was little lower than Degussa P25, although, as a magnetic photocatalyst it could be easily recovered while Degussa P25 cannot.

Gad-Allah et al. (2009) prepared the $TiO_2/SiO_2/Fe_3O_4$ photocatalyst and used it for the treatment of synthetic dyes, namely remazole golden yellow G, remazole brilliant blue R, and reactive orange 16 from wastewater. They studied different operating conditions such as catalyst load, pH, and dye concentration in order to determine the optimum operating conditions. The results of the optimization parameters suggested that the pH 3 and 2500 ppm TiO_2 photocatalyst dose was suitable for the treatment of dyes. However, increasing the dose of photocatalyst reduces the rate of reaction due to aggregation of photocatalyst as well as increasing of cloudiness inside the photoreactor. The equilibrium was established within 45 min, and complete decolorization and degradation of the organic dyes was obtained under optimized conditions. Kinetic studies revealed that the rate of reaction solely depends on dose of photocatalyst, dye concentration, and UV light irradiance inside the photoreactor. $TiO_2/SiO_2/Fe_3O_4$ photocatalysts can be used three times with nearly the same efficiency, after that the photocatalytic efficiency of nanocatalysts decrease. In general TiO_2 photocatalyst it is difficult to recover the used catalyst, however, in this case where the prepared photocatalyst is a magnetic nanophotocatalyst it can be easily separated from solution using an external magnet. The reuse of the photocatalyst is supposed to decrease the final cost of the dye wastewater treatment process. Based on the obtained result the authors proposed a flow chart for the treatment of dye-bearing wastewater (Fig. 9.6).

Hong et al. (2008) synthesized the Fe_3O_4 magnetic nanoparticles (MNPs) by a coprecipitation method and applied for the photocatalytic degradation of methyl orange from water. The synthesized catalyst was characterized by the XRD, SEM, and TEM, and for magnetic properties. TEM images show

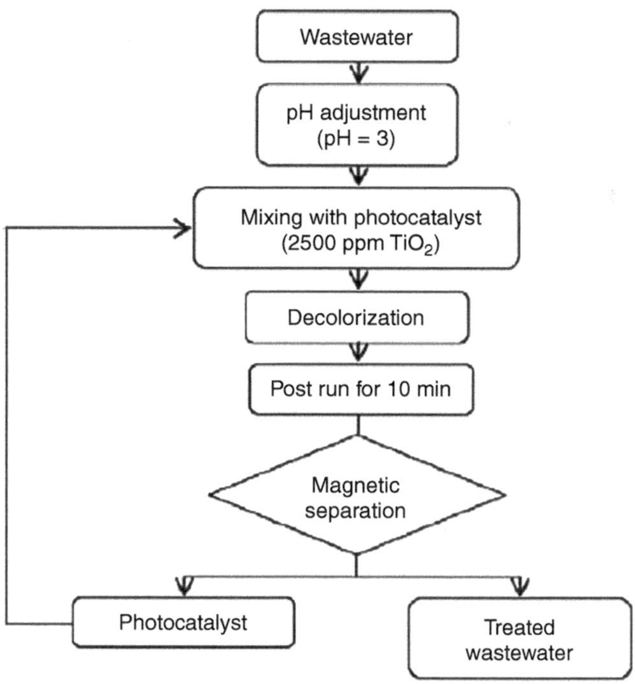

Figure 9.6 ***Proposed flowchart for the dye-bearing wastewater.*** *(Source: Gad-Allah et al., 2009; © Elsevier).*

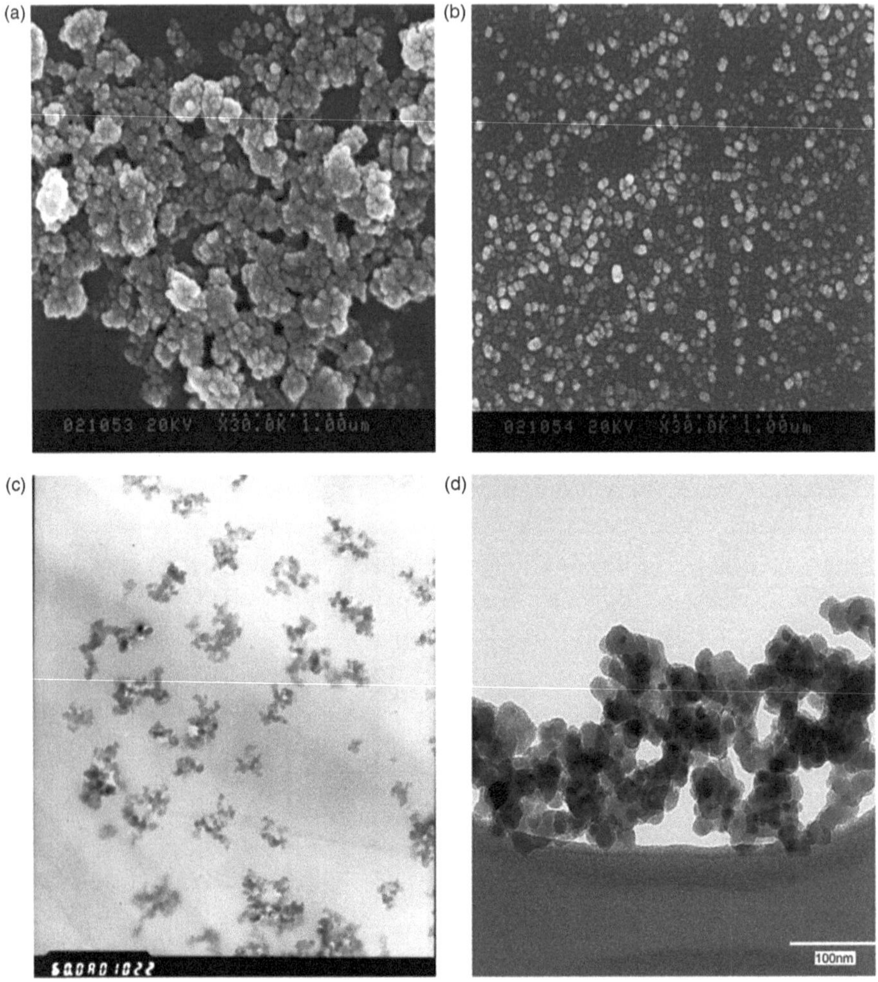

Figure 9.7 ***SEM images of Fe_3O_4 MNPs.*** (a) Before and (b) after surface modification with sodium citrate; (c) TEM images of unmodified Fe_3O_4 MNPs coated by ZnO, and (d) sodium citrate modified Fe_3O_4 nanoparticles coated by ZnO. *(Source: Hong et al., 2008; © Elsevier).*

that the average particle size was found to be 10 nm (Fig. 9.7). The synthesized magnetic nanoparticles were modified by sodium citrate (Fig. 9.8). The Fe_3O_4 MNPs were coated with zinc acetate and ammonium carbonate directly to obtain the Fe_3O_4/ZnO core/shell MNPs. The photocatalytic degradation of methyl orange revealed that the core/shell MNPs have higher photocatalytic activity than that of the ZnO nanoparticles, and the highest degradation percentage was found to be 83%. Because of the magnetic

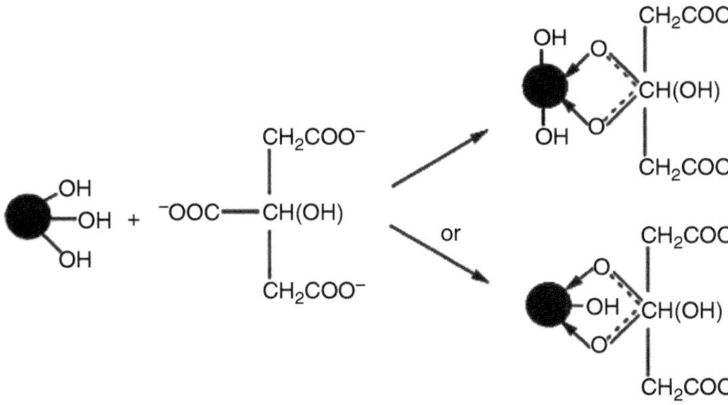

Figure 9.8 ***Schematic diagram of sodium citrate modification of magnetic nanomaterials.*** *(Source: Hong et al., 2008; © Elsevier).*

properties of the photocatalyst it was very easy to separate it from the liquid. The synthesized magnetic photocatalyst can be used up to four times, and the photocatalytic degradation percentage of the core/shell MNPs is about 70%, and after that the photocatalytic performance decreased.

A novel magnetic photocatalyst $TiO_2/SiO_2/\gamma\text{-}Fe_2O_3$ was prepared by Chen et al. (2001) and characterized for STM and XRD. The magnetic photocatalyst $TiO_2/SiO_2/\gamma\text{-}Fe_2O_3$ was used for the photocatalytic degradation of dyes namely, fluorescein, orange II, and red acid G from aqueous solutions. STM images show that the particle size was uniform and regular in diameter at 50 nm (Fig. 9.9). The photodegradation of fluorescein and orange II on $TiO_2/SiO_2/\gamma\text{-}Fe_2O_3$ was much faster than that of $TiO_2/\gamma\text{-}Fe_2O_3$ under both UV and visible light irradiation. Fig. 9.10 shows the photocatalytic degradation of dye on photocatalyst $TiO_2/\gamma\text{-}Fe_2O_3$ under UV and visible light.

Recently, Cao et al. (2015) synthesized a novel magnetically separable visible-light-induced photocatalyst Cu_2O/chitosan–Fe_3O_4 nanocomposite (Cu_2O/CS–Fe_3O_4NC) using magnetic chitosan chelating copper ions as a precursor, and the synthesized photocatalyst was evaluated by decolorization of reactive brilliant red X-3B (X-3B) under visible light irradiation. The characterization of synthesized Cu_2O/CS–Fe_3O_4NC exhibited relatively large specific surface areas and special dimodal pore structures because Cu_2O was wrapped in chitosan matrix embedded with Fe_3O_4 nanoparticles which facilitates the degradation of organic dyes under visible irradiance. The good combination of Fe_3O_4 and Cu_2O on chitosan made the catalyst superparamagnetic which can be easily separated from water by an external

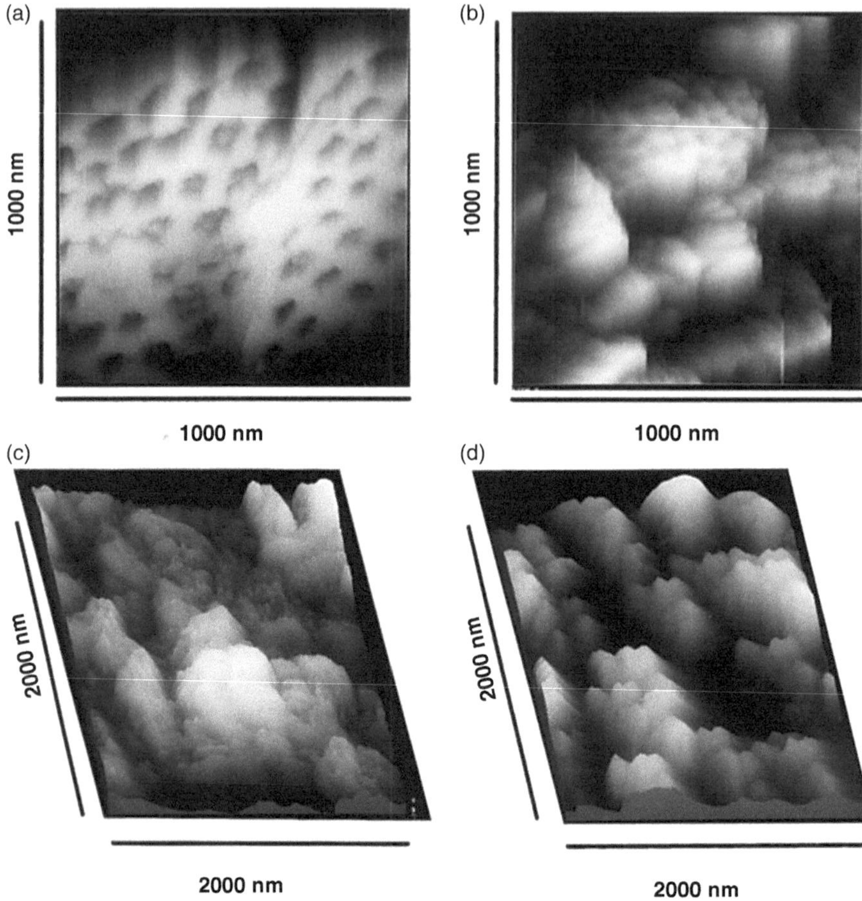

Figure 9.9 ***STM photographs of the photocatalysts.*** (a) γ-Fe_2O_3, (b) SiO_2/γ-Fe_2O_3, (c) TiO_2/γ-Fe_2O_3, and (d) TiO_2/SiO_2/γ-Fe_2O_3. *(Source: Chen et al., 2001; © Elsevier).*

magnet. Lower pH of the solution was favorable for the decomposition of reactive brilliant red X-3B, whereas the higher pH value inhibits the catalytic degradation of dye molecules. The decolorization of reactive brilliant red X-3B was ascribed to the synergistic effect of photocatalysis and adsorption. The synthesized Cu_2O/CS–Fe_3O_4 nanocatalyst was easily separated from the solution by an external magnet, and the decolorization rates of reactive brilliant red X-3B were above 87% after five reaction cycles, indicating that Cu_2O/CS–Fe_3O_4NC had excellent reusability and stability.

Gao et al. (2003) synthesized a photocatalyst of TiO_2/γ-Fe_2O_3 and applied for the degradation of dye contaminants in water. The synthesized product was characterized by XRD, AFM, and TEM for the phase identification and

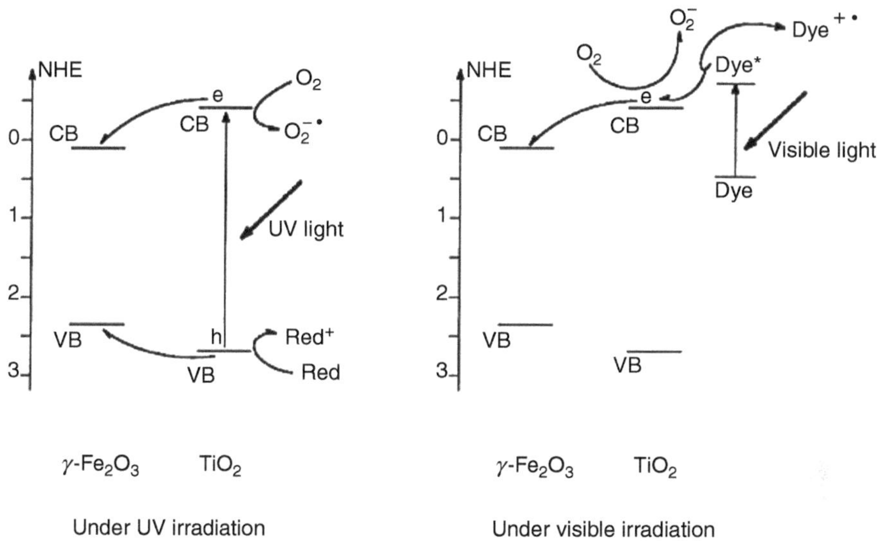

Figure 9.10 ***Photocatalytic degradation of dye on photocatalyst $TiO_2/\gamma\text{-}Fe_2O_3$ under UV and visible light.*** *(Source: Chen et al., 2001; © Elsevier).*

to decipher the crystallinity with sintering temperature of the catalyst. The motive of the studied materials depends on its photocatalytic and magnetic properties. As the $TiO_2/\gamma\text{-}Fe_2O_3$ photocatalyst was composed of two parts, the TiO_2 shell used for photocatalysis and the $\gamma\text{-}Fe_2O_3$ core for separation by the magnetic field. The photocatalytic degradation of acridine dye was carried out using the synthesized photocatalyst in water. Under UV or visible light irradiance the excited dyes inject electrons into the conduction band of the titania, then the injected electrons react with the preadsorbed oxygen to form oxidizing species ($O_2^{\bullet-}$, $^{\bullet}OOH$, and then $^{\bullet}OH$ radicals) that can bring about the photooxidiation of dyes. However, it has been observed that due to the strong light absorption by $\gamma\text{-}Fe_2O_3$, when the amount of the loaded TiO_2 content was under 30% in the catalyst, the photocatalytic activity of $TiO_2/\gamma\text{-}Fe_2O_3$ was significantly lower than that of the pure TiO_2. The sample, sintered at 500°C, showed the highest activity for the degradation of aqueous solution of acridine dye. To evaluate the influence of the amount of titania on the photocatalytic performance of magnetically separated photocatalyst in the dye degradation process, three samples were prepared at 500°C: $TiO_2:\gamma\text{-}Fe_2O_3 = 3:7$, $TiO_2:\gamma\text{-}Fe_2O_3 = 1:1$, and $TiO_2:\gamma\text{-}Fe_2O_3 = 7:3$, containing 30, 50, and 70% of TiO_2 nanoparticles, respectively. The obtained result revealed that the degradation rate increased with the increasing amount of supported TiO_2 in regard to the concentration range of acridine dye.

Chung et al. (2004) synthesized magnetically separable photocatalysts by a continuous multistep spray pyrolysis process, in which firstly $NiFe_2O_4$ core particles were prepared and tetraethyl orthosilicate and titanium isopropoxide were sequentially injected and coated on the surface of the core particles. Thus sequentially coated layers were decomposed to form silica and titania layers in high temperature in furnace. Catalytic results show that the titania–silica layered particles displayed higher photoactivity than particles coated only with titania and titania–silica mixture-coated particles. The synthesized magnetic photocatalyst shows very strong magnetic saturization, which was helpful in the separation of catalysts from aqueous solutions. This photocatalyst was applied for the degradation of methylene blue dye in aqueous solutions. Fig. 9.11 shows the graph methylene blue photocatalytic decomposition with respect to contact time of 90 min. Among the synthesized catalyst only titania-coated $NiFe_2O_4$ particles show strong photocatalytic activity, however, decreased by 30% after the first magnetic separation and recycle. This shows that the coating of titania particles was

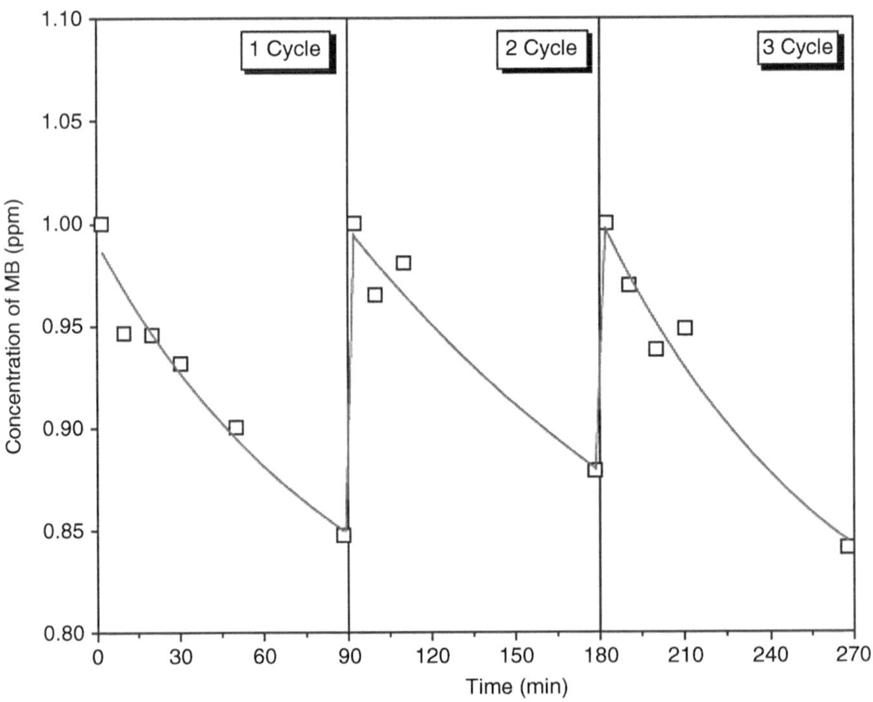

Figure 9.11 ***Photocatalytic degradation of methylene blue in continuous photodecomposition reaction [TiO_2 (3%)–SiO_2 ([10%) layer-coated $NiFe_2O_4$ particle].*** *(Source: Chung et al., 2004; © Elsevier).*

not completely strong, or it may be possible that there exist free titania particles that are not coated on the core particles.

Safavi and Momeni (2012) synthesized palladium/hydroxyapatite/Fe_3O_4 (Pd/HAP/Fe_3O_4) nanocatalyst and evaluated it for its catalytic activity toward the degradation of azo dyes (methyl red, methyl orange, and methyl yellow) from aqueous solutions. It has been reported that the acidic medium favors the catalytic degradation of azo dyes in the presence of synthesized Pd/HAP/Fe_3O_4. The catalytic degradation of azo dyes was monitored by UV–vis measurements during a certain time interval. The decay of the absorbance at 515 nm for methyl red, 505 nm for methyl orange, and 510 nm for methyl yellow was monitored over time (Fig. 9.12). It was proposed that the dye degradation takes place due to the reaction of Pd/HAP/Fe_3O_4 with dissolved oxygen with the assistance of acid to form a Pd hydroperoxide, which oxidizes azo dyes. As the solution pH changes from 1 to 7 the degradation percentage decreases significantly, showing that the prepared catalyst can only work in an acidic environment. It was also suggested that the hydroxyapatite plays a significant role in the support of the Pd nanocatalyst. However, the degradation of dye was initially low, which increases exponentially with time. Kinetic study shows that the degradation rate for dyes was fitted to pseudo-first-order reaction. Due to excellent stability and magnetic properties, it was easily recoverable from the liquid and recycled and further reused in the degradation processes.

Yao et al. (2015) has synthesized N-doped $TiO_2/ZnFe_2O_4$ catalysts by coupling nitrogen-modified TiO_2 with $ZnFe_2O_4$ via a one-pot vapor-thermal method. Three products, labeled as N–$TiO_2/ZnFe_2O_4$-1, N–$TiO_2/ZnFe_2O_4$-2, and N–$TiO_2/ZnFe_2O_4$-3 were obtained by varying the dosage of urea at 0.0537, 0.1073, and 0.1610 g, respectively. The synthesized photocatalyst was used for the catalytic degradation of dyes and various parameters such as effects of the amount of photocatalyst, different types of dyes, and catalyst stability on photodegradation of organic dyes were investigated. N-doped $TiO_2/ZnFe_2O_4$ shows good catalytic properties compared with $TiO_2/ZnFe_2O_4$ or $ZnFe_2O_4$ because of the formation of a heterostructure at the interface as well as the introduction of N species. Active species such as holes, electrons, hydroxyl radicals, and superoxide radicals involved in the photodegradation process were detected by using different types of scavengers. Due to the presence of $ZnFe_2O_4$ in the hybrid, the photocatalyst shows ferromagnetism and is easily separable from the liquid. Fig. 9.13 shows the photodegradation of rhodamine B dye in the presence of N–$TiO_2/ZnFe_2O_4$-2.

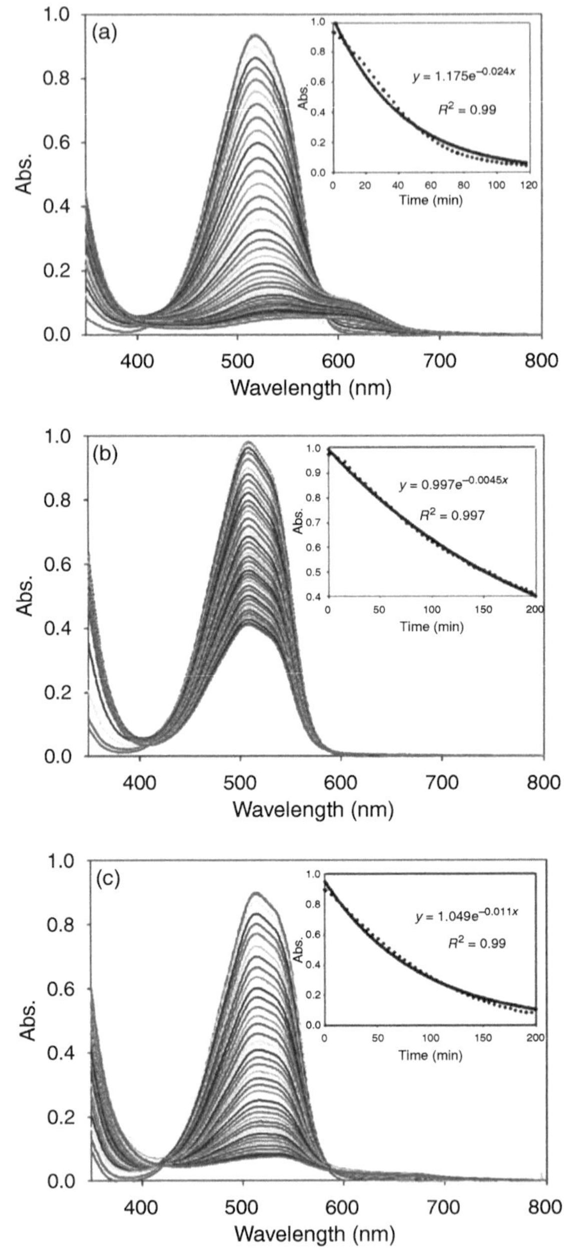

Figure 9.12 Change in the absorbance spectrum, with time, during the reaction of methyl red (a), methyl orange (b), and methyl yellow (c) solution in the presence of Pd/HAP/Fe_3O_4 catalyst in 0.1 M H_2SO_4; inset shows the absorbance decay of azo dyes with time in the presence of Pd/HAP/Fe_3O_4. *(Source: Safavi and Momeni, 2012; © Elsevier).*

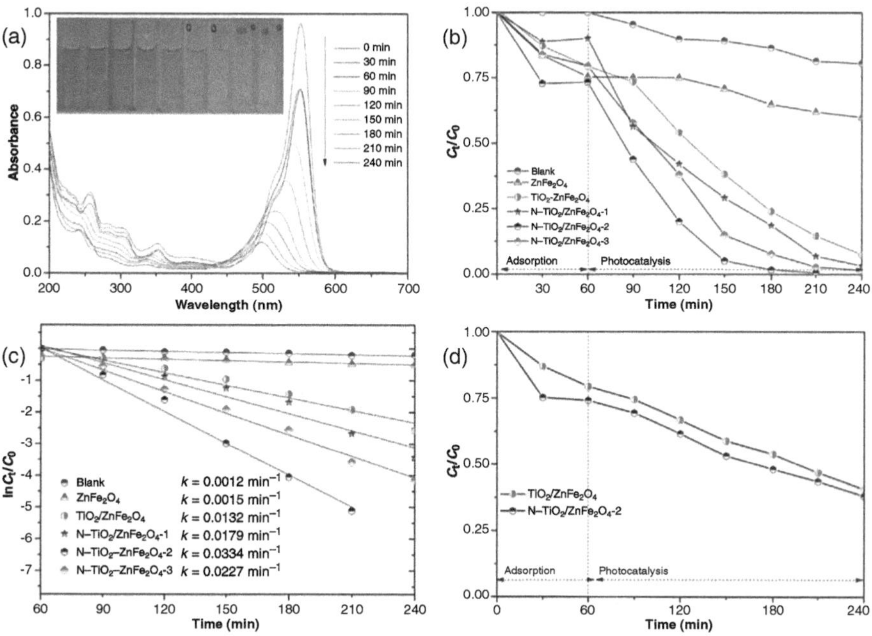

Figure 9.13 (a) Time-dependent optical absorbance spectra and photograph (inset) of the rhodamine B solution in the presence of N–TiO_2/$ZnFe_2O_4$-2 after exposure to UV–vis light. (b) Photocatalytic degradation and (c) kinetic data of rhodamine B over the different catalysts under UV–vis irradiation. (d) Photocatalytic activity over TiO_2/$ZnFe_2O_4$ and N–TiO_2/$ZnFe_2O_4$-2 hybrid under visible-light irradiation (reaction conditions: rhodamine B = 5 mg/L, catalyst = 0.40 g/L, T = 25°C). *(Source: Yao et al., 2015; © Elsevier).*

Anatase–titania coated nickel ferrite nanoparticles of size 10–20 nm was synthesized by combined reverse micelle and hydrolysis method (Rana et al., 2005b). The anatase TiO_2-coated $NiFe_2O_4$ nanoparticles show superparamagnetism and nonsaturation of magnetic moments at high field. The synthesized anatase TiO_2-coated $NiFe_2O_4$ nanoparticles were used for the catalytic degradation of methyl orange at room temperature under UV light irradiance. The photocatalytic experiment shows that the methyl-orange degradation was inversely proportional to the reaction time, and the concentration was decreased from ~25 to ~4 μmol on irradiation for 7 h. Kinetic study shows that the degradation process for methyl orange was best fitted with zero-order reaction. The anatase TiO_2-coated $NiFe_2O_4$ shows good magnetism, which was helpful in the separation of catalyst after the completion of degradation reaction from aqueous solution.

Xu et al. (2014) synthesized monodisperse nanostructured Fe_3O_4/ZnO microrods and applied for the removal of toxic metal ions from aqueous solutions; experiments display that Fe_3O_4/ZnO heterostructures show the best removal efficiency compared with pure ZnO and Fe_3O_4 structures, and the photocatalytic results show that the Fe_3O_4/ZnO heterostructures exhibit the excellent photocatalytic degradation of rhodamine B. The degradation was initially very fast, and 100% degradation was obtained within 40 min. Different added amounts (0, 0.125, 0.167, 0.25, 0.5 mmol) of $FeCl_3.6H_2O$ in the synthesized, nanostructured Fe_3O_4/ZnO microrods were denoted as ZF1, ZF2, ZF3, ZF4, and ZF5, respectively. The ZF3 photocatalyst shows the highest degradation of rhodamine B dye in 40 min under UV light irradiance.

Chen et al. (2015) prepared graphite carbon nitride hybridized chromium doping strontium titanate spheres (CrSTO/g-CN) nanocomposites and applied for the degradation of rhodamine b dye in aqueous solutions. It has been found that the CrSTO/g-CN hybrid nanocomposite was highly stable in aqueous environment and revealed significant photocatalytic activity for the degradation of rhodamine B under visible light irradiation. In addition, it has been found that the photocatalytic activity of CrSTO/g-CN-70% was almost 4.5 times higher than that of pure g-C_3N_4 and 3.5 times higher than that of pure Cr-doped $SrTiO_3$, respectively. The CrSTO/g-CN-70% hybrid nanocomposites exhibited the highest photocatalytic activity with a degradation rate of 97% for rhodamine B within 30 min in comparison with other pure samples. The high degradation of rhodamine B using the CrSTO/g-CN hybrid nanocomposites was due to a synergistic effect following the Z-scheme charge transfer mechanism to enhance charge-separation ability. Fig. 9.14 shows the transfer route of photoinduced charge carriers and photocatalytic degradation mechanisms of rhodamine B using CrSTO/g-CN-70% hybrid nanocomposites under visible light irradiation.

Li et al. (2015) synthesized Fe_3O_4@C@CdS magnetic photocatalysts and used the photocatalyst for the degradation of rhodamine B dye in aqueous medium. Due to the carbonization of ferrocene, the produced carbon-coated Fe_3O_4 nanocomposite had both magnetic cores and deposition substrates for in situ generating crystalline CdS nanocrystals on them by rapid microwave irradiation. Saturation magnetization of Fe_3O_4@C and Fe_3O_4@C@CdS was measured to be 36 and 30 emu/g, respectively. It was found that there was no hysteresis in both samples, showing that both of them exhibited superparamagnetic behavior. This was very helpful in the separation of magnetic photocatalysts from water. The synthesized product Fe_3O_4@C@CdS nanoparticles exhibit a uniform core-shell structure and display good photocatalytic activity and recyclability in the degradation of

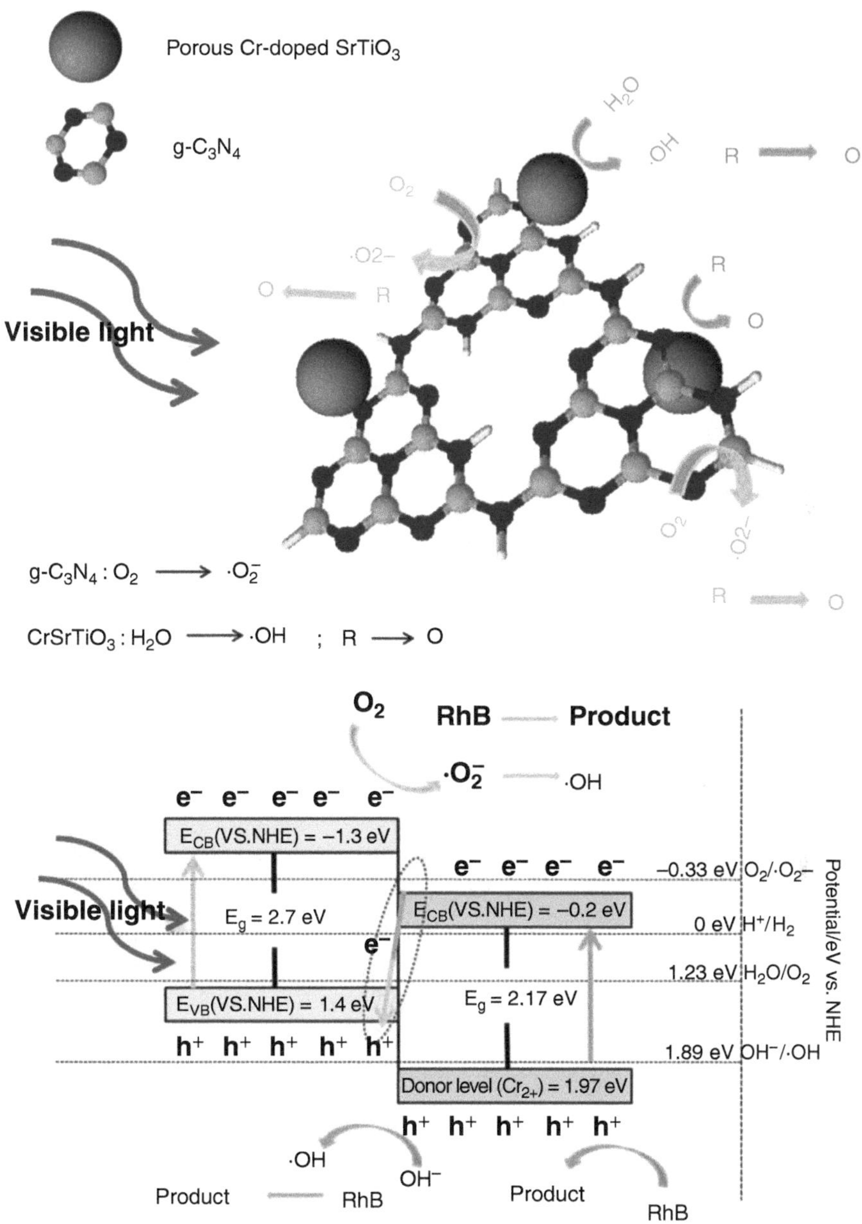

Figure 9.14 ***Schematic illustration of the transfer route of photoinduced charge carriers and photocatalytic degradation mechanisms of rhodamine B using CrSTO/g-CN-70% hybrid nanocomposites under visible light irradiation.*** *(Source: Chen et al., 2015; © Elsevier).*

rhodamine B dye. By using the 15 mg of the Fe_3O_4@C@CdS nanoparticle, most of the rhodamine B molecules in solution were quickly degraded by the photocatalyst within 50 min.

Maghemite nanoparticles (γ-Fe_2O_3) and maghemite/silica nanocomposite microspheres (γ-Fe_2O_3/SiO_2) have been evaluated as magnetic heterogeneous Fenton catalysts for the degradation of dye pollutants in aquatic systems (Ferroudj et al., 2013). It was found that the two materials differ in size, morphology, porosity, and microstructure, although the maghemite nanoparticles are not modified by their encapsulation into the silica. The BET surface area for γ-Fe_2O_3 was found to be 159 m^2/g. The synthesized magnetic nanoparticles possess good magnetic properties and were easily separable from the aqueous solutions by external magnetic field. Three model pollutants, namely methyl orange, methylene blue, and paranitrophenol, differing in structure and electrostatic charge were degraded by the magnetic catalyst in aqueous solutions. The type of contaminants and dose of catalyst highly influence the reaction rate and degradation processes. The synthesized magnetic catalysts showed quite rapid adsorption rate and reached equilibrium within 30 min. Thus, it was established that adsorption does not compete with the catalytic process, since H_2O_2 is added 2 h after the pollutants in the catalytic medium. The amounts of pollutant adsorbed at the equilibrium in ascending order are paranitrophenol $<$ methylene blue $<$ methyl orange for magnetic nanoparticles, and paranitrophenol $<$ methyl orange $\ll$ methylene blue for magnetic microsphere. The percentage of adsorbed pollutants ranges from less than 5 to 30%, except for methylene blue, for which a very strong adsorption occurs on the magnetic microspheres. It was also found that the mineralization rates were moderate for both catalysts, showing the larger stability toward oxidation of the uncolored organic intermediates resulting from the primary degradation of the model pollutants. Moreover, the magnetic microsphere materials were recovered and used for five consecutive cycles.

Recently, Zhang et al. (2014) synthesized CdS nanoparticles directly grown on the surfaces of chitosan microspheres via photochemical method with irradiation from an 8 W UV lamp. The nanoparticles showed almost spherical shape, were closely arranged, and were immobilized on the microspheres due to the orientation of the amino and hydroxyl groups of the chitosan. The chitosan microsphere-immobilized CdS nanoparticles were highly active and easily recyclable than pure CdS particles when used in the decolorization of methyl orange under UV–vis light. The photodegradation of methyl orange was due to the superoxide radicals generated in the reaction of electrons and adsorbed O_2. The functional groups of chitosan microspheres may prevent the recombination of electron/holes and effectively promote photocatalytic reactions. The decolorization tests were divided into two stages, that is, 30 min adsorption stage and 70 min photodegradation stage. The removal efficiency of methyl orange was only 20.9% in 100 min due to the photodegradation under UV light without any catalyst. This decolorization enhanced to 34.1% when pure CdS nanoparticles were applied, which constitute 7.8% adsorption and 26.3% photodegradation decolorization. When chitosan microspheres/CdS were used the decolorization efficiency reached to 22.4% within 30 min due to the adsorption. In addition, methyl orange decolorization reached 95.2% under light irradiance, and the COD decreased 71.6% in 70 min, which indicated that chitosan microsphere/CdS had a higher photocatalytic activity than pure CdS particles.

Fu et al. (2006a) synthesized $TiO_2/SrFe_{12}O_{19}$ composite nanoparticles with core-shell structure and applied for the removal of procion red MX-5B from aqueous solutions. The magnetic core $SrFe_{12}O_{19}$ nanoparticles were first prepared with Fe/Sr ratio of 10.8, and then the shell TiO_2 nanocrystals were derived via sol–gel technology. The presence of a small amount of polyethylenimine on the surface of the strontium ferrite nanoparticles facilitates this coating process. The synthesized $TiO_2/SrFe_{12}O_{19}$ composite has enough magnetic saturation and was able to be separated from liquid by external magnet. However, the result of the experiments showed that the saturation magnetization of titania-coated $SrFe_{12}O_{19}$ nanoparticles decrease with increasing the surface coating by titania. Photocatalytic activities of these powders increase with increasing the thickness of TiO_2 coating layer. The zeta potential of prepared $SrFe_{12}O_{19}$ nanoparticles and modified strontium ferrite nanoparticles was found to be ≈4.0 and ≈10.5, respectively. The photocatalytic degradation of procion red MX-5B by the prepared $TiO_2/SrFe_{12}O_{19}$ nanoparticles increases with increasing the content of TiO_2 in core-shell structure $TiO_2/SrFe_{12}O_{19}$ nanoparticles. When the content of TiO_2 is 30%, and there is enough UV illumination the photodegradation ratio of procion red MX-5B by the composite of $TiO_2/SrFe_{12}O_{19}$ is 98% similar to that of pure TiO_2 within 5 h.

Recently, Zhang et al. (2016) synthesized a novel three-component $Fe_3O_4/TiO_2/Ag$ composite mircosphere and prepared it via a facile chemical deposition route. The Fe_3O_4/TiO_2 microspheres were first prepared by the solvothermal method, and then Ag nanoparticles were anchored onto the external surface of TiO_2 by the tyrosine-reduced method. Photocatalytic degradation of methylene blue was carried out by the synthesized $Fe_3O_4/TiO_2/Ag$ composite.

The obtained results indicated that the photocatalytic activity of Fe_3O_4/TiO_2/Ag composite microspheres was better in comparison with Fe_3O_4/TiO_2 due to the dual effects of the enhanced light absorption and reduction of photoelectron-hole pair recombination in TiO_2 with the introduced Ag nanoparticles. Moreover, these magnetic Fe_3O_4/TiO_2/Ag composite microspheres were recycled from the liquid by external magnet and were further reused with little or no loss of catalytic properties.

Wang et al. (2010a) synthesized Fe_3O_4 MNPs with much improved peroxidase-like activity through an advanced reverse coprecipitation method under the assistance of ultrasound irradiation. The characterization of prepared material indicated that the ultrasound irradiation in the preparation induced the production of Fe_3O_4 MNPs possessing average particle sizes of 16.5 nm, and the BET surface area was 82.5 m^2/g. The catalytic property of Fe_3O_4 MNPs was evaluated for the degradation of rhodamine B in presence of H_2O_2. The obtained result showed that the degradation of rhodamine B was 90% within 60 min at pH 5.4 and 40°C by sonochemically synthesized Fe_3O_4 MNPs. This high performance of catalytic activity was attributed to smaller particle size, larger BET surface area, and higher dispersibility of the Fe_3O_4 MNPs, which were achieved by the ultrasonic irradiation during the preparation of Fe_3O_4 MNPs. The properties, including surface area and particle size along with chemical composition, were observed to be important in the catalytic activity of Fe_3O_4 MNPs.

Sono-enhanced degradation of dye pollutant rhodamine B was investigated by using H_2O_2 as a green oxidant and Fe_3O_4 MNPs as a peroxidase mimetic (Wang et al., 2010b). The result shows that the degradation of rhodamine B enhanced in the ultrasound irradiance in the presence of Fe_3O_4 MNPs catalyzes the breakdown of H_2O_2 in the liquid at pH ranges of 3 to 9. The degradation of dye was nearly 95% at pH 5 and 55°C under the ultrasound-assisted H_2O_2–Fe_3O_4 MNPs system. It was found that the synergistic behavior between Fe_3O_4 MNPs catalysis and ultrasonics was dependent on Fe_3O_4 MNPs dosage, H_2O_2 concentration, pH value, and temperature. The probable result was due to the synergistic effects attributed to the positive interaction between cavitation effect accelerating the catalytic breakdown of H_2O_2 over Fe_3O_4 MNPs, and the function of Fe_3O_4 MNPs providing more nucleation sites for the cavitation inception.

Wang et al. (2016) synthesized Fe_3O_4@ZnO binary nanoparticles and applied for the photocatalytic degradation of rhodamine B from aqueous solution under UV light irradiance. The synthesized Fe_3O_4@ZnO core/shell nanoparticles show enhanced catalytic degradation of dye in comparison

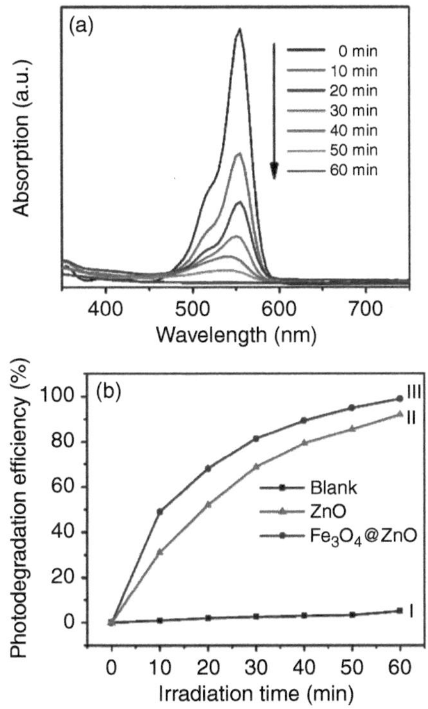

Figure 9.15 Adsorption changes of rhodamine B aqueous solutions in the presence of Fe_3O_4@ZnO core/shell nanoparticles (a), and degradation efficiency versus irradiance time for different photocatalysts (b). *(Source: Wang et al., 2016; © Elsevier).*

with ZnO nanoparticles. The enhanced catalytic degradation of dye by Fe_3O_4@ZnO core/shell nanoparticles was due to the higher concentration of surface oxygen vacancies and the suppressing effect of the Fe^{3+} ions on the recombination of photoinduced electron hole pairs. The Fe_3O_4@ZnO core/shell nanoparticles show good magnetization and were easily recovered from liquid with external magnet and reused six times in catalytic degradation of dye. Fig. 9.15 shows the adsorption changes and photocatalytic degradation of rhodamine B aqueous solutions in the presence of Fe_3O_4@ ZnO core/shell nanoparticles.

A magnetically reduced graphene oxide–bismuth oxybromide (MRGO–BiOBr) composite was successfully synthesized and was applied for the photocatalytic degradation of methylene blue in water (Janani et al., 2016). The characterization of synthesized materials shows the presence of Fe-O bond–stretching vibrations and bromo compounds using FT-IR, and the diffraction peaks of the composite obtained from XRD showed a combination of peaks of both MRGO and BiOBr. UV-vis DRS was applied to access the reduction in band gap energy from 2.76 eV for pure BiOBr to 1.91 eV for the composite. Photocatalytic degradation of methylene blue dye under visible light irradiation was examined using MRGO–BiOBr, and around 96.84% of dye was degraded in 2 h. The result revealed that MRGO–BiOBr showed the highest photocatalytic activity of 76.24% dye degradation as compared to 47.54 and 52.14% for pure MRGO and BiOBr, respectively. The percentage removal of the dye decreased from 69.89 to 49.12% with increasing initial concentration of dye. The degradation percentage was increased from 59.44 to 96.84% on increasing the catalyst loading from 0.1 to 0.8 g. Further increase in the catalyst loading does not affect the decrease in dye concentration. The photocatalyst was recovered by the external magnet and reused three times in the degradation reaction.

A magnetically separable photocatalyst $TiO_2/SiO_2/NiFe_2O_4$ nanosphere with egg-like structure was prepared and applied for high photocatalytic activity for the degradation of methyl orange in water (Xu et al., 2008). The synthesized magnetic nanoparticle shows a superparamagnetic nature which was helpful in the recovery and recycling of the used catalyst. The characterization of prepared nanosphere revealed that nickel ferrite core nanoparticles were completely encapsulated into monodisperse silica nanospheres as carriers, and titania nanoparticle aggregates were coated onto the surface of $SiO_2/NiFe_2O_4$ nanospheres, forming an imperfect TiO_2 shell for photocatalysis. The presence of the SiO_2 layer between the $NiFe_2O_4$ core and the TiO_2 shell effectively prevents the injection of charges from TiO_2 particles to $NiFe_2O_4$, which gives rise to an increase in photocatalytic activity.

Santala et al. (2009) synthesized the $CoFe_2O_4$ loaded TiO_2 nanotubes by atomic layer deposition and electrospinning and used them for the photocatalytic degradation of methylene blue in water. The experimental results showed that the $CoFe_2O_4$ loaded TiO_2 nanotubes were photocatalytically active in decomposing methylene blue. In addition, the photocatalytic efficiency of nanotubes remained the same in the repeated experiments that were made by collecting the $CoFe_2O_4$ loaded TiO_2 nanotubes with a magnet and then dispersing them into a fresh solution of methylene blue.

Jiang et al. (2014) synthesized multifunctional crumpled graphene-based ternary nanocomposite photocatalysts for the degradation of dye wastewater. The crumpled graphene oxide was subsequently encapsulated with commercially available TiO_2 and Fe_3O_4 nanoparticles. The synthesized crumpled graphene$-TiO_2-Fe_3O_4$ ternary core-shell nanostructures possess superior aqueous-based photocatalytic properties for methyl orange, and the degradation of dye was more than 20-fold as compared to TiO_2 alone. The degradation performance after a 90 min reaction in the case of 20 mg/L methyl orange and 16 mg/L catalysts showed ~41% and ~28% methyl orange remained in the solution for TiO_2 and graphene−TiO_2−Fe_3O_4, respectively. The photocatalytic degradation of methyl orange was fitted to pseudo-first order kinetic reaction. The synthesized graphene−TiO_2−Fe_3O_4 nanomaterial was recycled and reused, and the result of reused photocatalysts in photodegradation of methyl orange shows that they can be used in up to five consecutive cycles.

Li et al. (2011) synthesized a magnetic Fe_3O_4@C@Cu_2O composites with a beanlike core/shell nanostructure using a self-assembly approach. The average particle size of the prepared magnetic composites was ~420 nm, and the shells were made up of various nanoparticles in the range of 5–10 nm. The carbonaceous layer acted as both the linker and the stabilizer between Fe_3O_4 and Cu_2O. The Fe_3O_4@C@Cu_2O composites exhibit ferromagnetic behavior and were applied to the photocatalytic activity for the degradation of rhodamine-B, methyl orange, and alizarin red relative to commercial Cu_2O and Degussa P-25 powders under visible light irradiation. Approximately complete degradation of rhodamine B, methyl orange, and alizarin red was observed by magnetic Fe_3O_4@C@Cu_2O composites within 210, 180, and 150 min, respectively.

Liu et al. (2011) synthesized one-dimensional (1D) Fe_3O_4/C/CdS coaxial nanochains via a magnetic field–induced assembly and microwave-assisted deposition method. CdS nanoparticles were deposited onto the 1D Fe_3O_4/C nanochains via a rapid microwave-irradiation route to form 1D Fe_3O_4/C/CdS coaxial nanochains. Photocatalytic degradation of rhodamine B was undertaken by the 1D chainlike Fe_3O_4/C/CdS nanocomposites under visible light irradiation of a 500 W Xe lamp. The maximum decolorization efficiency for rhodamine B dye was 94.7% by 1D chainlike Fe_3O_4/C/CdS nanocomposites after visible light irradiation for 30 min. The occurrence of homogeneously deposited CdS nanoparticles and good dispersion of the magnetic nanocomposites in an aqueous solution allow more reactive sites for adsorption and degradation of rhodamine B. The photocatalytic activity of 1D chainlike Fe_3O_4/C/CdS nanocomposites was slightly higher than that of C/CdS composites. Due to excellent magnetic properties the synthesized magnetic photocatalyst was easily recovered and reused for multiple degradation cycles. After 12 photocatalytic degradation cycles there was a slight decrease in the photocatalytic efficiency of the 1D chainlike Fe_3O_4/C/CdS nanocomposites.

The photocatalytic activity of Gd-doped $BiFeO_3$ to degrade rhodamine B under visible light irradiation was studied by Guo et al. (2010), and the result shows that the degradation of rhodamine B enhanced with the changing of the polar behavior of the nanoparticles. Blank test for degradation of rhodamine B under visible light irradiation exhibited a little photolysis, which shows that the self degradation of rhodamine B was extremely slow. The degradation of rhodamine B with the $BiFeO_3$ nanoparticles in the dark condition for 15 h was demonstrated for the absorption of rhodamine B. The obtained result shows that after 15 h irradiation with $BiFeO_3$ nanoparticles, the rhodamine B degradation rate was about 25%. In addition, the photodegradation efficiency of $BiFeO_3$ nanoparticles was extremely lower than that of other visible-light photocatalysts containing bismuth element under the same conditions, such as $BiVO_4$ and Bi_2WO_6. However, rhodamine B dye has a negative charge, whereas $BiFeO_3$ nanoparticles have a positive charge in acid medium, which was beneficial for promoting the adsorption and photocatalytic efficiency.

Kumar et al. (2013) prepared $g\text{-}C_3N_4-Fe_3O_4$ nanocomposites by an in situ growth mechanism, and the results show that monodisperse Fe_3O_4 nanoparticles with diameters ~8 nm were uniformly deposited on $g\text{-}C_3N_4$ sheets. The synthesized $g\text{-}C_3N_4-Fe_3O_4$ nanocomposites were applied for the catalytic degradation of rhodamine B under visible-light irradiation. The $g\text{-}C_3N_4-Fe_3O_4$ nanocomposites were highly stable and revealed good recyclability without loss of apparent photocatalytic activity even after six cycles. Further, the used magnetic nanocomposites were recycled with the external handheld magnet. The photodegradation of the rhodamine B by $g\text{-}C_3N_4-Fe_3O_4$ followed the pseudo-first-order kinetic model. It has also been observed that the optimum photocatalytic activity of $g\text{-}C_3N_4-Fe_3O_4$ at a Fe_3O_4 mass content of 15.2% under visible-light irradiation was almost seven times higher than that of pure $g\text{-}C_3N_4$. The recycled $g\text{-}C_3N_4-Fe_3O_4$ was reused for the degradation of rhodamine B, and the degradation was over 90% of its original photocatalytic activity after six successive experimental runs. The enhanced degradation of rhodamine B by the $g\text{-}C_3N_4-Fe_3O_4$ photocatalysts was due to a synergistic effect including the large surface-exposure area, high visible-light-absorption efficiency, and enhanced charge-separation properties.

9.3.2 Degradation of Organochlorine Pollutants DDT and DDE

1,1,1-Trichloro-2,2-bis(4-chlorophenyl) ethane (DDT) is an organochlorine pollutant that is highly persistent in aquatic environments and responsible for many ecological and health consequences. In the last few years,

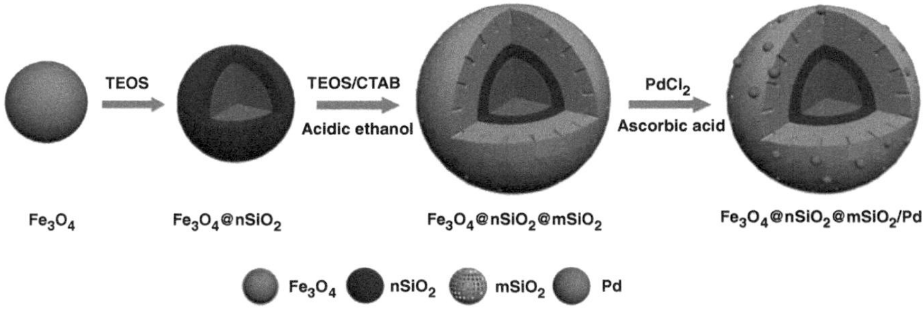

Figure 9.16 ***Schematic illustration of the synthesis process of Fe_3O_4@$nSiO_2$@$mSiO_2$/Pd composites.*** *(Source: Tian et al., 2015; © Elsevier).*

various effective measures have taken place for the degradation and remediation of DDT from soil and water. Recently, Tian et al. (2015) synthesized a triple-functional nanocomposite by integrating superparamagnetic Fe_3O_4 and palladium (Pd) nanoparticles onto mesoporous Fe_3O_4@$nSiO_2$@$mSiO_2$ nanospheres. Fig. 9.16 shows the synthesis process for the Fe_3O_4@$nSiO_2$@ $mSiO_2$/Pd nanocomposites. Fig. 9.17 shows the TEM image of synthesized nanomaterials, which indicated the Fe_3O_4 is in spherical shape. In a typical TEM image of Fe_3O_4@$nSiO_2$@$mSiO_2$/Pd nanocomposites, the Pd

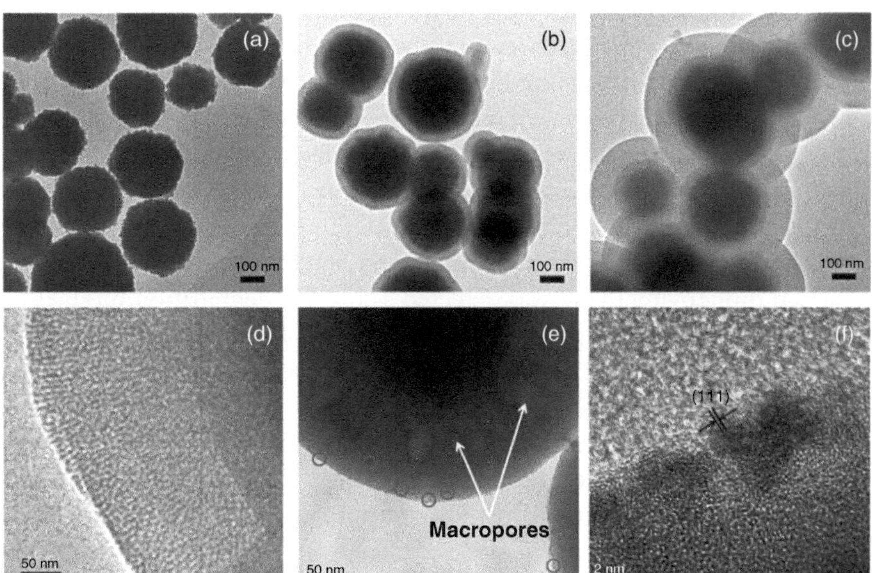

Figure 9.17 ***TEM images of magnetic mesoporous nanoparticles.*** (a) Fe_3O_4, (b) Fe_3O_4@ $nSiO_2$, (c) Fe_3O_4@$nSiO_2$@$mSiO_2$, and (e) 1.0Pd/Fe_3O_4@$nSiO_2$@$mSiO_2$; (d) and (f) are high-resolution TEM images of (c) and (e), respectively. Circles in (e) highlight Pd nanoparticles. *(Source: Tian et al., 2015; © Elsevier).*

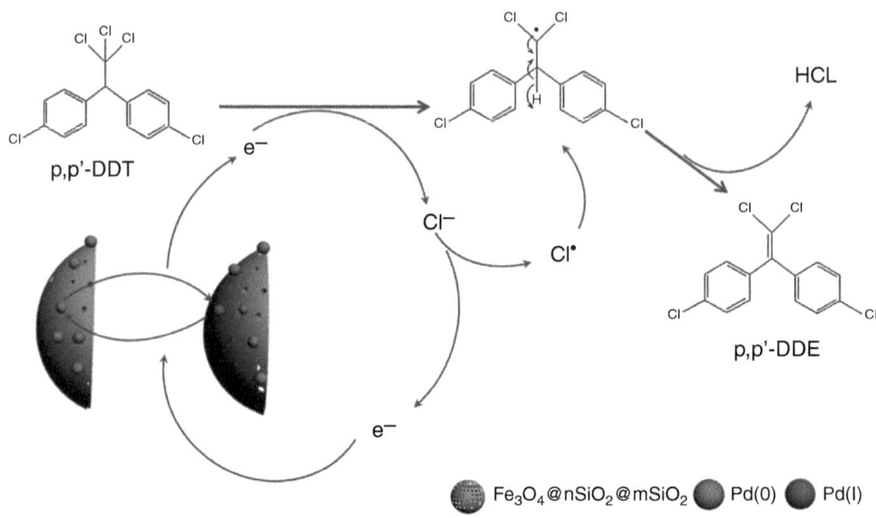

Figure 9.18 ***Proposed mechanisms for DDT degradation on Pd/Fe_3O_4@nSiO_2@mSiO_2 nanocomposites.*** *(Source: Tian et al., 2015; © Elsevier).*

nanoparticles were clearly formed and were well dispersed on the mesoporous silica shell. Some Pd nanoparticles tended to agglomerate with each other, covering some surface area or mesopores of Fe_3O_4@nSiO_2@mSiO_2 nanospheres. These magnetic mesoporous materials were used for capturing and catalytically degrading DDT in water. Over these nanocomposites, DDT and its metabolite, 1,1-dichloro-2,2-bis(4-chlorophenyl) ethylene (DDE), could be quickly enriched and completely degraded at as low as 150°C. Fig. 9.18 shows the proposed mechanisms for DDT degradation on Fe_3O_4@nSiO_2@mSiO_2/Pd nanocomposites. Fig. 9.19 displays the proposed degradation pathways of DDT over magnetic mesoporous catalysts. Bibenzene and diphenylmethane was detected as the last degradation product of DDT. Due to the magnetic nature of the catalysts it can be easily separated and reused further in the degradation process, which is convenient, energy-saving, and just requires ambient pressure and mild reaction conditions.

Tian et al. (2014) successfully synthesized a novel multifunctional microsphere with an iron oxide-improved mesoporous silica shell and a Fe_3O_4@SiO_2 core by a hydrothermal method and impregnation process. Fig. 9.20 depicts the schematic presentation of the synthesis process of Fe_3O_4@SiO_2@mSiO_2–Fe core-shell microspheres. Fig. 9.21 shows the representative TEM images of prepared magnetic microspheres. TEM images of Fe_3O_4@SiO_2@mSiO_2–Fe samples show that the functionalized microspheres have the sandwich structure, and a few magnetic nanoparticles were encapsulated

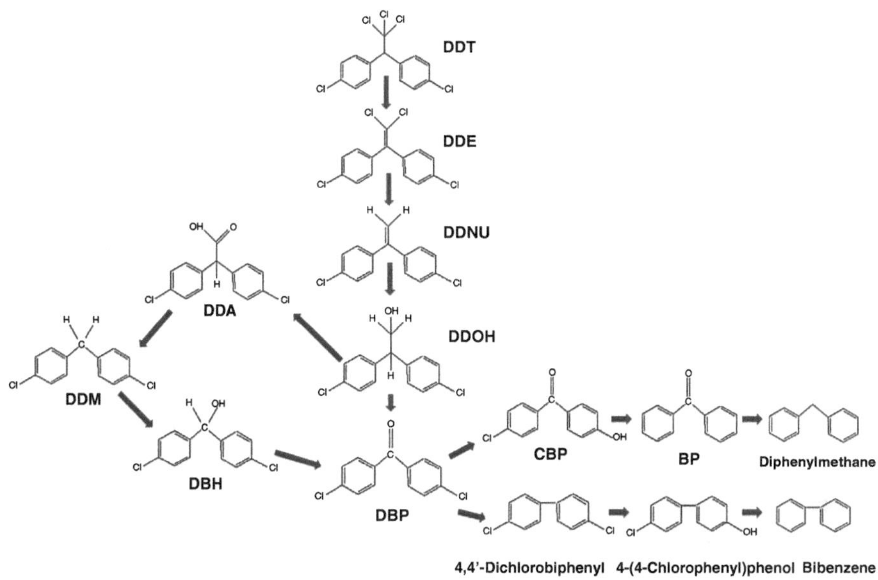

Figure 9.19 ***Proposed degradation pathways of DDT over magnetic mesoporous catalysts.*** Bibenzene and diphenylmethane was detected as the last degradation product of DDT. *(Source: Tian et al., 2015; © Elsevier).*

on the surface pore. The synthesized Fe_3O_4@SiO_2@mSiO_2–Fe core-shell microspheres were applied as a catalyst for the removal of 1,1,1-trichloro-2,2-bis(4-chlorophenyl) ethane (DDT) and its derivatives, ie, 1,1-dichloro-2,2-bis(4-chlorophenyl)ethane (DDD) and 1,1-dichloro-2,2-bis(4-chlorophenyl) ethylene (DDE) from aqueous solutions, and the operating conditions were investigated to fully degrade the pollutants. The result shows that the DDT, DDD, and DDE could be quickly and effectively removed from aqueous media in 60 min and completely dechlorinated at 350°C by Fe_3O_4@SiO_2@mSiO_2–Fe. Fig. 9.22 shows the photographs of a magnetic separation and redispersion process using Fe_3O_4@SiO_2@mSiO_2–5%Fe microspheres during the degradation of DDT in aqueous medium. Fig. 9.23 demonstrates the removal efficiencies of DDT, DDD, and DDE on Fe_3O_4@SiO_2@mSiO_2 materials with different iron contents. In addition, due to the superparamagnetic nature of the catalyst Fe_3O_4@SiO_2@mSiO_2–Fe microspheres could be separated and collected easily and rapidly using an external magnet.

9.3.3 Degradation of Pharmaceuticals

Magnetic TiO_2/Fe_3O_4 and TiO_2/SiO_2/Fe_3O_4 nanoparticles were prepared by an ultrasonic-assisted sol–gel method (Álvarez et al., 2010). Magnetic

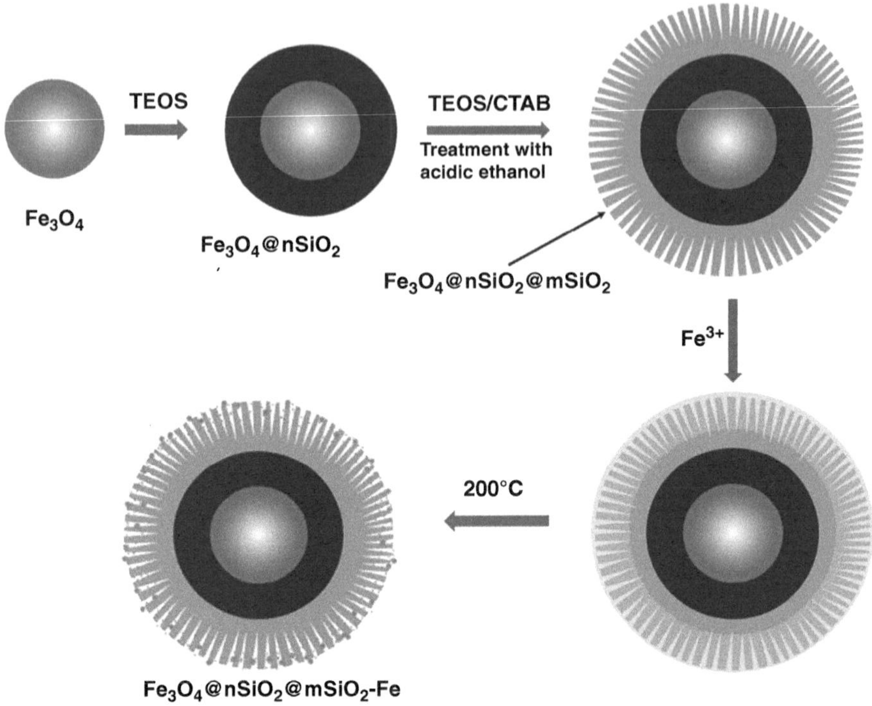

Figure 9.20 ***Schematic illustration of the synthetic procedure of Fe_3O_4@SiO_2@mSiO_2–Fe core–shell microspheres.*** *(Source: Tian et al., 2014; © Elsevier).*

nanocomposites were characterized by XRD, TEM, and BET surface areas as well as the SQUID magnetometer. Structure analyses indicated that TiO_2/Fe_3O_4 nanoparticles presented a core-shell structure with an anatase TiO_2 coating wrapped around the Fe_3O_4 surface. Nanoparticles of $TiO_2/SiO_2/Fe_3O_4$ showed a ternary structure with a core of Fe_3O_4, a SiO_2 mesosphere, and an anatase TiO_2 crust. The surface area of magnetic iron oxide, Degussa P25 TiO_2, TiO_2/Fe_3O_4, and $TiO_2/SiO_2/Fe_3O_4$ nanoparticles were 42, 47, 27, and 22 m^2/g, respectively. TiO_2/Fe_3O_4 and $TiO_2/SiO_2/Fe_3O_4$ nanoparticles exhibited magnetic properties with saturation magnetization about 40 emu/g and low remnant magnetization and coercivity, meaning it can be easily separated from aqueous solutions. Both TiO_2/Fe_3O_4 and $TiO_2/SiO_2/Fe_3O_4$ catalysts showed the ability to catalyze the photodegradation of acetaminophen and four other pharmaceutical and personal care products (antipyrine, caffeine, metoprolol, and bisphenol A) from an aqueous solution.

Teixeira et al. (2016) studied the heterogeneous photocatalytic degradation of 14 pharmaceuticals present in a wastewater treatment plant effluent.

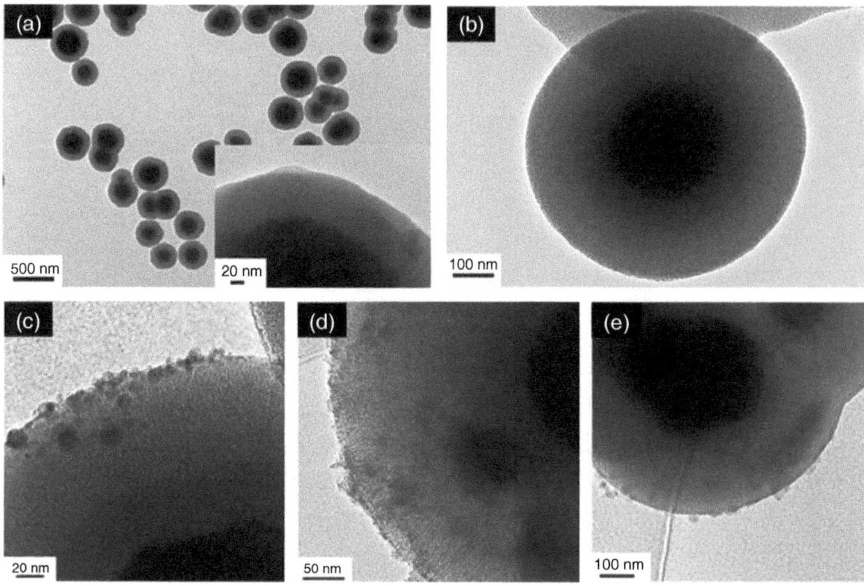

Figure 9.21 ***Representative TEM images of prepared magnetic microspheres.*** (a) Fe_3O_4@ $nSiO_2$ (inset: TEM image at high magnification), (b) Fe_3O_4@SiO_2@$mSiO_2$, (c) Fe_3O_4@SiO_2@$mSiO_2$–5%Fe, (d) Fe_3O_4@SiO_2@$mSiO_2$–10%Fe, and (e) Fe_3O_4@SiO_2@$mSiO_2$–15%Fe. *(Source: Tian et al., 2014; © Elsevier).*

Figure 9.22 ***Photographs of a magnetic separation and redispersion process using Fe_3O_4@SiO_2@$mSiO_2$–5%Fe microspheres.*** *(Source: Tian et al., 2014; © Elsevier).*

They studied the effect of various parameters such as UVA irradiation, dose of TiO_2 P25, or ZnO nanoparticles on the degradation of pharmaceuticals. The result showed that with ZnO nanoparticles more than 95% of pharmaceuticals were degraded within 40 min UVA irradiation. However, it was very tough to degrade using TiO_2 P25 and took more than six times longer to reach the same level.

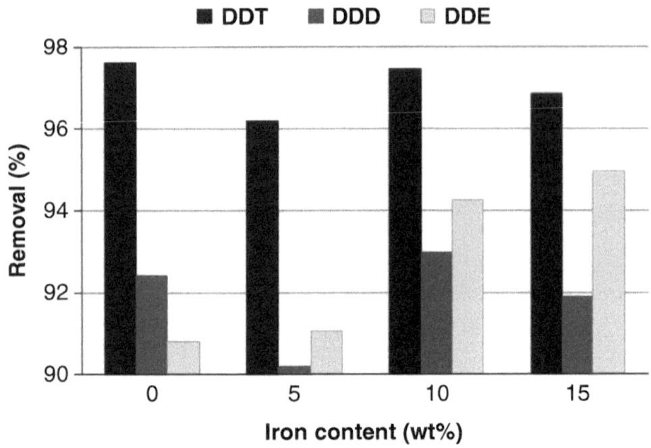

Figure 9.23 ***Removal efficiencies of DDT, DDD, and DDE on Fe_3O_4@SiO_2@mSiO_2 materials with different iron contents.*** *(Source: Tian et al., 2014; © Elsevier).*

9.3.4 Reduction of Chromium [Cr(VI)]

Idris et al. (2012) synthesized magnetic separable photocatalyst beads containing maghemite nanoparticles (γ-Fe_2O_3) in polyvinyl alcohol (PVA) polymer and used them in the reduction of Cr(VI) to Cr(III) in an aqueous solution under sunlight. The synthesized magnetic photocatalyst shows superparamagnetic properties due to the core γ-Fe_2O_3, which was easily recyclable from water by external magnet, and after washing it was regenerated and reused seven times in the photocatalytic reduction of Cr(VI). Photocatalysis over a semiconductor oxide such as γ-Fe_2O_3 was initiated by the absorption of a photon, producing electron–hole pairs, so this electron–hole pair was responsible for the Cr(VI) reduction to Cr(III). Under exposure to sunlight, the electron and hole generated were capable of initiating the triple-electron reaction of Cr(VI) to stable Cr(III). Various doping effects of γ-Fe_2O_3 was studied on the reduction of Cr(VI) by magnetic beads. Hence the concentration of γ-Fe_2O_3 was varied from 8 to 27% (v/v) and the obtained values show that the 8% (v/v) γ-Fe_2O_3 containing beads have highest catalytic properties within 30 min under sunlight. It was also proposed that the addition of polyvinyl alcohol has improved the bead properties and life cycle of the beads.

Idris et al. (2011) studied the physical adsorption and photocatalytic reduction of Cr(VI) in magnetic separable beads. The kinetics of the photocatalytic degradation was studied by optimizing operating parameters such as photocatalyst dose and initial Cr(VI) concentration. The catalytic reduction of Cr(VI) was initially fast with increase in catalyst dose as it translated

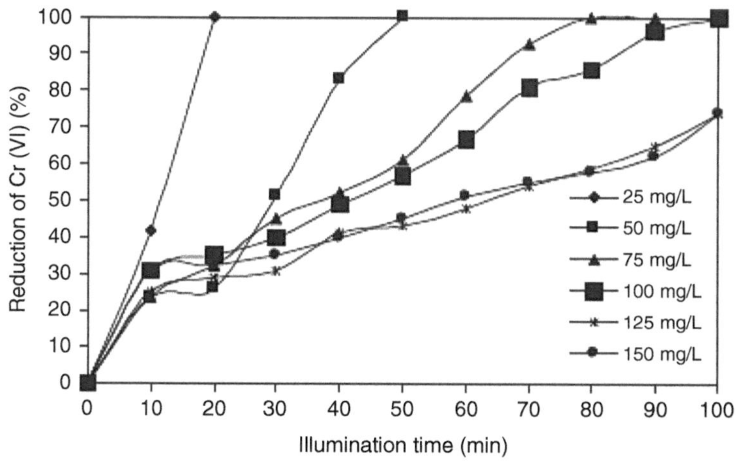

Figure 9.24 ***Effect of time and initial concentration on photocatalytic reduction of Cr(VI) (catalyst dosage 16% (v/v); pH 1).*** *(Source: Idris et al., 2011; © Elsevier).*

into an increase in the number of available active sites. However, it was observed that the increasing initial concentration of Cr(VI) in solution reduces the rate of reduction of Cr(VI) into Cr(III) (Fig. 9.24). The kinetic study shows that the reduction was followed to pseudo-first-order reactions, and magnetic bead surface was the rate-controlling step in the entire reduction process. The maximum removal efficiency of Cr(VI) at initial concentrations of 25, 50, 75, 100, 125, and 150 mg/L was found to be 100, 100, 100, 100, 70, and 70%, respectively, at equilibrium. Fig. 9.25 shows the effect of catalyst concentration on photoreduction of Cr(VI). The effect of photocatalyst loading on the photoreduction of Cr(VI) was studied by varying catalyst concentrations from 8 to 25% (v/v). The 100% reduction of Cr(VI) was obtained within 50 min by 16% (v/v) loading of photocatalyst. Further increase in photocatalyst loading did not cause much change in the reduction of pollutants in aqueous medium. In addition, the magnetic beads were recovered by elution with sodium hydroxide, and they were used for three adsorption–desorption cycles.

In an attempt to reduce the Cr(VI) contamination in water, Idris et al. (2010) synthesized magnetically separable photocatalyst beads containing nano-sized iron oxide in alginate polymer. The magnetic photocatalyst beads were used in slurry-type reactors to decontaminate the Cr(VI) from aqueous solutions. These synthesized magnetic photocatalyst beads were sunlight-driven photocatalysts. Fig. 9.26 shows the photograph of magnetic photocatalyst alginate beads attracted by a magnet. The photocatalytic

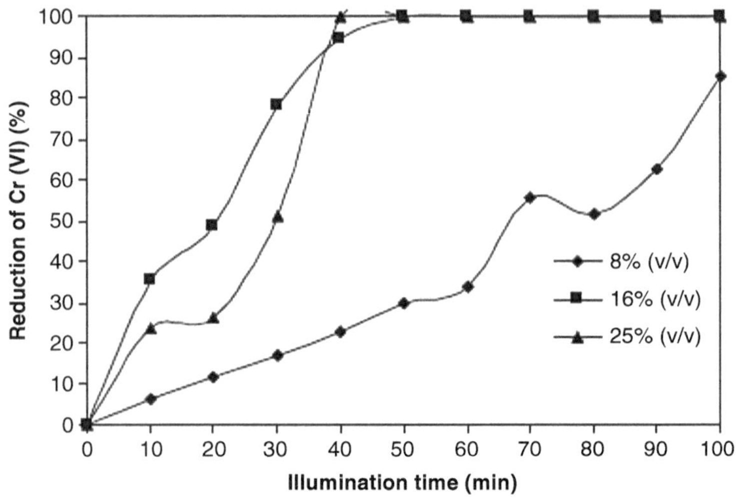

Figure 9.25 ***Effect of catalyst concentration on photoreduction of Cr(VI) [initial Cr(VI) concentration is 50 mg/L; pH 1].*** *(Source: Idris et al., 2011; © Elsevier).*

Figure 9.26 ***Photograph of magnetic photocatalyst alginate beads attracted by a magnet.*** *(Source: Idris et al., 2010; © Elsevier).*

reduction of Cr(VI) was completed within just 50 min under sunlight irradiation due to strong photocatalytic activity of magnetic beads. However, the reduction rate for Cr(VI) was just about 10% when the sunlight was removed from the system. Fig. 9.27 shows the photocatalytic activity of magnetic photocatalyst beads and alginate beads under sunlight irradiation. In addition, there was no catalytic reduction of Cr(VI) without the magnetic beads, as the Cr(VI) has strong stability in sunlight. The highest reduction of

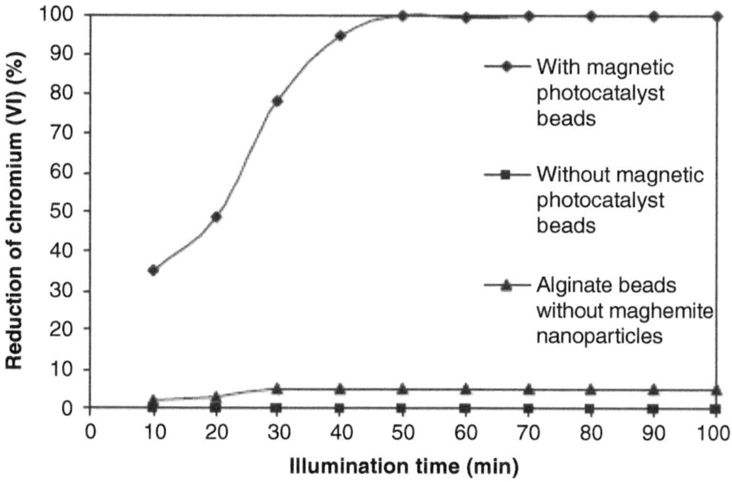

Figure 9.27 ***The photocatalytic activity of magnetic photocatalyst beads and alginate beads under sunlight irradiation.*** *(Source: Idris et al., 2010; © Elsevier).*

Cr(VI) onto magnetic beads in the liquid phase was due to the adsorption on the active surface of beads and simultaneous reduction. Furthermore it has been reported that the reduction rate was very fast in highly acidic medium at pH 1–2. The obtained result of photocatalytic reduction shows that the sunlight irradiance plays a crucial role in the reduction of Cr(VI) and suggests that the photocatalytic process for the removal of notorious Cr(VI) may occur via direct reduction by the photogenerated electrons. In a typical photocatalysis reaction over a semiconductor, an oxide such as γ-Fe_2O_3 is initiated by the absorption of a photon, producing electron–hole pairs. In this photocatalytic reduction process, the Cr(VI) reduction probably occurred via three subsequent one-electron transfer processes, ending in Cr(III), which is the stable final product. Due to the magnetic nature of the photocatalytic beads, it was easily separated from the water, making the entire treatment process cost effective.

9.3.5 Oxidation of Arsenic [As(III)]

Yu et al. (2013) prepared a novel γ-Fe_2O_3–TiO_2 binary oxide magnetic nanoparticle by a simple coprecipitation method. The photocatalytic oxidation property of TiO_2 and the high adsorption capacity and magnetic property of γ-Fe_2O_3 was very helpful in the oxidation and adsorption of As(III) from aqueous solutions. The oxidation of As(III) to As(V) was investigated by varying the solution pH, common coexisting ions and initial concentration of As(III). It has been found that the γ-Fe_2O_3–TiO_2 nanoparticle has efficient capacity to oxidize the As(III) to As(V) by dissolved O_2 under UV light, and at the same time the As(V) was adsorbed on the surface of nanomaterials. As pH of the solution increases from 3 to 9, the removal percentage increased from 35.56 to 64.38% for arsenic by the γ-Fe_2O_3, while the percent removal of arsenic by the γ-Fe_2O_3–TiO_2 nanoparticles decreased from 99.82 to 91.45% with increasing solution pH. The γ-Fe_2O_3–TiO_2 nanoparticles show relatively high removal efficiency for As(III) in comparison with γ-Fe_2O_3 nanoparticles. This difference in the removal efficiency for As(III) of γ-Fe_2O_3 and γ-Fe_2O_3–TiO_2 nanoparticles was due to the different removal mechanism for As(III). The result shows that the phosphate was the greatest competitor with arsenic for the active adsorption sites on the γ-Fe_2O_3–TiO_2 nanoparticles. The effect of anions for arsenic removal was phosphate > silicate > bicarbonate in the aqueous solutions (Fig. 9.28). The recycling of the loaded γ-Fe_2O_3–TiO_2 nanoparticles was carried out with 0.1 M NaOH by stirring the solution for 3 h. The regenerated γ-Fe_2O_3–TiO_2 nanoparticles were capable of removal efficiency up to five adsorption–desorption cycles, without losing the photocatalytic property, and the removal percentage was above 84%.

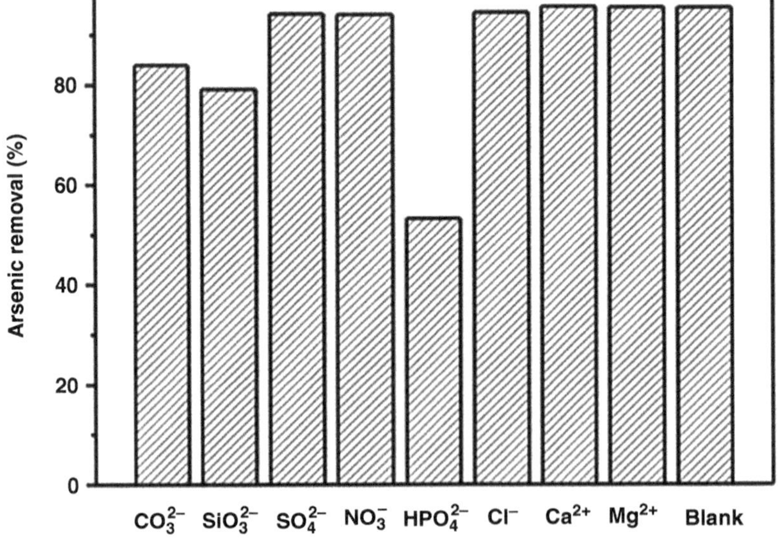

Figure 9.28 ***Effects of coexisting ions (1 mM) on As(III) removal at fixed initial As(III) concentration 1.01 mg/L, adsorbent dosage 0.5 g/L, solution pH 7.0 ± 0.1, and reaction time 3 h.*** *(Source: Yu et al., 2013; © Elsevier).*

9.3.6 Degradation of Phenol

Zhang et al. (2009) synthesized Fe_3O_4 magnetic nanoparticles (MNPs) with diameters about 10 nm and used them to remove phenol and aniline from aqueous solutions. The study shows that the acidic or neutral pH favors the decomposition of phenol and aniline in the presence of MNPs and H_2O_2. Phenol and aniline were completely degraded in 6 h at 308 K, and the total organic carbon abatement efficiency for phenol and aniline were 42.79 and 40.38%, when the concentration of Fe_3O_4 MNPs and H_2O_2 were 5 g/L and 1.2 M, respectively. It was proposed that phenol and aniline might be degraded by the hydroxyl free radicals (·OH) released from H_2O_2 in the presence of Fe_3O_4 MNPs as catalysts. Highest degradation of phenol and aniline were observed under acidic and neutral conditions, and the removal efficiency was decreased with an increase in the solution pH. The total organic carbon abatement efficiency was about 35% under acidic and neutral conditions, and when pH was above 7, the total organic carbon abatement decreased sharply to about 5% at pH 11. It was suggested that under the acidic or neutral condition, protonated surface of iron oxide was beneficial to the formation of metal oxide–phenol/aniline inner-sphere complexes, which promoted the degradation of pollutants; while under the alkaline condition the combination of OH^- with Fe^{3+} and Fe^{2+} on the surface of the MNPs might separate the reactants from surface metal centers. This leads to the decrease in the degradation activity of the oxides at a higher pH, thus making harsh conditions for the formation of the protonated inner-sphere complex. Formic acid, acetic acid, fumaric acid, and hydroquinone were detected as intermediate products in the decomposition reactions. The synthesized Fe_3O_4 MNPs revealed superparamagnetism and were readily separated from solution by applying an external magnetic field.

Recently, Wang et al. (2015) synthesized iron–copper bimetallic nanoparticles embedded within ordered mesoporous carbon composite catalysts (CuFe-MC) via one-pot block-copolymer self-assembly strategy. The characterization of prepared CuFe-MC composites showed the catalyst was ordered 2D hexagonal mesostructures and iron–copper nanoparticles highly dispersed in the matrix of ordered mesoporous carbon. The catalyst was applied for the degradation of recalcitrant organic pollutants from water as a heterogeneous Fenton catalyst and showed a promising application in the degradation of nonbiodegradable organic contaminants. Namely eight organic pollutants were chosen for the degradation analysis, namely, phenol; benzoic acid; bisphenol A; 2,4,6 trichlorophenol; imidacloprid; ketoprofen; methylene blue; and methyl orange from aqueous solutions. High-efficiency removal of total organic carbon (TOC) for each organic contaminant was achieved by using CuFe-MC as a catalyst, which was higher than that by iron oxide at the same reaction condition. Because of the high BET surface area (639 m^2/g) and mesostructure, the CuFe-MC showed high adsorption of bisphenol A from aquatic media than Fe-MC and Cu-MC. Fig. 9.29 shows the possible pathways of bisphenol A degradation induced by $^{\bullet}OH$ radicals in the CuFe-MC/H_2O_2 system. Further analysis by electron spin resonance (ESR) and high performance liquid chromatography (HPLC) indicated that the concentration of generated hydroxyl radicals ($^{\bullet}OH$) with CuFe-MC was much higher than Fe-MC and Cu-MC.

Nouri and Sargolzaei (2014) synthesized SnO_2/Fe_3O_4 nanoparticles and studied the removal of phenol red in a photocatalytic reactor. The degradation kinetic for phenol red in water followed the pseudo-second order model. In order to determine the optimal values of each of the parameters and the optimal conditions of the process, parameters were introduced to response surface methodology. The absorption was stronger at basic medium due to the intensive anionic feature of phenol red molecules, thus increasing absorption efficiency and consequently photocatalytic efficiency of the process. Other factors were the efficiency increase in basic pH and in the amount of OH^- ions that were the precursor of $^{\bullet}OH$ radicals. The greater radiation intensity of UV light causes more $^{\bullet}OH$ radical production and thus the greater the removal efficiency of phenol red.

Heterogeneous 3D magnetic $ZnFe_2O_4/MnO_2$ hybrid catalysts with two hierarchical nanostructures were synthesized by Wang et al. (2014), and their catalytic activities were investigated in terms of peroxymonosulfate activation for phenol degradation. Kinetics of the degradation of phenol were fit to first-order kinetic models, and the activation energies for microsphere/nanosheet hierarchical structure and urchin-like structure were calculated to be 41.7 and 49.4 kJ/mol, respectively. The mechanism of catalytic reaction for activation of peroxymonosulfate was illustrated by competitive radical tests and electron paramagnetic resonance spectra. Catalytic degradation of phenol occurred by the formation of sulfate radicals as the major reactive species, and the generation of sulfate radicals was induced by hydroxyl radicals produced at the initial stage.

9.3.7 Degradation of Chlorophenol

Xu and Wang (2012) applied Fe_3O_4 nanoparticles as a catalyst for the Fenton-like degradation of 2,4-dichlorophenol (2,4-DCP) in aqueous solution. The catalytic performance of Fe_3O_4 nanoparticles revealed good properties

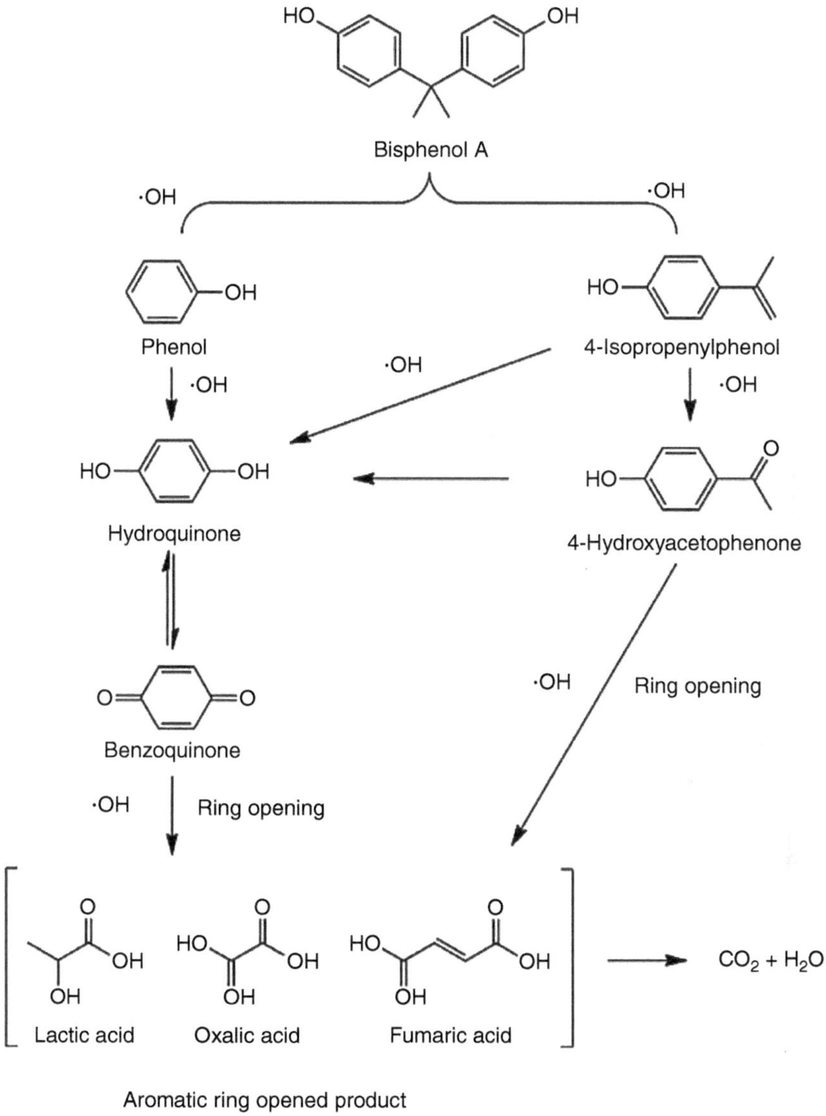

Figure 9.29 ***Possible pathways of bisphenol A degradation induced by •OH radicals in the CuFe–MC/H_2O_2 system.*** The catalyst CuFe–MC was heated at 800°C. *(Source: Wang et al., 2015; © Elsevier).*

for the degradation and mineralization of 2,4-DCP, achieving complete decomposition of 2,4-DCP and 51% of TOC removal after 180 min at reaction conditions of H_2O_2 12 mM; Fe_3O_4 nanoparticles 1.0 g/L; 2,4-DCP 100 mg/L; pH 3.0; and T 30°C. It was observed that within 180 min reaction the concentration of 2,4-DCP diminished very slowly with only

30 mM H_2O_2, and even in 24 h only 13% removal was obtained, which is ascribed to the weak oxidation potential of H_2O_2 compared with hydroxyl and perhydroxyl radicals. Less than 15% degradation of 2,4-DCP was observed in control reactions with 1.5 g/L Fe_3O_4 nanoparticles, and about 0.027 mM free chloride was detected in the solution after 24 h reaction, which was mainly expected to surface adsorption as well as reductive dechlorination. The effect of different reaction parameters such as initial pH; H_2O_2 dosage; Fe_3O_4 nanoparticle addition; initial concentration of 2,4-DCP; and temperature on two-stage first-order kinetics of 2,4-DCP degradation was studied. A high utilization efficiency of H_2O_2 calculated as 73% was observed. According to the analyses of iron leaching, reactive oxidizing species, and degradation intermediates, a possible mechanistic step of 2,4-DCP degradation dominated by •OH reactions (especially by free ˙OH in the bulk solution) were proposed. Besides, stability and reusability of Fe_3O_4 nanoparticles were tested.

Choi et al. (2011) reported a novel photofunctional magnetic nanoparticle that generates singlet oxygen in high quantum yield. The photocatalytic degradation of 2,4,6-trichlorophenol with the photofunctional magnetic nanoparticles was carried out. In order to check the application possibility for water purification, the photocatalytic oxidation reaction of 2,4,6-trichlorophenol in an aqueous solution was conducted at pH 10 under visible light irradiation of the Xe lamp. Only the photofunctional magnetic nanoparticles efficiently degraded the 2,4,6-trichlorophenol solution under the visible light, while the light-induced decomposition of 2,4,6-trichlorophenol without the magnetic nanoparticles and the decomposition of 2,4,6-trichlorophenol induced by the photofunctional magnetic nanoparticles without the light were not observed.

Rashid et al. (2015) reported facile synthesis of stable magnetic core-shell-shell $Fe_3O_4/SiO_2/TiO_2$ nanoparticles and their effectiveness for photocatalysis. The photocatalytic degradation of 2-chlorophenol was investigated, and reaction parameters for best catalyst performance were optimized to obtain maximum removal efficiency. It has been reported that the complete degradation of 25 mg/L 2-chlorophenol was achieved by a catalyst dose of 0.5 g/L within 130 min under UV irradiation, while 97.2% degradation of 50 mg/L 2-chlorophenol was achieved within 3 h. The degradation kinetics was modeled by pseudo-first-order kinetics and Hugul's kinetic equations. The obtained result shows that the Hugul's kinetics was best fitted to the present catalysis system than the pseudo-first-order reaction kinetics. Maximum photocatalytic efficiency was observed at pH 3, which was due to the electrostatic binding of 2-chlorophenol molecules with a positively charged surface of TiO_2 particles. However, the low degradation rate in basic medium (pH 9) was due to the electrostatic repulsion between negatively charged TiO_2 surface and phenolate ions.

Zeng et al. (2015) synthesized yolk–shell Co_3O_4@MOFs and applied for the catalytic degradation of 4-chlorophenol by activated peroxymonosulfate, which was selected as a model reaction, and various control experiments were conducted. The adsorption and degradation profiles of 4-chlorophenol against the reaction time in various situations was investigated. The degradation rate was not observed when peroxymonosulfate alone was applied for the catalytic degradation of 4-chlorophenol. The degradation rate was increased when the Co_3O_4 nanoparticles were applied in the reaction, which means that the combination of Co oxides with peroxymonosulfate did possess a certain degradability for 4-chlorophenol. The degradation rate of 4-chlorophenol was highly increased when yolk–shell Co_3O_4@MOFs catalysts and peroxymonosulfate was applied. The combination of yolk–shell Co_3O_4@MOFs catalysts and peroxymonosulfate shows that almost 100% decomposition was achieved within 60 min, while only 59.6% decomposition was observed for bare Co_3O_4 nanoparticles under the same conditions. However, in the absence of peroxymonosulfate, the degradation of 4-chlorophenol was only 19% by the yolk–shell Co_3O_4@MOFs composite even after 150 min, which was much slower than that in the presence of yolk–shell Co_3O_4@MOFs together with peroxymonosulfate.

9.3.8 Degradation of Nitrophenol

P-nitrophenol ($C_6H_5NO_3$) is a weak acid (pKa ~ 7.2) with a molecular weight equal to 139.11 g/mol, with a colorless appearance in aqueous solutions, whereas its phenolic salt is bright yellow. Recently, Obeid et al. (2015) synthesized a magnetic adsorbent by encapsulation of magnetic functionalized nanoparticles using alginate as a green biopolymer matrix and used for the adsorption of p-nitrophenol. The effect of a cationic surfactant, cetylpyridinium chloride, was studied on the adsorption behavior of magnetic beads and thus to promote their adsorption efficiency. Fig. 9.30 shows that the magnetic adsorbent has efficient saturation magnetization that can be easily recovered by external magnet. The effect of different optimization parameters such as initial concentrations of p-nitrophenol, cetylpyridinium chloride, and contact time and solution pH value on the adsorption of p-nitrophenol was investigated. The maximum removal of p-nitrophenol was found to be 140 mg/L. The beads were regenerated by 10 mL ethanol/

Figure 9.30 ***Photograph of the magnetic alginate beads attracted with a small permanent NdFeB magnet (diameter, 2.5 cm; height, 1 cm).*** *(Source: Obeid et al., 2015, © Elsevier).*

water solution and separated by external handheld magnet. It is a known fact that combination of magnetic and adsorption properties in the same material is an interesting challenge which could overcome the recovery problems of pollutant-loaded adsorbents.

Recently, Dong et al. (2015) synthesized a novel fibrous nano-silica (KCC-1)-based nanocatalyst (Ni@Pd/KCC-1) by modifying KCC-1 using Ni@Pd core-shell nanoparticles and applied for the catalytic reduction of 4-nitrophenol and hydrodechlorination of 4-chlorophenol. The Ni@Pd NPs possessed a magnetic nickel (Ni) core with palladium (Pd) shell structural composite. Fig. 9.31 shows the representative TEM images of the synthesized magnetic nanocatalysts. The use of Ni in the catalyst makes the nanocatalyst easy to recycle and regenerate after the completion of reaction without reducing the catalytic performance. The coating of Ni@Pd nanocatalysts KCC-1 nanospheres was helpful to check the aggregation and to enhance the reactive surface sites. The Ni@Pd/KCC-1 nanocatalyst displayed excellent catalytic activities for both the reduction of 4-nitrophenol and hydrodechlorination of 4-chlorophenol under green conditions. The UV-visible spectra for the reduction of 4-nitrophenol to 4-aminophenol over the prepared Ni@Pd/KCC-1 nanocatalyst and its kinetics have been represented in Fig. 9.32.

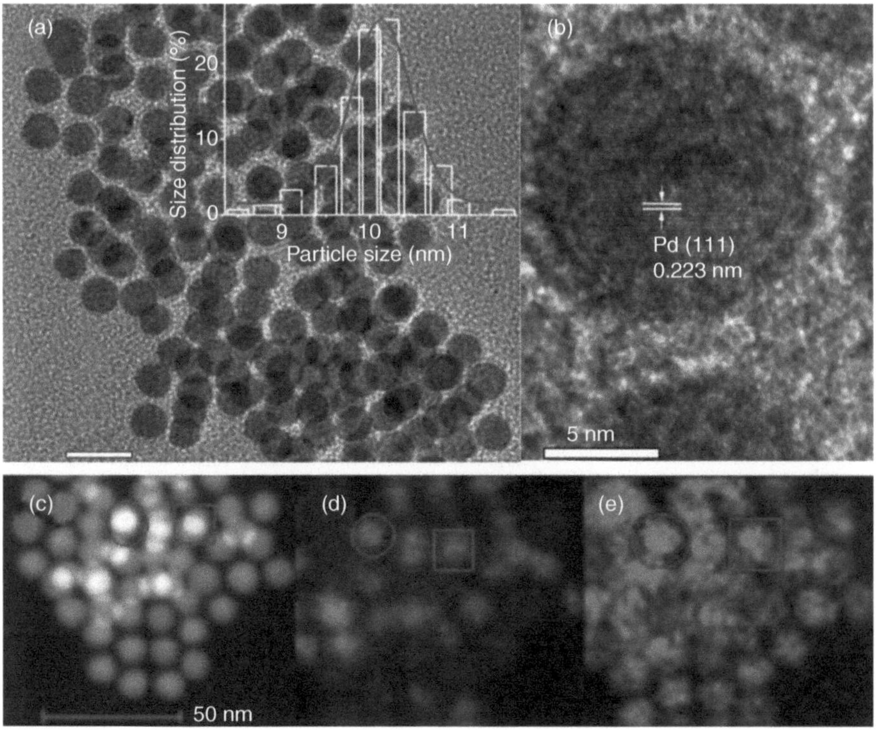

Figure 9.31 (a) TEM image, (b) HRTEM image, (c) the HAADF–STEM image, (d) the corresponding high resolution Ni mapping, and (e) the corresponding high-resolution Pd mapping of Ni@Pd NPs. *(Source: Dong et al., 2015; © Elsevier).*

9.3.9 Photocatalytic Degradation of the Insecticide

Yang et al. (2010) synthesized the recyclable Fe_3O_4/hydroxyapatite composite nanoparticles via the homogeneous precipitation method. The characterization of particles revealed that the nanoparticles were spherical in shape and had a unique size of ~25 nm. The synthesized Fe_3O_4/hydroxyapatite nanocomposite applied for the catalytic degradation of the diazinon insecticide under UV irradiation. At solution pH above 2.6 the diazinon was negatively charged, whereas hydroxyapatite was positively charged at a pH below 6.8. Thus, it leads to the strong interaction between hydroxyapatite and insecticide at pH 5.5, which was between 2.6 and 6.8, exhibiting a high degradation rate. The photocatalytic degradation rate was higher in the presence of Fe_3O_4/hydroxyapatite nanoparticles, and it was found that 75% removal of diazinon was achieved in 60 min of contact time. The degradation process was due to the band gap values of Fe_3O_4 and hydroxyapatite which is reported as 5.5 eV and 3.9 eV, respectively. Thus, hydroxyapatite

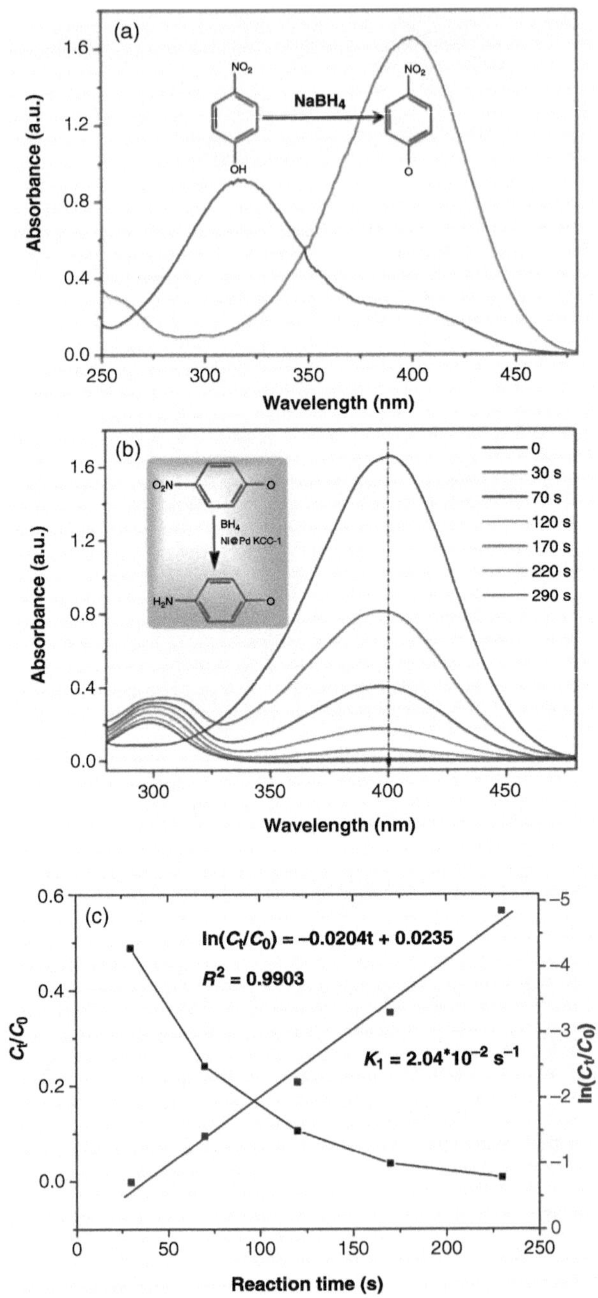

Figure 9.32 (a) UV–vis spectra of 4-nitrophenol before and after adding $NaBH_4$ solution, (b) the successive reduction of 4-nitrophenol to 4-aminophenol over the prepared Ni@Pd/KCC-1 nanocatalyst, and (c) plots of C_t/C_0 and $\ln(C_t/C_0)$ versus reaction time for the reduction of 4-nitrophenol over Ni@Pd/KCC-1 nanocatalyst. *(Source: Dong et al., 2015; © Elsevier).*

with a low band gap was a suitable candidate to be used as a photocatalyst. The superparamagnetic nature of the photocatalyst was suitable for its separation from liquid phase. The recycled magnetic nanocomposite was reused seven times, and the degradation capacity was higher than 95%.

9.3.10 Degradation of Oil

The photocatalytic degradation of organic compounds of olive mill wastewater was investigated by using core-shell-shell $Fe_3O_4/SiO_2/TiO_2$ nanoparticles as a catalyst (Ruzmanova et al., 2013). The preparation of nanoparticles was performed by coating onto Fe_3O_4 nanoparticles with SiO_2 and TiO_2 layers. The photocatalytic degradation of olive mill oil was conducted in a batch photoreactor irradiated by a UV lamp of 45 W. The catalytic degradation of organic content of olive mill wastewater was evaluated by COD measurements. The obtained results showed a high activity of synthesized nanoparticles for photocatalytic degradation of the olive mill wastewater organic compounds. Moreover, the recovery of the magnetic photocatalyst by using a handheld magnet was easy and cost effective.

Peng et al. (2014) synthesized the $TiO_2/SiO_2/Fe_3O_4$ core-shell nanoparticles with good photocatalytic activity and applied for the degradation of crude oil wastewater. Fig. 9.33 shows the TEM micrographs of the synthesized nanoparticles. The average particle size of the synthesized materials was 10–30 nm. When the $TiO_2/SiO_2/Fe_3O_4$ core-shell nanoparticles are added to the crude oily wastewater, the oil concentration was reduced to lower than 30 ppm within 20 min for the case of initial oil concentration less than 350 ppm. Fig. 9.34 shows the absorption spectra of crude oil wastewater exposed to UV light, photocatalytic performance of $TiO_2/SiO_2/Fe_3O_4$ nanoparticles in comparison with pure titania P25, and kinetics of degradation of crude oil. It was reported that that the $TiO_2/SiO_2/Fe_3O_4$ nanoparticles show superparamagnetic properties and were easily recyclable from water by external magnet. The recovered $TiO_2/SiO_2/Fe_3O_4$ nanoparticles were reused in the degradation process and were highly effective up to three catalytic cycles for crude oil wastewater degradation.

9.3.11 Disinfection of Pathogenic Bacteria

Disinfection of toxic bacteria under solar light by photocatalysis is a reliable technique which can be used in drinking water treatment. Semiconductor photocatalysts based on TiO_2 nanoparticles when suspended in water and irradiated with UV light generates •OH free radicals. The •OH free radicals have been proven to be highly toxic toward microorganisms.

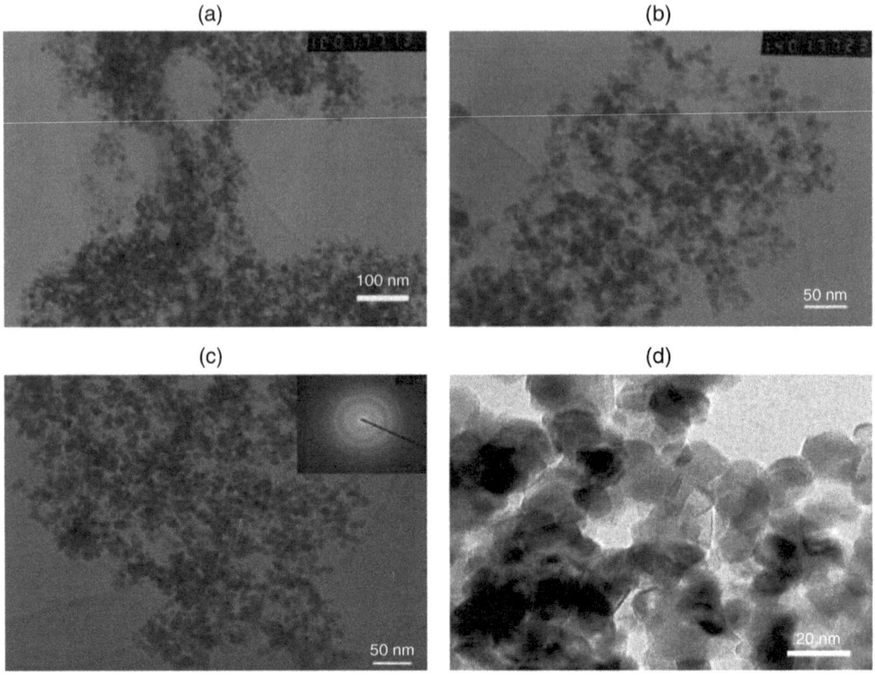

Figure 9.33 TEM micrographs of (a) Fe_3O_4, (b) SiO_2/Fe_3O_4 nanoparticles, (c) $TiO_2/SiO_2/Fe_3O_4$ nanoparticles (inset is the electronic diffraction pattern), and (d) HRTEM images of $TiO_2/SiO_2/Fe_3O_4$ nanoparticles. *(Source: Peng et al., 2014, under the Creative Commons Attribution License).*

Rawat et al. (2007) harnessed this idea for the inhibition of the growth rate of *Escherichia coli* bacteria in the presence of $TiO_2/NiFe_2O_4$ nanoparticles under UV light irradiation. The result shows that at a constant irradiation time both heat treated and non–heat treated $TiO_2/NiFe_2O_4$ nanoparticles decreased the concentration of *E. coli* bacteria, but the decrease in concentration was more for heat treated $TiO_2/NiFe_2O_4$ nanoparticle system compared to non–heat treated $TiO_2/NiFe_2O_4$ nanoparticle systems. The disinfection of bacteria was directly controlled by the UV light absorption by the DNA of the bacteria. During the UV light irradiation, bacterial chromosome form thymine dimers, a cross-link between the thymine bases within the same strand of DNA. The formation of thymine dimers barricade the conformation of the double helix and interfere with normal DNA replication. In addition, photocatalytic inactivation of bacteria in the presence of TiO_2 has been explained by the attack of $O_2^{\bullet -}$, $^{\bullet}HO_2$, and $^{\bullet}OH$ radicals photogenerated at the surface of the catalyst.

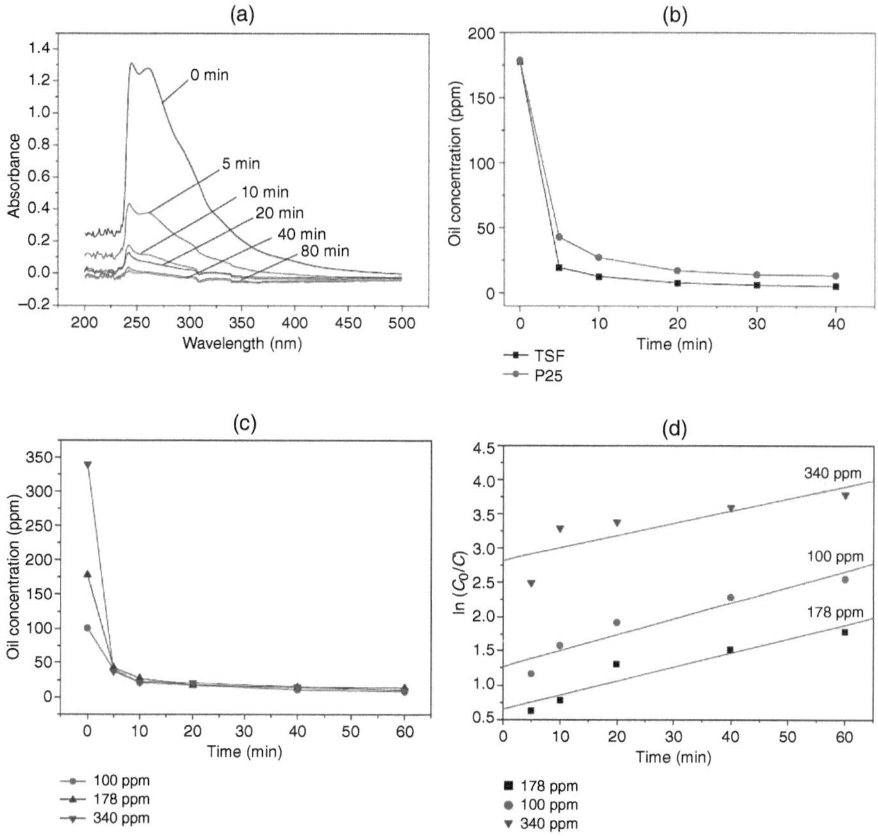

Figure 9.34 (a) Absorption spectra of crude oil wastewater exposed to UV light, (b) photocatalytic performance of $TiO_2/SiO_2/Fe_3O_4$ nanoparticles and P25, (c) the effect of initial oil concentration on the photocatalysis process of $TiO_2/SiO_2/Fe_3O_4$ nanoparticles, and (d) pseudo-first-order model fitting of (c) in degrading crude oily wastewater with the oil concentration less than 350 ppm. *(Source: Peng et al., 2014, under the Creative Commons Attribution License).*

Ma et al. (2015) synthesized the Fe_3O_4-TiO_2 nanosheets by means of lamellar reverse micelles and solvothermal method and applied for the photocatalytic antibacterial activity toward Gram-negative (G-) *E. coli* and Gram-positive (G+) *Staphylococcus aureus*. Fe_3O_4-TiO_2 nanosheets show better performance than pure Fe_3O_4 and TiO_2, and the antibacterial efficiency reached 87.2 and 93.7% toward *E. coli* and *S. aureus* with 100 μg/mL Fe_3O_4-TiO_2 nanosheets after 2 h of simulated solar light irradiation, respectively. The Fe_3O_4-TiO_2 nanosheets inactivated G- *E. coli* and G+ *S. aureus* by different mechanisms: the destruction of outer membranes and ruptured cell bodies were responsible for the bactericidal effect against *E. coli*, while the antibacterial effect toward *S. aureus* were due to the fact that the cells were adsorbed in the form of clusters by massive Fe_3O_4-TiO_2 nanosheets, which could restrict their activities and cause malfunction of the selective permeable barriers. The magnetic nanoparticles were recycled from the water by external magnets and applied for five cycles, and the result indicated that Fe_3O_4-TiO_2 nanosheets still retained more than 90% bacteria removal efficiency.

Rana et al. (2005a) synthesized composite nanoparticles consisting of a photocatalytic shell of titania and a magnetic core of nickel ferrite by reverse micelle and chemical hydrolysis techniques. The photocatalytic response of composite anatase- and brookite-titania-coated nickel ferrite nanoparticles in the degradation of methyl orange was evaluated in batch mode, and it was found that the methyl-orange photodegradation was enhanced with the increase in irradiation time. The methyl-orange concentrations were inversely proportional to the reaction time, and the concentration decreased from ~25 to ~4 μmol on irradiation for 7 h. The synthesized material was also applied for the decontamination of *E. coli* bacteria in water. The obtained result showed that at a constant irradiation time both anatase and brookite titania decrease the concentration of *E. coli* bacteria, but the decrease in concentration was more for the anatase-titanium–nickel ferrite system compared to the brookite-titania–nickel ferrite system.

9.4 CONCLUSIONS

Magnetically recyclable photocatalysts have been gaining much interest in the photocatalytic degradation of environmental contaminants from aqueous solutions due to the integration of magnetic materials as easy recycling and semiconductors material as highly responsive under photo systems. In the pure form neither any magnetic material nor semiconductor materials have been found suitable for the catalytic degradation of pollutants because of the stability issue, low photocatalytic efficiency, high leaching rate, lower formation of electron–hole pair, and tough regeneration and recyclability. Hence, extensive focus has been given to the development of core-shell nanocomposites, which show high catalytic efficiency and easy recyclability from water.

Photocatalytic efficiency of the magnetically separable photocatalysts shows that these nanomaterials can be effectively utilized in the treatment of wastewater. High removal rate has been obtained in the treatment of organic dyes, organophosphates, chlorophenols, nitrophenols, arsenic, and chromium from wastewater. Due to their magnetic nature the used photocatalyst can be easily separated in external magnetic fields. Various studies show that magnetic photocatalysts have high efficiency and were used up to 7–10 cycles without any loss of catalytic properties. Still there needs to be more economical photocatalysts developed of highly stable nature in harsh physicochemical environments. However, this technology has been extensively applied in the batch treatment mode on laboratory scale, and its practical application is shortly reported. Efforts should be made to commercialize the product, and its application in real wastewater treatment should be carried out.

REFERENCES

Álvarez, P.M., Jaramillo, J., López-Pinero, F., Plucinski, P.K., 2010. Preparation and characterization of magnetic TiO_2 nanoparticles and their utilization for the degradation of emerging pollutants in water. Appl. Catal. B 100, 338–345.

Beydoun, D., Amal, R., Low, G., McEvoy, S., 2002. Occurrence and prevention of photodissolution at the phase junction of magnetite and titanium dioxide. J. Mol. Catal. A Chem. 180, 193–200.

Cao, C., Xiao, L., Chen, C., Cao, Q., 2015. Magnetically separable Cu_2O/chitosan–Fe_3O_4 nanocomposites: preparation, characterization and visible-light photocatalytic performance. Appl. Surf. Sci. 333, 110–118.

Chen, F., Xie, Y., Zhao, J., Lu, G., 2001. Photocatalytic degradation of dyes on a magnetically separated photocatalyst under visible and UV irradiation. Chemosphere 44, 1159–1168.

Chen, X., Tan, P., Zhou, B., Dong, H., Pan, J., Xiong, X., 2015. A green and facile strategy for preparation of novel and stable Cr-doped $SrTiO_3$/g-C_3N_4 hybrid nanocomposites with enhanced visible light photocatalytic activity. J. Alloy. Compd. 647, 456–462.

Choi, K.H., Wang, K.K., Shin, E.P., Oh, S.L., Jung, J.S., Kim, H.K., Kim, Y.R., 2011. Water-soluble magnetic nanoparticles functionalized with photosensitizer for photocatalytic application. J. Phys. Chem. C 115, 3212–3219.

Chung, Y.S., Park, S.B., Kang, D.W., 2004. Magnetically separable titania-coated nickel ferrite photocatalyst. Mater. Chem. Phys. 86, 375–381.

Dong, Z., Le, X., Dong, C., Zhang, W., Li, X., Ma, J., 2015. Ni@Pd core–shell nanoparticles modified fibrous silica nanospheres as highly efficient and recoverable catalyst for reduction of 4-nitrophenol and hydrodechlorination of 4-chlorophenol. Appl. Catal. B 162, 372–380.

Ferroudj, N., Nzimoto, J., Davidson, A., Talbot, D., Briot, E., Dupuis, V., Bee, A., Medjram, M.S., Abramson, S., 2013. Maghemite nanoparticles and maghemite/silica nanocomposite microspheres as magnetic Fenton catalysts for the removal of water pollutants. Appl. Catal. B 136–137, 9–18.

Fu, W., Yang, H., Li, M., Li, M., Yang, N., Zou, G., 2005. Anatase TiO_2 nanolayer coating on cobalt ferrite nanoparticles for magnetic photocatalyst. Mater. Lett. 59, 3530–3534.

Fu, W., Yang, H., Chang, L., Hari-Bala, Li, G., Zou, M., 2006a. Anatase TiO_2 nanolayer coating on strontium ferrite nanoparticles for magnetic photocatalyst. Colloids Surf. A 289, 47–52.

Fu, W., Yang, H., Li, M., Chang, L., Yu, Q., Xu, J., Zou, G., 2006b. Preparation and photocatalytic characteristics of core-shell structure $TiO_2/BaFe_{12}O_{19}$ nanoparticles. Mater. Lett. 60, 2723–2727.

Gad-Allah, T.A., Kato, S., Satokawa, S., Kojima, T., 2009. Treatment of synthetic dyes wastewater utilizing a magnetically separable photocatalyst ($TiO_2/SiO_2/Fe_3O_4$): parametric and kinetic studies. Desalination 244, 1–11.

Gao, Y., Chen, B., Li, H., Ma, Y., 2003. Preparation and characterization of a magnetically separated photocatalyst and its catalytic properties. Mater. Chem. Phys. 80, 348–355.

Guo, R., Fang, L., Dong, W., Zheng, F., Shen, M., 2010. Enhanced photocatalytic activity and ferromagnetism in Gd doped $BiFeO_3$ nanoparticles. J. Phys. Chem. C 114, 21390–21396.

Hong, R.Y., Zhang, S.Z., Di, G.Q., Li, H.Z., Zheng, Y., Ding, J., Wei, D.G., 2008. Preparation, characterization and application of Fe_3O_4/ZnO core/shell magnetic nanoparticles. Mater. Res. Bull. 43, 2457–2468.

Idris, A., Hassan, N., Ismail, N.S.M., Misran, E., Yusof, N.M., Ngomsik, A.F., Bee, A., 2010. Photocatalytic magnetic separable beads for chromium (VI) reduction. Water Res. 44, 1683–1688.

Idris, A., Hassan, N., Rashid, R., Ngomsik, A.F., 2011. Kinetic and regeneration studies of photocatalytic magnetic separable beads for chromium (VI) reduction under sunlight. J. Hazard. Mater. 186, 629–635.

Idris, A., Misran, E., Hassan, N., Jalil, A.A., Seng, C.E., 2012. Modified PVA-alginate encapsulated photocatalyst ferro photo gels for Cr(VI) reduction. J. Hazard. Mater. 227–228, 309–316.

Janani, S., Rani, K.S.S., Ellappan, P., Miranda, L.R., 2016. Photodegradation of methylene blue using magnetically reduced graphene oxide bismuth oxybromide composite. J. Environ. Chem. Eng. 4, 534–541.

Jiang, Y., Wang, W.N., Biswas, P., Fortner, J.D., 2014. Facile aerosol synthesis and characterization of ternary crumpled graphene−TiO_2−magnetite nanocomposites for advanced water treatment. ACS Appl. Mater. Interfaces 6, 11766–11774.

Kumar, S., Surendar, T., Kumar, B., Baruah, A., Shanker, V., 2013. Synthesis of magnetically separable and recyclable g-C_3N_4−Fe_3O_4 hybrid nanocomposites with enhanced photocatalytic performance under visible-light irradiation. J. Phys. Chem. C 117, 26135–26143.

Li, S.K., Huang, F.Z., Wang, Y., Shen, Y.H., Qiu, L.G., Xie, A.J., Xu, S.J., 2011. Magnetic Fe_3O_4@C@Cu_2O composites with bean-like core/shell nanostructures: synthesis, properties and application in recyclable photocatalytic degradation of dye pollutants. J. Mater. Chem. 21, 7459–7466.

Li, D., Xie, J., Zhang, Y., Qiao, R., Li, S., Li, Z., 2015. Convenient synthesis of magnetically recyclable Fe_3O_4@C@CdS photocatalysts by depositing CdS nanocrystals on carbonized ferrocene. J. Alloy. Compd. 646, 978–982.

Liu, Y., Zhou, L., Hu, Y., Guo, C., Qian, H., Zhang, F., (David) Lou, X.W., 2011. Magnetic-field induced formation of 1D Fe_3O_4/C/CdS coaxial nanochains as highly efficient and reusable photocatalysts for water treatment. J. Mater. Chem. 21, 18359–18364.

Ma, S., Zhan, S., Jia, Y., Zhou, Q., 2015. Superior antibacterial activity of Fe_3O_4–TiO_2 nanosheets under solar light. ACS Appl. Mater. Interfaces 7, 21875–21883.

Nouri, A., Sargolzaei, J., 2014. Removal of phenol dyes using a photocatalytic reactor with SnO_2/Fe_3O_4 nanoparticles. J. Disper. Sci. Technol. 35, 1031–1039.

Obeid, L., Kolli, N.E., Talbot, D., Welschbillig, M., Bée, A., 2015. Influence of a cationic surfactant on adsorption of p-nitrophenol by a magsorbent based on magnetic alginate beads. J. Colloid Interface Sci. 457, 218–224.

Peng, Z., Tang, H., Tang, Y., Yao, K.F., Shao, H.H., 2014. Synthesis and photocatalytic activity of magnetically recoverable core-shell nanoparticles. Int. J. Photoenerg. 2014, 1.

Rana, S., Rawat, J., Misra, R.D.K., 2005a. Anti-microbial active composite nanoparticles with magnetic core and photocatalytic shell: TiO_2–$NiFe_2O_4$ biomaterial system. Acta Biomaterialia 1, 691–703.

Rana, S., Srivastava, R.S., Sorensson, M.M., Misra, R.D.K., 2005b. Synthesis and characterization of nanoparticles with magnetic core and photocatalytic shell: anatase TiO_2–$NiFe_2O_4$ system. Mater. Sci. Eng. B 119, 144–151.

Rashid, J., Barakat, M.A., Ruzmanova, Y., Chianese, A., 2015. $Fe_3O_4/SiO_2/TiO_2$ nanoparticles for photocatalytic degradation of 2-chlorophenol in simulated wastewater. Environ. Sci. Pollut. Res. 22, 3149–3157.

Rawat, J., Rana, S., Srivastava, R., Misra, R.D.K., 2007. Antimicrobial activity of composite nanoparticles consisting of titania photocatalytic shell and nickel ferrite magnetic core. Mater. Sci. Eng. C 27, 540–545.

Ruzmanova, Y., Stoller, M., Chianese, A., 2013. Photocatalytic treatment of olive mill wastewater by magnetic core titanium dioxide nanoparticles. Chem. Eng. Trans. 32, 2269–2274.

Safavi, A., Momeni, S., 2012. Highly efficient degradation of azo dyes by palladium/hydroxyapatite/Fe_3O_4 nanocatalyst. J. Hazard. Mater. 201–202, 125–131.

Santala, E., Kemell, M., Leskela, M., Ritala, M., 2009. The preparation of reusable magnetic and photocatalytic composite nanofibers by electrospinning and atomic layer deposition. Nanotechnology 20, 035602, doi:10.1088/0957-4484/20/3/035602.

Tanaka, K., Capule, M.F.V., Hisanaga, T., 1991. Effect of crystallinity of TiO_2 on its photocatalytic action. Chem. Phys. Lett. 187, 73–76.

Teixeira, S., Gurke, R., Eckert, H., Kuhn, K., Fauler, J., Cuniberti, G., 2016. Photocatalytic degradation of pharmaceuticals present in conventional treated wastewater by nanoparticle suspensions. J. Environ. Chem. Eng. 4, 287–292.

Tian, H., Liu, F., He, J., 2014. Multifunctional Fe_3O_4@SiO_2@mSiO_2–Fe core–shell microspheres for highly efficient removal of 1,1,1-trichloro-2,2-bis(4-chlorophenyl) ethane (DDT) from aqueous media. J. Colloid Interface Sci. 431, 90–96.

Tian, H., Chen, J., He, J., Liu, F., 2015. Pd-loaded magnetic mesoporous nanocomposites: a magnetically recoverable catalyst with effective enrichment and high activity for DDT and DDE removal under mild conditions. J. Colloid Interface Sci. 457, 195–202.

Wang, N., Zhu, L., Wang, D., Wang, M., Lin, Z., Tang, H., 2010a. Sono-assisted preparation of highly-efficient peroxidase-like Fe_3O_4 magnetic nanoparticles for catalytic removal of organic pollutants with H_2O_2. Ultrason. Sonochem. 17, 526–533.

Wang, N., Zhu, L., Wang, M., Wang, D., Tang, H., 2010b. Sono-enhanced degradation of dye pollutants with the use of H_2O_2 activated by Fe_3O_4 magnetic nanoparticles as peroxidase mimetic. Ultrason. Sonochem. 17, 78–83.

Wang, Y., Sun, H., Ang, H.M., Tade, M.O., Wang, S., 2014. Facile synthesis of hierarchically structured magnetic $MnO_2/ZnFe_2O_4$ hybrid materials and their performance in heterogeneous activation of peroxymonosulfate. ACS Appl. Mater. Interfaces 6, 19914–19923.

Wang, Y., Zhao, H., Zhao, G., 2015. Iron-copper bimetallic nanoparticles embedded within ordered mesoporous carbon as effective and stable heterogeneous Fenton catalyst for the degradation of organic contaminants. Appl. Catal. B 164, 396–406.

Wang, J., Yang, J., Li, X., Wang, D., Wei, B., Song, H., Li, X., Fu, S., 2016. Preparation and photocatalytic properties of magnetically reusable Fe_3O_4@ZnO core/shell nanoparticles. Physica E 75, 66–71.

Xu, L., Wang, J., 2012. Fenton-like degradation of 2,4-dichlorophenol using Fe_3O_4 magnetic nanoparticles. Appl. Catal. B 123–124, 117–126.

Xu, M.W., Bao, S.J., Zhang, X.G., 2005. Enhanced photocatalytic activity of magnetic TiO_2 photocatalyst by silver deposition. Mater. Lett. 59, 2194–2198.

Xu, S., Shangguan, W., Yuan, J., Chen, M., Shi, J., Jiang, Z., 2008. Synthesis and performance of novel magnetically separable nanospheres of titanium dioxide photocatalyst with egg-like structure. Nanotechnology 19, 095606.

Xu, M., Li, Q., Fan, H., 2014. Monodisperse nanostructured Fe_3O_4/ZnO microrods using for waste water treatment. Adv. Powder Technol. 25, 1715–1720.

Yang, Z., Gong, X., Zhang, C., 2010. Recyclable Fe_3O_4/hydroxyapatite composite nanoparticles for photocatalytic applications. Chem. Eng. J. 165, 117–121.

Yao, Y., Qin, J., Chen, H., Wei, F., Liu, X., Wang, J., Wang, S., 2015. One-pot approach for synthesis of N-doped TiO_2/$ZnFe_2O_4$ hybrid as an efficient photocatalyst for degradation of aqueous organic pollutants. J. Hazard. Mater. 291, 28–37.
Yu, L., Peng, X., Ni, F., Li, J., Wang, D., Luan, Z., 2013. Arsenite removal from aqueous solutions by γ-Fe_2O_3–TiO_2 magnetic nanoparticles through simultaneous photocatalytic oxidation and adsorption. J. Hazard. Mater. 246–247, 10–17.
Zeng, T., Zhang, X., Wang, S., Niu, H., Cai, Y., 2015. Spatial confinement of a Co_3O_4 catalyst in hollow metal–organic frameworks as a nanoreactor for improved degradation of organic pollutants. Environ. Sci. Technol. 49, 2350–2357.
Zhang, S., Zhao, X., Niu, H., Shi, Y., Cai, Y., Jiang, G., 2009. Superparamagnetic Fe_3O_4 nanoparticles as catalysts for the catalytic oxidation of phenolic and aniline compounds. J. Hazard. Mater. 167, 560–566.
Zhang, L., Sun, F., Zuo, Y., Fan, C., Xu, S., Yang, S., Gu, F., 2014. Immobilisation of CdS nanoparticles on chitosan microspheres via aphotochemical method with enhanced photocatalytic activity in thedecolourisation of methyl orange. Appl. Catal. B 156–157, 293–300.
Zhang, L., Wu, Z., Chen, L., Zhang, L., Li, X., Xu, H., Wang, H., Zhu, G., 2016. Preparation of magnetic Fe_3O_4/TiO_2/Ag composite microspheres with enhanced photocatalytic activity. Solid State Sci. 52, 42–48.

CHAPTER 10

Alumina Nanoparticles and Alumina-Based Adsorbents for Wastewater Treatment

Sushmita Banerjee
Department of Chemistry, University of Allahabad, Allahabad, India

10.1 INTRODUCTION

During the last few years, increasing attention has been focused on the development of nanosized alumina powders for advanced engineering applications such as catalysts, electronics, metals, and alloy manufacturing, as well as separation and purification (Shojaie-Bahaabad and Taheri-Nassaj, 2008). Alumina exists in various metastable structures including γ-, η-, δ-, θ-, κ- and χ-aluminas, as well as its stable α-alumina phase (Wang et al., 2008). Among all the metastable structures of alumina, γ-alumina is one kind of extremely important nanosized material. It has been used as a catalyst and catalyst substrate in automotive and petroleum industries, structural composites for spacecraft, abrasive and thermal wear coatings, and separation and purification of wastewater (Banerjee et al., 2015). Recently, it has been established that γ-alumina is thermodynamically stable relative to α-alumina when a critical surface area is achieved, and that nano γ-alumina powder can promote the sintering behavior of alumina and silicon carbide fibers (Castro et al. 2006). Various conventional processes have been developed for synthesizing alumina nano powder, which includes mechanical synthesis, vapor phase reaction, precipitation, combustion, and sol–gel methods. All of them have their own benefits and drawbacks.

10.2 SYNTHESIS

The sol–gel method is the most promising and thus has been widely investigated for nanomaterials because it produces solid particles of high purity and high specific surface area.

Shojaie-Bahaabad and Taheri-Nassaj (2008) synthesized a nano Al_2O_3 powder by a simple aqueous sol–gel method using inexpensive $AlCl_3{\cdot}6H_2O$ and Al powder as raw materials. It was shown that the gel, calcined at

Nanomaterials for Wastewater Remediation
http://dx.doi.org/10.1016/B978-0-12-804609-8.00010-8

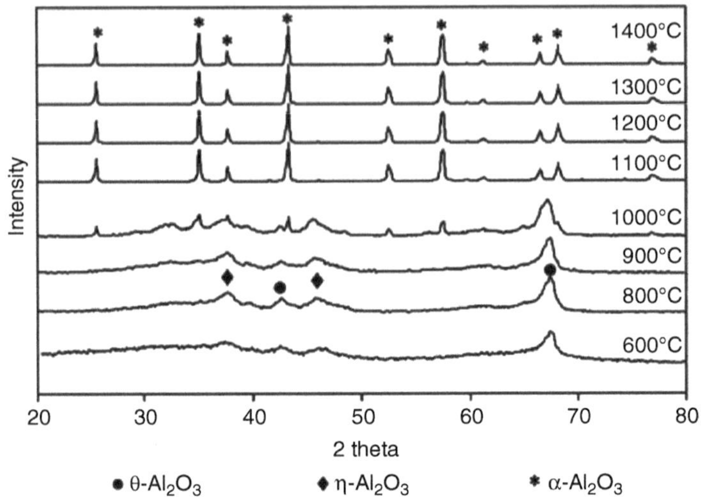

Figure 10.1 ***XRD pattern of gel heat treated at various temperatures for 4 h.*** *(Source: Shojaie-Bahaabad and Taheri-Nassaj, 2008; © Elsevier).*

1100°C, resulted in the formation of a crystalline α-Al2O3 nano powder. It had a particle size distribution ranging from 32 to 100 nm after heat treatment at 1200°C. Fig. 10.1 shows the XRD patterns of gel heat treated at various temperatures for 4 h.

Recently, nanosized γ-alumina powder was prepared via sol–gel precipitation in ethanol followed by a washing drying treatment and calcination (Wang et al., 2008). The starting materials $AlCl_3·6H_2O$ and conc. $NH_3·H_2O$ were both dissolved in C_2H_5OH to form certain concentration solutions. The $NH_3·H_2O$ solution was then added dropwise to the vigorously stirred $AlCl_3·6H_2O$ solution to produce a white, gelatinous $Al(OH)_3$ precipitate. After that the obtained material was filtered and washed with C_2H_5OH, dried at 100°C for 2 h, and calcined at certain temperatures for 2 h; the $Al(OH)_3$ gel was transformed into Al_2O_3. When the calcination temperatures were controlled between 686 and 1029°C, the nanosized γ-alumina particles with a size range of 5–9 nm were obtained. The highest surface area of 313.5 m^2/g was reported for alumina nanoparticles calcined at 600°C.

Coprecipitation is another most widely used method for the synthesis of alumina nanoparticles. Parida et al. (2009) reported synthesis of nanosized alumina particles using the control precipitation method. The effect of different precipitating agents (ammonium bicarbonate, ammonium carbonate, sodium bicarbonate, and sodium carbonate) on the size and surface area of the synthesized sample were also investigated. The chemical reactions during synthesis of alumina nanoparticles, by using ammonium bicarbonate as a precipitating agent and aluminium nitrate as precursors, can be represented as follows (Parida et al., 2009):

$$NH_4HCO_3 + HOH \rightarrow NH_4OH + H_2O + CO_2 \tag{10.1}$$

$$Al(NO_3)_3 + 3NH_4OH \xrightarrow{pH\sim7.5-8.5} Al(OH)_3 \downarrow + 3NH_4NO_3 \tag{10.2}$$

$$Al(OH)_3 \rightarrow AlOOH \downarrow + H_2O \tag{10.3}$$

$$AlOOH \rightarrow Al_2O_3 + H_2O \tag{10.4}$$

The results demonstrated that the samples synthesized using ammonium bicarbonate exhibited nanosized spherical-shaped particles with crystallite size in the range 4.7–5.7 nm and a high surface area of 190 m^2/g. Banerjee et al. (2015) also reported synthesis of nanosized alumina particles using the coprecipitation route. However in the coprecipitation method it is not possible to control the size of the crystallites, therefore, in several cases in order to avoid such problems coprecipitated samples were further subjected to hydrothermal treatment to obtain well-crystallized samples. Recently, Zhou et al. (2013) reported hydrothermally synthesized alumina hollow microspheres using urea and trisodium citrate (CTC) as precipitating and mediating agents, respectively. CTC exhibited remarkable effects on the morphology as well as the microstructure of the sample. It was observed that with the increase in CTC concentration from 0.125 mmol to 1.0 mmol the specific surface area and pore volume of the sample increases from 160.3 to 255.0 m^2/g and 0.30 to 0.52 cm^3/g, respectively. Moreover the effect of CTC concentration on the morphology of the synthesized material can be interpreted from the micrographs as given in Fig. 10.2. It can be observed in Fig. 10.2a and b that the irregular aggregated dense particles were obtained in the absence of CTC. An increase in CTC to 0.125 mmol resulted in hollow microspheres and loosely connected nanoflakes (Fig. 10.2c and d). Fig 10.2e and f and associated insets revealed intensifying of the hollow structures and the presence of smaller pore sizes when CTC further increased to 0.25 mmol. At 0.5 mmol CTC concentration, the morphology of the sample demonstrated well-defined microspheres (Fig. 10.2g). Finally at 1 mmol CTC the microspheres appeared as collapsed form, given in Fig. 10.2h.

The combustion method was also reported by several researchers for the synthesis of alumina nanoparticles as the process is considered energy saving and eco-friendly. Rahmani et al. (2010) prepared nanostructured alumina

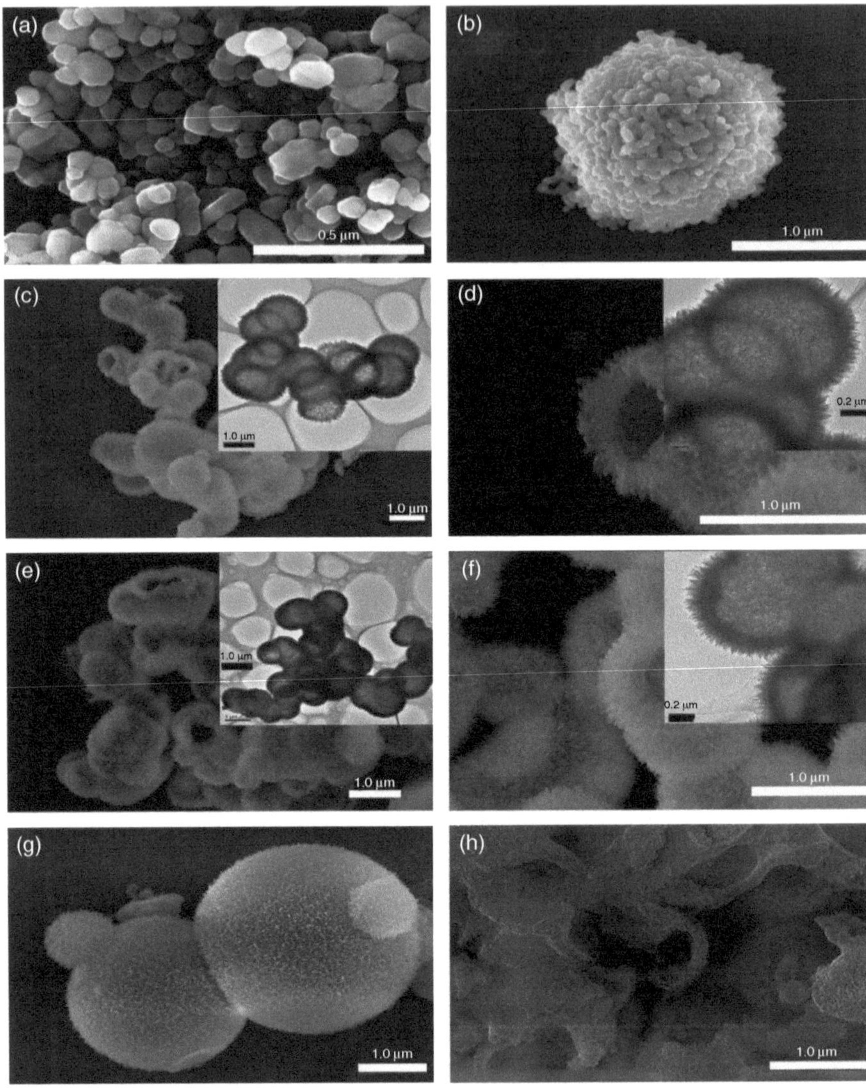

Figure 10.2 ***SEM and TEM images of alumina microspheres.*** *(Source: Zhou et al., 2013; © Elsevier).*

with mean pore diameter of 7.21 nm and surface area of 206 m^2/g by the combustion method using ammonium acetate as fuel. Bhargavi et al. (2015) synthesized alumina-based nanoparticles using similar techniques. Glycine and ammonium nitrate along with aluminium nitrate were the basic raw materials used in the synthesis process. The oxidizer to fuel ratio was maintained at 1:1. The solution combustion reaction is carried out in a muffle furnace at temperatures <45°C. As the combustion reaction completed, alumina nanoparticles were produced in the form of the resultant product. The theoretical equation for the combustion-based synthesis of alumina nanoparticles was represented as:

$$\begin{aligned} &4\,Al(NO_3)_3 + 10\,C_2H_5O_2N + 15\,NH_4NO_3 \\ &\rightarrow 2\,Al_2O_3 + 20\,CO_2 + 55\,H_2O + 26\,N_2 \end{aligned} \tag{10.5}$$

The synthesis of alumina nanoparticles via chemical templating method also has been intensively researched (Ogihara et al., 1991; Kim et al., 2004; Renuka et al., 2012; Patra et al., 2012), as the method permits the formation of porous material with a controlled porosity and high surface area. Kim et al. (2004) synthesized mesoporous alumina using the template method. The process involved post-hydrolysis of aluminum sec-butoxide (aluminium precursors) along with stearic acid as a chemical template, and water and sec-butyl alcohol were used as solvents. The synthesized sample revealed a narrow pore size distribution with a pore diameter of 3.5 nm, high surface area, and a high pore volume of 307 and 0.39 m^3/g, respectively. Patra et al. (2012) also reported template-assisted synthesis of alumina nanoparticles by using sodium salicylate as the template. The technique involves precipitation of aluminum precursor followed by its hydrolysis along with sodium salicylate that results in aggregation and its subsequent calcinations. The proposed method for the synthesis is given in Fig. 10.3.

In addition to the aforementioned methods, some other techniques such as thermal decomposition of aluminium sulfate (Kato et al., 1981), laser ablation of aluminium (Johnston et al., 1992), ball milling of Al powder (Wu, 2001), chemical vapor decomposition of organic aluminium (Noda et al., 2003), ultrasonic flame pyrolysis (Varatharajan et al., 2003), flux method (Park et al., 2003), and flame spray pyrolysis (Tok et al., 2006) were used for the synthesis alumina nanoparticles.

10.3 APPLICATION

High surface area, good porosity, thermal resistivity, and rich surface chemistry make alumina an ideal adsorbent material for wastewater treatment. Many investigations (Somasundaran and Fuerstenau, 1966; Parida et al., 2009; Tang et al., 2009; Tarun et al., 2009; Srivastava et al., 2011; Renuka et al., 2012; Bhargavi et al., 2015) were reported in which alumina nanoparticles were employed for the sequestration of various organic and inorganic pollutants from the wastewater system. However it was investigated that for some

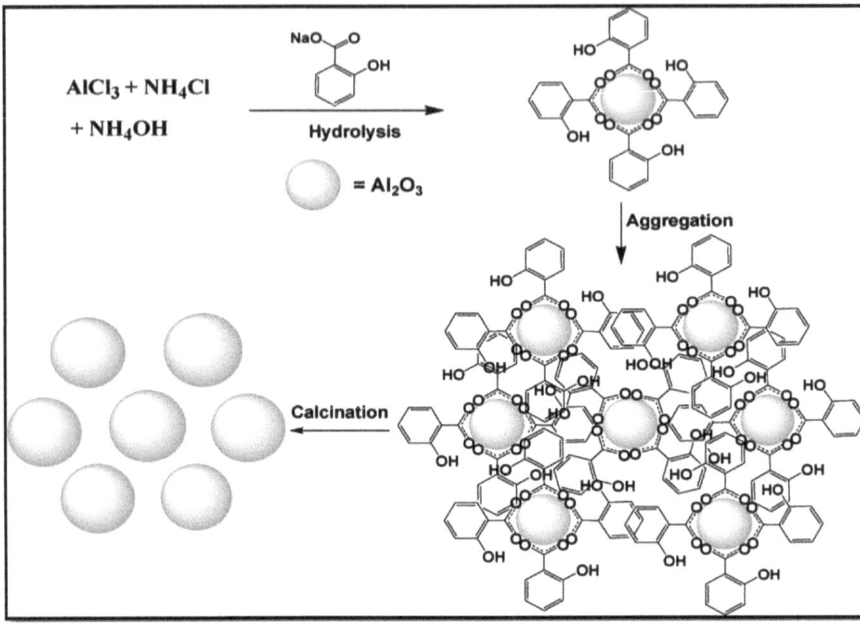

Figure 10.3 ***Schematic representation of the template-assisted synthesis of mesoporous alumina.*** *(Source: Patra et al., 2012; © Elsevier).*

pollutant species the scavenging performance of alumina nanoparticles was not very satisfying; therefore, in order to enhance the adsorption efficiency the alumina was further modified through physical or chemical treatment. Various treatment studies were reported (Adak and Pal, 2006a; Saha et al., 2008a; Afkhami et al., 2010; Zolgharnein et al., 2015) using modified alumina particles, and results demonstrated enhanced decontamination performance. In the subsequent sections we briefly discuss the application of bare and modified alumina nanoparticles for the treatment of different classes of pollutant species from aquatic environments.

10.3.1 Removal of Phenolic Compounds

Phenol and its derivatives are some of the most prominent water pollutants that are usually discharged as waste effluents from a number of industrial factories (Li-nan et al., 2007). Phenolic compounds are considered priority pollutants, as they are harmful to organisms at low concentrations and many of them have been classified as hazardous pollutants because of their potential toxicity to humans and aquatic life (Zhu et al. 2007). According to the U.S. Environmental Protection Agency (USEPA) regulations the prescribed limit of phenolic content in wastewater must be less than 1 mg/L. Hence, its elimination from industrial effluents is necessarily required. Many studies based on phenol adsorption using alumina-based materials has been reported by researchers; for instance, Danis et al. (1998) reported removal of 2,4-dichlorophenol; 2,4,6-trichlorophenol; and pentachlorophenol from aqueous solutions using alumina-pillared montmorillonite (AlPMt) and mesoporous alumina aluminium phosphates (AAP). The adsorption of chlorophenols directly depends on the number of chlorine present in the molecules, thus pentachlorophenol was adsorbed at higher rates. AlPMt material adsorbs 26.3% of 2,4-dichiorophenol; 75.6% of 2,4,6-trichlorophenol; and 95.2% of pentachlorophenol at equilibrium condition. However, the adsorption performance of mesoporous AAP for chlorophenols was not comparable to AlPMt. The uptake was found to increase with the increase of P/Al ratio in AAP. The maximum removal was 14.8% for 2,4-dichlorophenol; 27.1% for 2,4,6-trichlorophenol; and 58.3% for pentachlorophenol when the AAP mixture with ratio P/Al = 0.6 was used. Adak and Pal (2006b) reported surfactant-modified alumina for the removal of phenol from aqueous solutions. Conceptually, the surfactant-modified metal oxides tend to form micelle-like structures on their surface, thus they possess great potential in solubilizing organic molecules within the structure formed. More than 90% of phenol removal was investigated at within 120 min of contact time. Solution pH greatly influenced the phenol sorption performance as the removal of phenol decreases with the increase of pH. At lower pH, phenol is adsorbed as molecular form. As pH increases phenol exists as phenolate ion, and the negatively charged head of SDS molecules adsorbed on the alumina surface repel these ions. At pH > 9.15, SDS molecules were desorbed from the alumina surface and caused reduction in the phenol removal. The maximum removal of phenol was observed at pH 6.7. Moreover, the presence of interfering cation and anions was also investigated, and results demonstrated that the removal efficiency was increased in the presence of anions and decreased a little in the presence of cations. However, temperature played no significant role in the sorption of phenolate ions. Zhou et al. (2013) investigated adsorption performances for p-nitrophenol from aqueous solutions by using alumina hollow microspheres. The adsorption kinetics investigation suggested that the removal process was clearly time dependent, as the majority of p-nitrophenol adsorption from aqueous solutions was completed within 180 min of contact time. Moreover, pseudo-second order kinetics better described the adsorption process. For adsorption isotherm study the equilibrium adsorption data of p-nitrophenol on Al_2O_3 adsorbent were further analyzed using Langmuir and Freundlich models. High R^2 values obtained for the Langmuir model indicated suitability of this model; furthermore, the q_{max} values calculated from the Langmuir model demonstrated that the p-nitrophenol adsorption capacity for Al_2O_3 samples were significantly improved from 137.0 to 217.4 mg/g with increasing CTC concentration from 0 to 1 mmol. Thus the results suggested that alumina microspheres are powerful adsorbents for the removal of p-nitrophenol from water. However, in some studies (Roostaei and Tezel, 2004; Blomqvist and Salo, 2009) it was reported that alumina exhibited poor affinity for phenol.

10.3.2 Removal of Dyes

Color is the most obvious indicator of water pollution. The presence of colored materials or dyes and pigments can be easily discernable even at minute levels (1.0 mg/L). Removal of such compounds at low levels is indeed a cumbersome task. A survey of the literature suggested that alumina or modified alumina-based adsorbents demonstrated tremendous potential for the removal of dye from wastewater. Adak et al. (2005) investigated the removal of crystal violet dye from aqueous solutions using surfactant-modified alumina. The dye uptake was rapid and more than 80% of removal was noticed within 5 min of contact time. However, the adsorption process reached equilibrium condition within 1 hr of contact time. The pH of the solution also influenced the dye adsorption process; maximum removal was observed at pH 8. The removal efficiency of the adsorbent in the presence of interfering ions was also interpreted, and it was found that the removal percent of dye was increased in the presence of anions and humic acid and was decreased in the presence of cations. Thus, surfactant-modified alumina was found to be a very effective adsorbent, and ~99% of removal was achieved under optimized conditions, even at a high dye concentration of 200 ppm. Khosla et al. (2013) examined the adsorption behavior of acid orange-7 (AO-7) on aluminum oxide nanoparticles (ANP). The dye adsorption was strongly dependent on adsorbent dose, temperature, and pH. The adsorption capacity of the adsorbent increases from 52 to 65 mg/g with the rise of adsorption temperature from 303 K to 323 K, which suggests the endothermic nature of the adsorption process. The pH-dependent removal study revealed the maximum dye uptake (>98%) observed in acidic pH condition at pH value 2.0. The maximum adsorption capacity of ANP for AO-7 was reported as 0.9766 mg/g at 303 K. Moreover, results of thermodynamic investigation indicated the feasible and spontaneous nature of the sorption process. A similar study for removal of color black G (CBG) by alumina nanoparticles was reported by Bhargavi et al. (2015). The maximum adsorption capacity of the alumina nanoparticles was reported as 263.16 mg/g. Adsorption of reactive red 120 by adsorption on nano alumina was reported by Nadafi et al. (2014). Maximum removal of 83% was achieved at initial dye concentration of 25 mg/L with adsorbent dose of 1.2 g/L at pH 3.0. Zolgharnein et al. (2014) investigated removal of binary mixture of dyes alizarin red and alizarin yellow from aqueous solutions using nano-alumina particles. The adsorption results suggested that under optimal conditions (pH = 5.0; adsorbent dose = 10 g/L; initial dye concentration = 0.3 mmol/L), the maximum removal percentages of 99 and 78.5%, and the capacity uptake of 54.4 and 39.0 mg/g, were obtained for both alizarin red and alizarin yellow, respectively. The kinetic behavior of alizarin red and alizarin yellow and their binary mixture during adsorption onto alumina nanoparticles was illustrated in Fig. 10.4. It can be clearly seen that kinetics of adsorption was very rapid, and for all component systems the removal achieved was more than 80%. A higher adsorption capacity was observed for alizarin red than alizarin yellow in both single and binary solutions, and in binary solutions the presence of alizarin red restricted the adsorption of alizarin yellow. The difference in adsorption capacity between the two dyes may be due to the chemical structure of each dye. Hence the

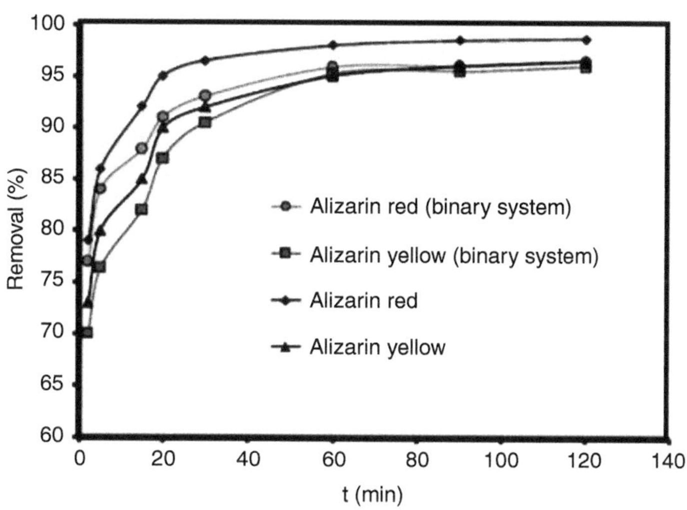

Figure 10.4 ***Adsorption kinetics for alizarin red and alizarin yellow in single and binary component systems.*** *(Source: Zolgharnein et al., 2014; © Elsevier).*

dye alizarin red revealed higher adsorption capacity and faster kinetics due to smaller molecular size as compared to alizarin yellow.

Recently, Zolgharnein et al. (2015) reported a study on simultaneous removal of binary mixtures of brilliant green (BG) and crystal violet (CV) dyes on surfactant-modified alumina nanoparticles. The maximum removal of 96 and 78% was observed for BG and CV at pH 8, respectively. Equilibrium modeling studies were performed for single and binary dye systems, and results suggested that single component system Sip's isotherm model and binary component extended Sip's isotherm model suitably represent the adsorption equilibrium of dyes. The maximum adsorption capacity obtained for CV and BG in single component systems was reported as 254 mg/g and 168 mg/g and for binary systems as 200 and 143, respectively. Adsorption kinetics investigation revealed that the adsorption of BG and CV was rapid, and within 20 min of interaction period more than 50% of the dye was removed (Fig. 10.5). Moreover, the mechanism of the adsorption process was determined by analyzing the kinetic data through various kinetic models and, on the basis of coefficient determination values, pseudo-second order kinetics were found to describe the adsorption process more effectively.

Some of the dye removal studies were reported using alumina-based nanocomposites; for instance, Zhang et al. (2012) have synthesized chitosan-alumina (CA) composites and evaluated their potential in the removal of methyl orange (MO) dye from aqueous solutions. The principle idea behind the synthesis of chitosan-alumina composites was to derive the benefit of both materials in its composite form. The pH study suggested that

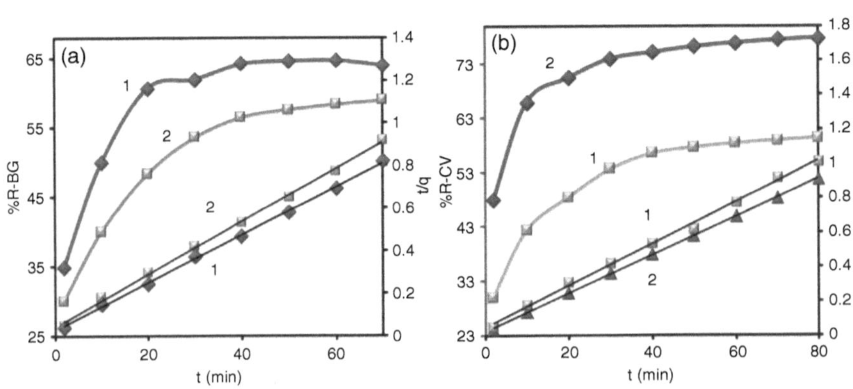

Figure 10.5 ***The effect of contact time on removal of BG and CV in (1) single and (2) binary systems.*** *(Source: Zolgharnein et al., 2015; © Elsevier).*

maximum dye removal of 98% was achieved at pH 6.0. The probable removal mechanism of MO was given as:

$$MO - SO_3^- + CA - NH_3^+ \xrightarrow{pH\,6.0} MO - SO_3^- \equiv NH_3^+ - CA \quad (10.6)$$

Adsorbent dose also influenced the removal process as the percent of MO removal increases from 92.84 to 99.23% for an increase in adsorbent dose from 1 to 12 g/L. The temperature exhibited adverse effects on the dye removal percent, because with increased temperature surface activity is decreased, which indicates that adsorption of MO on CA is an exothermic process. Desorption experiments indicated that MO was effectively desorbed in the presence of deionized water at 373 K with desorption rate of 84.91%. Likewise, Raut et al. (2013) investigated the alumina–carbon composite for the removal of methylene blue and alizarin red from aqueous solutions. The maximum adsorption capacity for methylene blue and alizarin red were reported as 1152.30 mg/g and 522.81 mg/g at pH 8 and pH 5, respectively.

10.3.3 Removal of Hydrophobic Organic Compounds (HOCs)

The removal of hydrophobic organic compounds from aqueous solutions employing alumina particles was extensively researched. However, in nearly all studies chemically modified alumina were reported for the adsorption of HOCs, whereas alumina particles in their bare form were rarely investigated, which indicates poor scavenging efficiency of bare alumina particles for HOCs. Sun and Jaffe (1996) investigated sorption of phenanthrene from the aqueous phase using a combination of dianionic surfactants and alumina. The results demonstrated that the dianionic surfactants (commercial DOWFAX) adsorbed on alumina are effective in sorption of phenanthrene, as the apparent distribution coefficient of pheanthrene on the solid phase was enhanced by up to three orders of magnitude. Further, the sorption performance of phenanthrene as a function of the surfactant loading rate and the surfactant alkyl chain length was interpreted. The sorption of phenanthrene on the alumina-coated surfactant increased with the increase in alkyl chain length. Further, it was also found that the partitioning of the sorbate between the adsorbed dianionic surfactant phases and water is 5-7 times more effective than that between the surfactant's phases in the aqueous phase and water. In another study, partitioning of naphthalene was attempted by Neupane and Park (2000) using three anionic surfactants with different molecular structures, such as conventional [sodium dodecylbenzene sulfonate (SDDBS)], gemini [dialkylated disulfonated diphenyl oxide with alkyl chain length of 12 (DADS-C12)], and dianionic [monoalkylated disulfonated diphenyl oxide with alkyl chain length of 12 (MADS-C12)]. The results of partition coefficients of naphthalene between surfactant-treated alumina and the water phase sorption with the change of adsorbed surfactant concentration in moles suggested that the partition coefficients increased with the increase in adsorbed surfactant molecules adsorbed onto aluminum oxide, and uptake of naphthalene by gemini surfactant-treated alumina was relatively higher than by dianionic surfactant-treated alumina and by conventional surfactant-treated alumina at a given adsorbed surfactant moles.

Fig. 10.6 clearly demonstrated that the slope representing DADS-C12 was nearly twice as steep as the slopes of the MADS-C12 and SDDBS at a sorbed surfactant concentration of 50 mmol/kg. This may be due to the presence of one or more alkyl chains in a single molecule of DADS-C12 in comparison to the other two surfactants. Further beyond this concentration the partition was not very prominent, which indicated that surfactants like DADS-C12 with comparatively more organic carbon content offered more efficient partitioning medium for naphthalene per mole of surfactant. However at lower surfactant concentrations, the uptake of naphthalene by the gemini surfactant-adsorbed alumina was similar to that of the adsorbed conventional surfactant for equivalent masses adsorbed. Gawade et al. (2005) reported removal of an herbicide (Isoproturon) from water using alumina and surfactant-treated alumina. It was found that surfactant-treated alumina exhibited far better sorption capacity than alumina. This

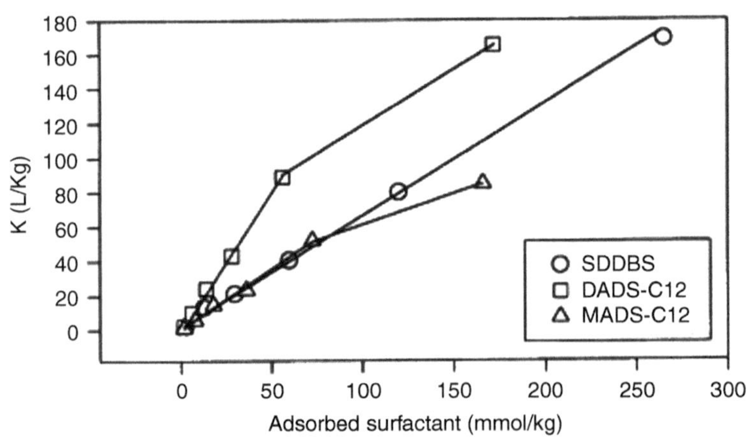

Figure 10.6 ***Partition coefficients (K) of naphthalene with adsorbed surfactant.*** *(Source: Neupane and Park, 2000; © Elsevier).*

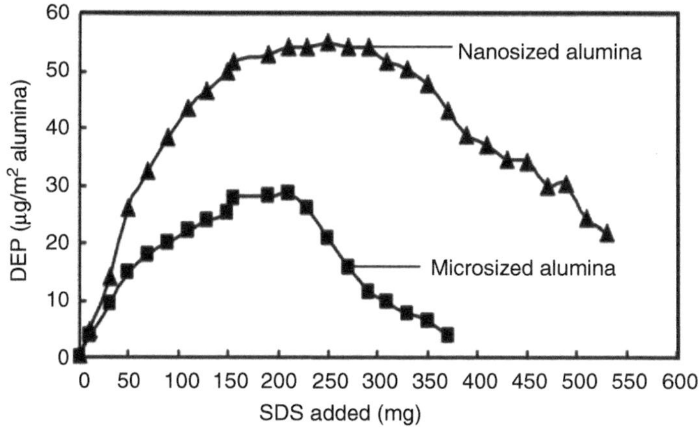

Figure 10.7 ***Effect of SDS concentration on DEP adsorption at pH 3.0.*** *(Source: Li et al., 2008; © Elsevier).*

enhanced herbicide removal by surfactant-modified alumina was due to the formation of surfactant aggregates at the solid–liquid interface that led to solubilization of herbicide within the aggregates. Li et al. (2008) studied the adsorption characteristics of sodium dodecyl sulfate (SDS)-coated nano/microsized alumina systems for removal of di-ethyl-phthalate (DEP) from the aqueous phase. The research results suggested that adsorption of DEP was highly dependent on SDS concentration, pH value, ionic strength, and temperature. The adsorption of DEP increased with the increasing SDS concentration (Fig. 10.7). The increase in adsorption of DEP may be attributed to the gradual formation of mixed hemimicelles on the alumina surface. Moreover the sorption capacity of the SDS-coated nanosized alumina began to exceed SDS-coated microsized alumina when the SDS amount added was above 25 mg. However it was also investigated that after a certain value of SDS concentration the adsorption amount began to decrease, which may be due to formation of micelles in bulk aqueous solution, and these micelles distributed the DEP into the bulk solution again.

pH of the solution also influenced the sorption process. The uptake was highest at acidic pH condition. This was explained on the basis of pH_{PZC} of alumina (8.5), as the surface of alumina was positively charged when the pH was below its PZC; that is, at low pH condition the anionic SDS molecules easily adsorbed to the alumina surface by a strong electrostatic attraction that led to the formation of mixed hemimicelles on the alumina surface, and DEP was captured by the mixed hemimicelles through hydrophobic actions. Furthermore, the presence of interfering ions also affects the sorption process that the DEP uptake amount reduced from 61 to 47 μg/m^2 for nanosized alumina when NaCl concentration in the adsorption solution increased from 0.005 to 0.4 M. The effect of temperature on DEP suggested that the adsorption systems were exothermic. With the increasing temperature the adsorbate–adsorbent bonds were weakened, and DEP exhibited desorption trends from the solid phase to the solution phase. In another investigation, removal of formaldehyde (FA) was accomplished by Afkhami et al. (2011) using alumina nanoparticles modified with 2,4-dinitrophenylhydrazine (DNPH). The chemical modification with DNPH results in the addition of amine groups over the adsorbent surface that further facilitates removal of formaldehyde from the bulk through physical interaction. The tentative mechanism for FA adsorption was given as:

$$\mathrm{HCHO + R - NH_2 \rightarrow H_2C = N - R + H_2O} \qquad (10.7)$$

According to Afkhami et al. (2011) the efficient removal of FA may be interpreted in two ways: first, due to the reaction between active hydrogen atoms of the adsorbent and FA that results in the formation of alcohol or, second, due the strong electrostatic interaction that was built up between the adsorbent (which possessed a partially negative charge due to the presence of nitrogen atoms in DNPH-alumina) and FA (in which the carbon atom showed a slightly positive charge). Moreover, the ratios of DNPH/alumina used in adsorbent synthesis were found to influences the uptake efficiency of FA. It was investigated that the adsorption was increased by increasing DNPH/alumina mass ratio up to 0.45. The mass ratio lower than 0.45 resulted in lower adsorption due to the presence of insufficient amine adsorption sites that lead to poor adsorption of FA. The effect of pH on FA removal was also studied by varying the pH condition from 2.0 to 10.0. The pH values of the FA aqueous solutions decreased slightly when they flowed through the adsorbent as the pH value of the solution dropped from ~7.0 to ~6.1–6.3 and from ~5.5 to ~4.2–4.5. The experimental results (Fig. 10.8) revealed that the maximum removal of 98% was achieved in between pH 4.5–8.5.

The effect of initial FA concentration on the removal efficiency suggested that with the increase in initial analyte concentration the time period to achieve equilibrium condition gets extended. For instance, from Fig. 10.9 it can be seen that at initial FA concentrations of 80, 200, and 300 mg/L the time required to attain equilibrium conditions were 45, 60 and 120 min, respectively. This may be due to the saturation of the binding sites that results

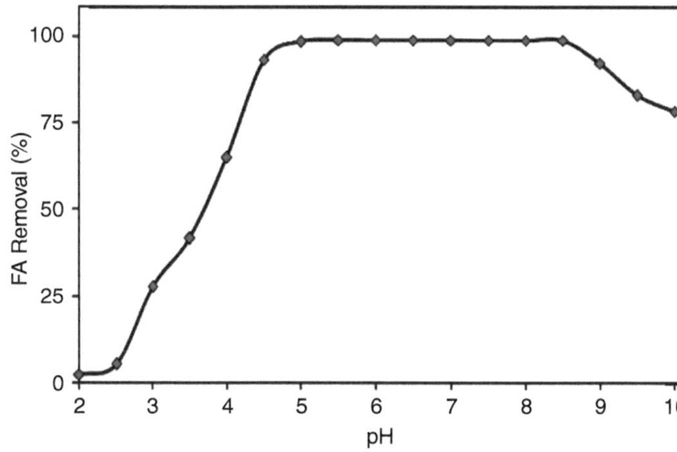

Figure 10.8 ***Effect of pH on removal of formaldehyde using DNPH-modified alumina.*** *(Source: Afkhami et al., 2011; © Elsevier).*

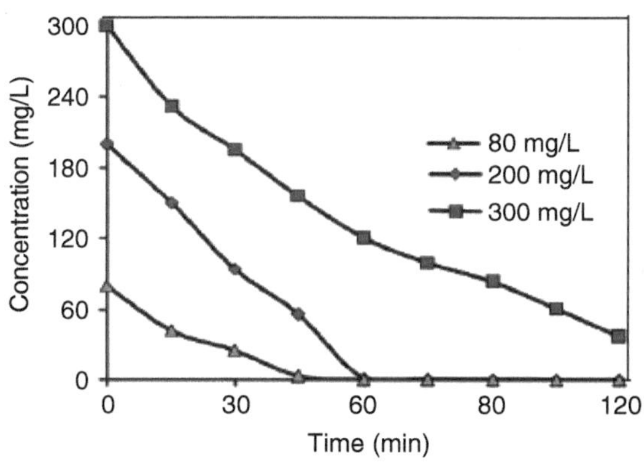

Figure 10.9 ***Effect of various initial concentrations of formaldehyde on removal efficiency at pH 7.0 and adsorbent dosage of 100 mg/100 mL.*** *(Source: Afkhami et al., 2011; © Elsevier).*

in FA molecules remaining unadsorbed at increased analyte concentration. The maximum adsorption capacity of the adsorbent was evaluated from the Langmuir isotherm parameter as 411.83 mg/g.

10.3.4 Removal of Surfactants

A large number of studies were reported on adsorption of ionic surfactants on alumina nanoparticles; for instance; Huang et al. (1996), investigated adsorption behavior of surfactant mixtures by using cationic surfactant; tetradecyl trimethyl ammonium chloride (TTAC); and nonionic surfactant, pentadecylethoxylated nonyl phenol (NP-15) mixtures at the alumina water interface. It was reported that NP-15 alone does not adsorb at the alumina water interface; as for adsorption on the polar surface, nonionic surfactants should have a moiety with sufficient adsorption force that can displace water molecules from the alumina. Conversely, cationic surfactant TTAC was adsorbed easily on alumina at pH 10 due to electrostatic attraction. However, in the presence ofTTAC the adsorption behavior of NP-15 was changed and adsorption of the same was reported due to hydrophobic interactions between adsorbed surfactant chains at the alumina–water interface. On the contrary, adsorption ofTTAC suddenly declined at 1:4 TTAC:NP-15 ratio (Fig. 10.10), which was mainly attributed to the competition of the NP-15 with TTAC for the adsorption sites under saturated adsorption conditions.

The adsorption of NP-15 in the presence ofTTAC was also examined (Fig. 10.11), and it was investigated that the presence ofTTAC causes significant adsorption of NP-15. With an increase in TTAC content in the mixtures, the adsorption of NP-15 was increased considerably due to the formation of large numbers ofTTAC aggregates at the alumina–water interface that results in effective coadsorption of NP-15.

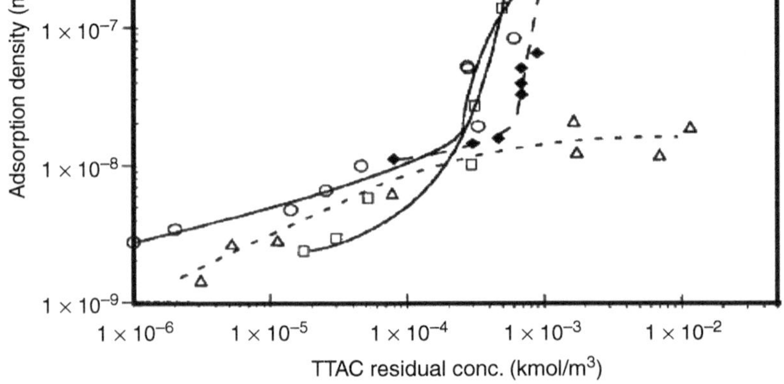

Figure 10.10 ***Adsorption of tetradecyl trimethyl ammonium chloride (TTAC) on alumina in the presence and absence of pentadecylethoxylated nonyl phenol (NP-15) at pH 10.*** *(Source: Huang et al., 1996; © Elsevier).*

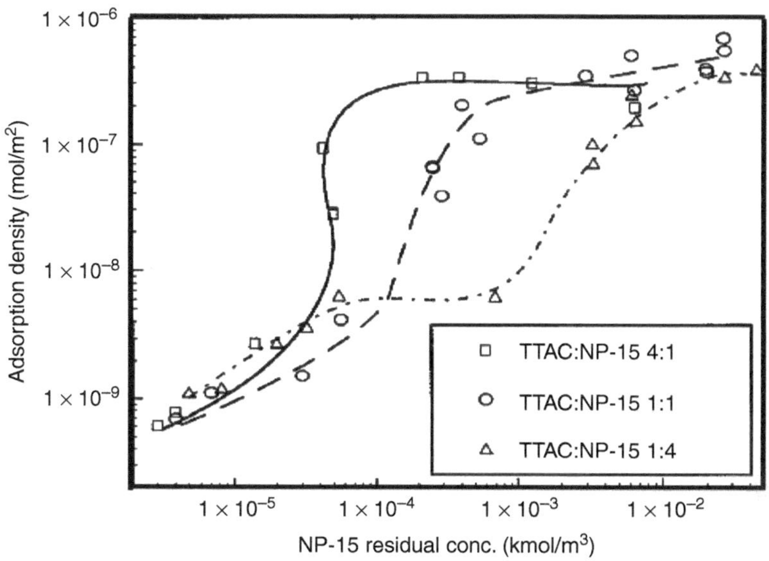

Figure 10.11 ***Adsorption of pentadecylethoxylated nonyl phenol (NP-15) on alumina in the presence of tetradecyl trimethyl ammonium chloride (TTAC) at pH 10.*** *(Source: Huang et al., 1996; © Elsevier).*

Fan et al. (1997) reported an in-depth mechanism of adsorption of cationic surfactant alkyltrimethylammonium bromides onto negatively charged alumina particles. It was assumed that the adsorption isotherm of the surfactant-alumina system was divided into four distinct regions. The first region corresponds to the adsorption of surfactant as individual ions onto the solid surface electrostatically. The second region indicated formation of hemimicelles. As adsorption proceeds, the surfactant starts aggregating, and molecules arranged themselves in an opposite orientation due to surface neutralization by the oppositely charged surfactant, and this phase corresponds to third region. Finally, the adsorbed layer possesses the structure of a bilayer and was indicative of the fourth adsorption region. However, Fan et al. (1997) detected three regions in the adsorption isotherm, which indicated loose and interpenetrating configurations of the surface colloids. The results also conveyed the nature of the surface colloids as the dominating factor that decides the shape of the adsorption isotherm. Neupane and Park (1999) reported the adsorption of a gemini surfactant, namely, dialkylated disulfonated diphenyl oxide, with alkyl chain lengths of twelve (DADSC12) onto positively charged alumina in water. Further, the adsorption behavior of the surfactant was also compared with

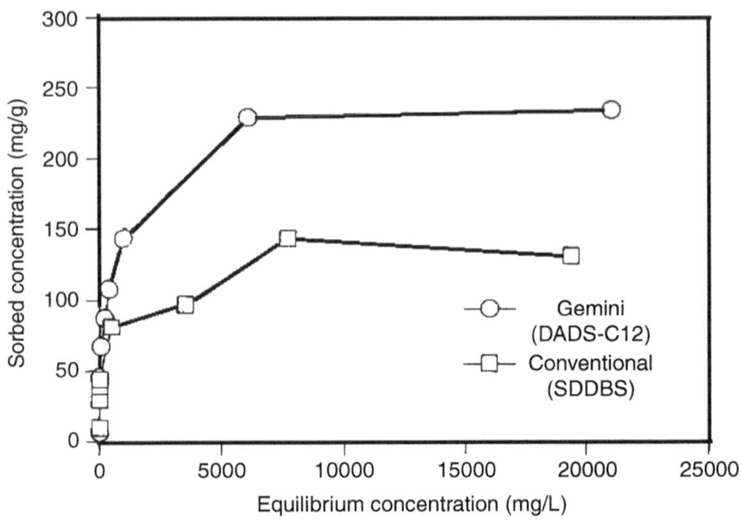

Figure 10.12 ***Adsorption of two surfactants onto alumina.*** *(Source: Neupane and Park, 1999; © Elsevier).*

another anionic surfactant (SDDBS). The results (Fig. 10.12) suggested that the uptake amount of gemini surfactant over alumina was higher than that of SDDBS. As at analyte equilibrium concentration of 20,000 mg/L the sorbed concentration reported for SDDBS was 136 mg/g and for gemini was 235 mg/g.

The kinetics for surfactant adsorption was also investigated at two different initial concentrations, such as 0.5 times CMC and 1.5 times CMC. The adsorption of each surfactant onto oxide particles attained equilibrium within 20 minutes at both concentrations. A shorter equilibrium time indicates quick formation of sorptive phase onto treated oxide particles, which makes the surfactant-treated oxide particles more suitable for practical applications in removing contaminants from the aqueous phase. Thus no difference in binding kinetics was noticed between gemini and SDDBS, which indicated that the adsorption of the two surfactants onto aluminum oxide follow the same mechanism, and the large molecular structure of the gemini surfactant does not hinder the adsorption of SDDBS. It was also anticipated that the gemini surfactant-treated solid particles will provide the stronger sorbent for the hydrophobic organic compounds.

Zhang et al. (2002) reported an interesting study on adsorption behavior of a sugar-based surfactant, *n* dodecyl-β-D-maltoside (DM), on alumina. The adsorption results as a function of pH were investigated and they found that the surfactant adsorption was not affected by the change in pH, which suggested that surface charge of alumina does not affect the adsorption of DM under the test conditions. Further, the adsorption performance of alumina in the presence of calcium ions was explored, and the results demonstrated that adsorption was not affected by the presence of calcium ions. The effect of temperature on the adsorption of *n*-dodecyl-*β*-D-maltoside on alumina was tested, and results revealed that temperature has almost no effect on the adsorption of *n*-dodecyl-*β*-D-maltoside. Zhang et al. (2002) proposed that hydrogen bonding is the most probable driving force of the adsorption of *n*-dodecyl-*β*-D-maltoside; therefore, in order to verify this hypothesis, adsorption of DM in solvents such as DMSO and urea were carried out.

From Fig. 10.13 it can be seen that the DMSO not only decreased the adsorption density but also yielded a very different isotherm shape, as the hydrogen bonds between alumina and surfactant wereweakened in the presence of DMSO. Moreover DMSO influenced the adsorption process by changing the solvent properties. It was found that in the presence of DMSO no CMC (critical micelles concentration) was detected, which indicated that aggregation of surfactant was disrupted in the presence of DMSO.

From Fig. 10.14 it was quite clear that urea also influences the adsorption of surfactant over alumina as urea exhibited profound effects on the surface tension of the surfactant. The presence of urea facilitates solvation of

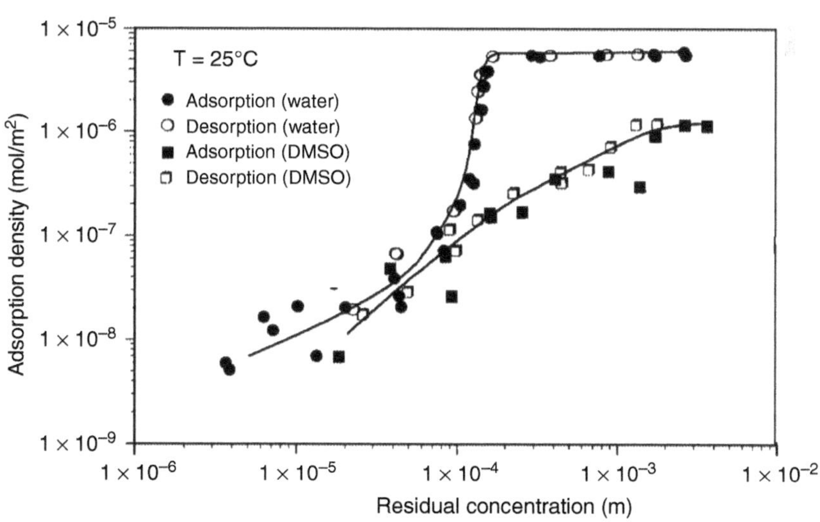

Figure 10.13 ***Adsorption and desorption of n-dodecyl-β-d-maltoside in the presence of 85% DMSO.*** *(Source: Zhang et al., 2002; © Elsevier).*

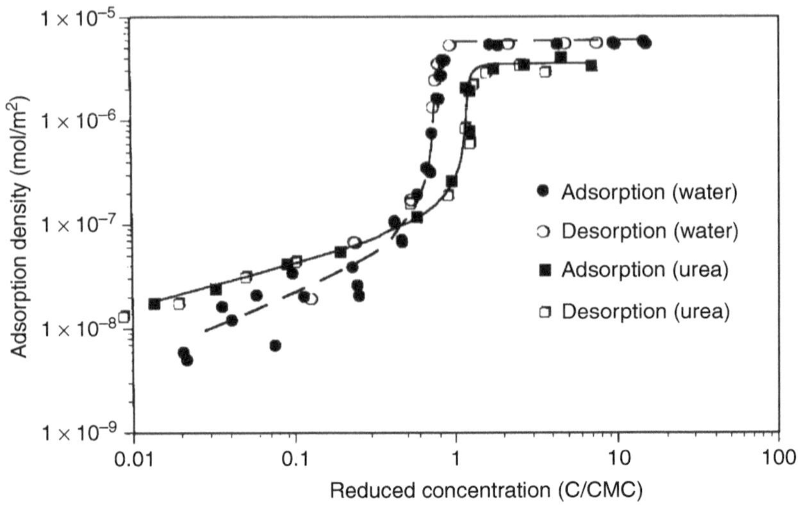

Figure 10.14 ***Adsorption and desorption of n-dodecyl-β-d-maltoside in urea solution as a function of reduced concentrations.*** *(Source: Zhang et al., 2002; © Elsevier).*

surfactant by replacing water molecules in the hydration shell of the solute. Thus, urea depressed the adsorption process of *n*-dodecyl-*β*-D-maltoside.

10.3.5 Removal of Inorganic Species

Alumina in its bare form as well as in modified form was widely investigated for the removal of various metallic species from aqueous solutions; however, arsenic and fluoride are the two most prominent species whose adsorption properties using alumina-based adsorbent materials were extensively explored by the researchers. For example, Kim et al. (2004) reported removal of arsenic ions using mesoporous alumina (MA). The adsorption experiments were conducted at different pH conditions to study the effect of arsenic adsorption on MA. Maximum uptake of arsenate ions was observed at pH 5. The maximum adsorption capacity for As(V) and As(III) was evaluated from Langmuir isotherm parameter as 121 and 47 mg/g, respectively. The sorption kinetics of As(V) and As(III) demonstrated that for both analyte species the adsorption was rapid in the first 5 h and then slowed considerably as the reaction approached equilibrium within 35 h. The mechanism for arsenic adsorption was interpreted through intraparticle diffusion plots. The resultant plots suggested that the arsenic was initially adsorbed by the exterior surface of the adsorbent. When the adsorption at the exterior surface became saturated, the arsenic ions moved into the pores within the particles and were adsorbed by the interior surfaces of the adsorbent particles. When the arsenic ion diffuses into the pores of the MA, the diffusion resistance becomes increased and overall diffusion rate was decreased. The diffusion rate for As(V) was two times faster than that of As(III), indicating that arsenate ions are more easily diffused and transported into MA pores than arsenite. Desorption experiments were also conducted using 0.05 M NaOH solution, and results revealed that more than 85% of arsenic was desorbed from MA within 1 h. Thus fast desorption kinetics recommended continuous use of MA adsorbents for treating arsenic-contaminated water. Similarly, Li et al. (2011) also reported removal of arsenic ions by highly ordered mesoporous alumina. The adsorption performance and stability of the adsorbent was tested as a function of pH at initial pH values 2.0–13.0.

Fig. 10.15 clearly demonstrates that adsorption of arsenic is strongly dependent on solution pH. The adsorbent efficiently removes arsenate ions in the initial pH range of 4–9. Meanwhile, the concentration of aluminum ion in the solution is very low, indicating that the synthesized mesoporous alumina is stable in this pH range. Moreover the adsorbent displayed remarkable arsenic scavenging ability, as only 1.0 g of adsorbent was able to treat 200 kg of arsenic-contaminated water with a pH value of 7, reducing the concentration of arsenate from 100 ppb to 1 ppb.

Patra et al. (2012) reported that self-assembled mesoporous gamma alumina nanoparticles were highly efficient in removing dissolved arsenic ions

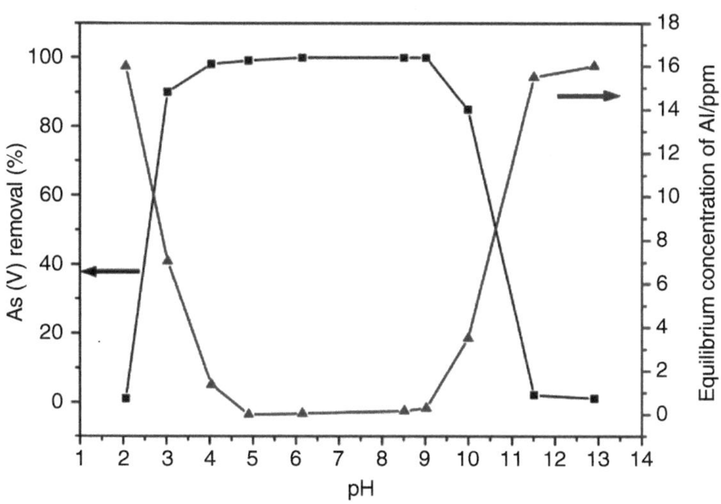

Figure 10.15 ***Effect of initial pH values on arsenate removal with mesoporous alumina and the dissolution amount of mesoporous alumina.*** *(Source: Li et al., 2011; © Elsevier).*

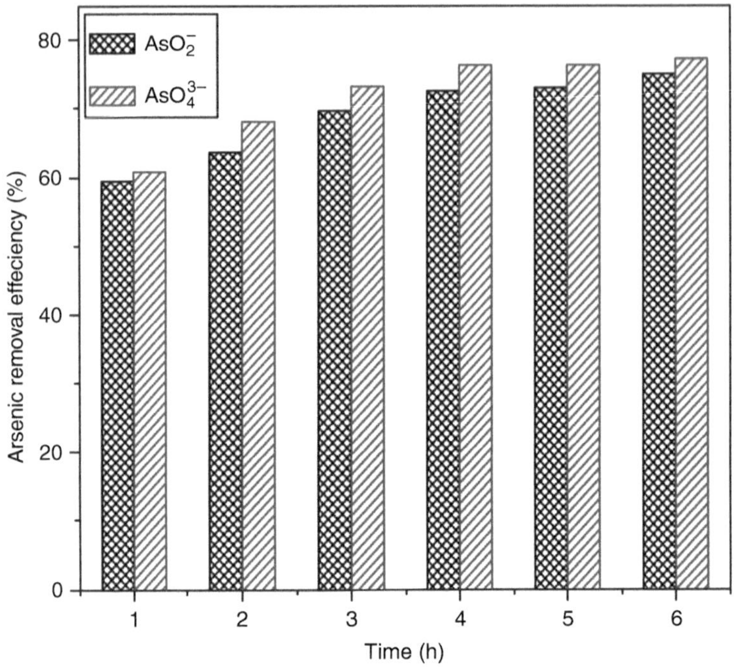

Figure 10.16 ***Adsorption kinetics of both As(III) and As(V) as a function of time over mesoporous gamma alumina nanoparticles.*** *(Source: Patra et al., 2012; © Elsevier).*

from the contaminated aqueous solutions. The kinetics study for the adsorption of As(III) and As(V) was also performed, and results demonstrated (Fig. 10.16) that more than 65% of removal was observed within 2 h of contact time, which was reportedly due to monolayer adsorption on the mesoporous surface of alumina nanoparticles, as the reaction time proceeds multilayered adsorption was reported. Moreover it was also observed that throughout the adsorption studies the removal efficiency remained higher for As(V) over As(III), suggesting higher affinity for the former over the latter at the mesoporous alumina surface. A desorption study was also performed and was found that mesoporous adsorbents could be effectively used for repeated adsorption studies, as arsenic removal efficiency remained at more than 70% even during the fifth recycling experiment.

In another study Saha and Sarkar (2012) investigated the potential of nanocomposites in the removal of As(V) by nano-alumina dispersed in chitosan-grafted polyacrylamide (CTS-g-PA) matrix. The removal of As(V) as a function of pH was investigated over a pH range of 4–9. The removal of arsenic was increased from 45 to 99.5% as pH condition varies from 4.0 to 7.2, and then decreased with further increase in pH. The arsenic removal efficiency of the adsorbent increased significantly from pH 6.8 due to the fact that the surface of nano-alumina and amino group of CTS-g-PA both became positively charged and exhibited high affinity for arsenate anions. The decrease in removal efficiency at pH greater than 7.2 may be attributed to the competition for the active sites by OH^- ions and the electrostatic repulsion of anionic As(V) by the negatively charged Al_2O_3 surface. An adsorption isotherm study revealed that the Freundlich isotherm model was more appropriate in describing adsorption of As(V). Dubinin-Radushkevich model suggested that chemisorption played an important role in the sorption process. The maximum adsorption capacity of the adsorbent was also evaluated using the Langmuir isotherm parameter as 4.23 mg/g.

The adsorption mechanism for the removal of As(V) was also proposed (Fig. 10.17), and it was anticipated that the mechanism of arsenic removal by alumina-loaded polymer bead was governed by both electrostatic adsorptions and complexation mechanisms. As it was examined that As(V) removal was maximum at pH 7.2, thus the presence of a free amino group (NH_2) in CTS-g-PA might exist in equilibrium with the protonated amino group in acidic aqueous solution. Thus arsenate ions might interact with the polymer through electrostatic interactions or complexation and hydrogen bonds. The positively charged Al^{3+} at the surface of chitosan would attract negatively charged As(V) species by electrostatic attraction. The regeneration study of the adsorbent was also investigated and results indicated a retention of 94% capacity in the fifth cycle.

Tang et al. (2009) reported adsorption characteristics of fluoride onto activated alumina. The adsorption kinetics for fluoride ions suggested that the adsorption process attained equilibrium within 10 h. The effect of pH on fluoride removal was also examined and maximum adsorption occurred between pH 5.5 and 7.5 (Fig. 10.18).

From Fig. 10.18, it was also observed that at lower initial fluoride concentrations (10 mg/L) the maximum adsorption of fluoride was reported at broader pH range (5.5–9.0) while at higher initial fluoride concentration (106 mg/L) the maximum adsorption was reported at narrow pH range (5.5–6.5). This may be due to saturation of some surface sites due to higher surface loading. When pH approached pH_{zpc} (8.8), fluoride adsorption decreased sharply with the increase of pH because the concentration of protonated surface sites, which served as fluoride binding sites, decreased significantly with the increase in pH.

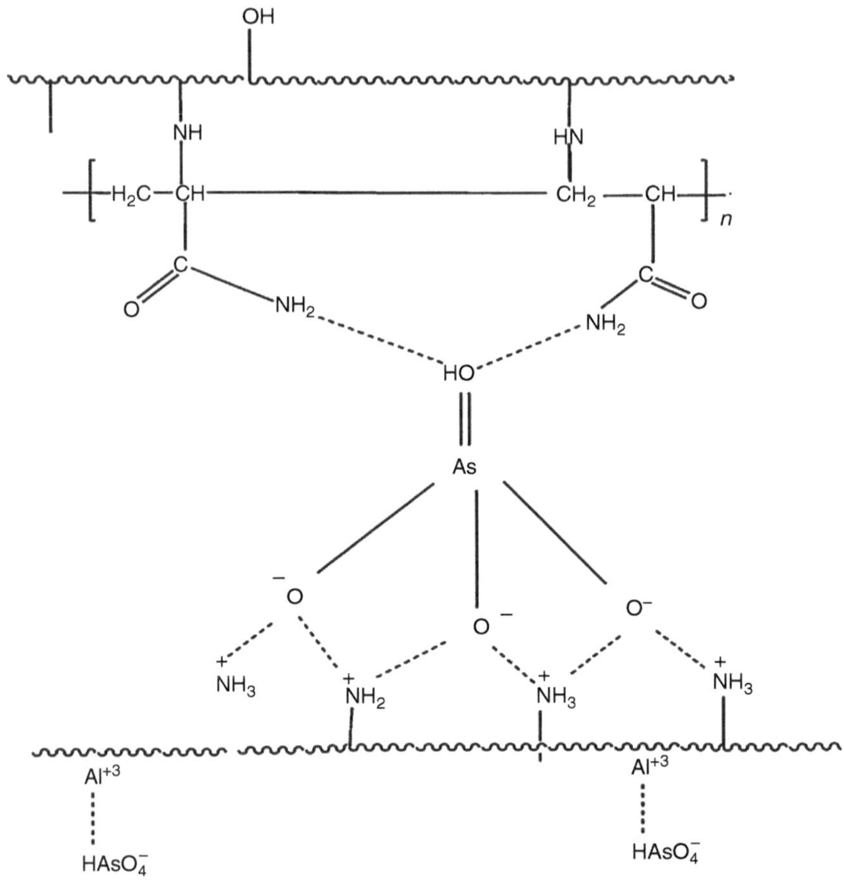

Figure 10.17 ***Schematic diagram representing possible mechanism of As(V) adsorption on nano-alumina particles dispersed in chitosan-grafted polyacrylamide.*** *(Source: Patra et al., 2012; © Elsevier).*

The effect of interfering anions on fluoride adsorption was also investigated. Fluoride adsorption was found to decrease substantially when bicarbonate and sulfate were present in the system however chloride anions had less impact on fluoride adsorption (Fig. 10.19). Chloride ions formed the outer-sphere complex, while sulfate ions formed both outer-sphere and inner-sphere complexes with surface active sites which therefore decline removal of fluoride ions.

Karthikeyan et al. (2009) synthesized polyaniline/alumina (PANi-AlO) and polypyrrole/alumina (PPy–AlO) composites for the removal of fluoride ions from aqueous solutions. It was observed that adsorption of fluoride

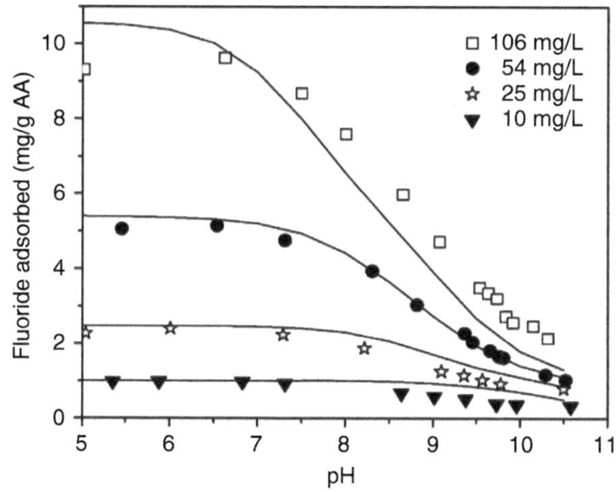

Figure 10.18 ***Effect of initial concentration on fluoride adsorption on activated alumina as a function of pH.*** *(Source: Tang et al., 2009; © Elsevier).*

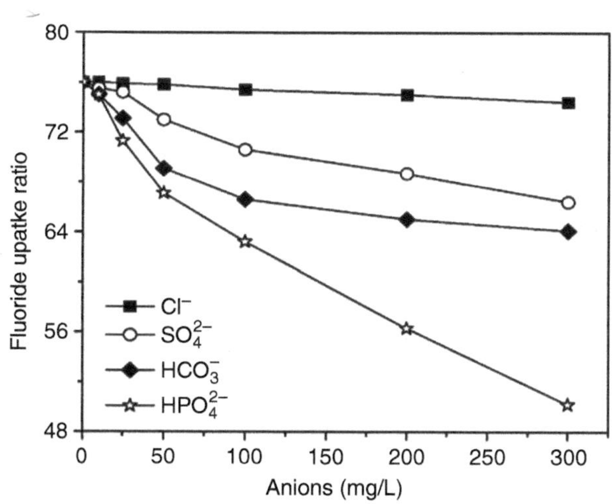

Figure 10.19 ***Effect of interfering anions on fluoride adsorption on activated alumina.*** *(Source: Tang et al., 2009; © Elsevier).*

ions per unit mass of the adsorbents was substantially higher than that of individual constituents. Fluoride ion removal was favorable at acidic pH conditions (pH 3.0) and higher temperature ranges (323 K). The maximum fluoride adsorption capacity for PANi-AlO and PPy–AlO was reported as 13.0 mg/g and 12.0 mg/g, respectively. Mechanism of fluoride adsorption was also proposed, and it was suggested that removal by composites involves interaction of fluoride ions with both the polymer and alumina moieties.

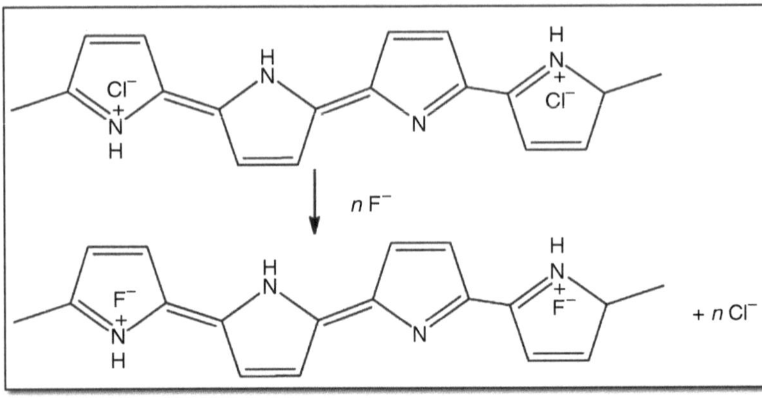

Figure 10.20 ***Plausible mechanism for removal of fluoride ions in the presence of polypyyrole.*** *(Source: Karthikeyan et al., 2009; © Elsevier).*

Usually in the presence of alumina, removal of fluoride occurred via formation of aluminium-fluoro complexes is given as follows:

$$Al_2O_3 + nF^- \xrightarrow{mH_2O} AlF_n^{3-n} + Al(OH)_{3-m}F_m \quad (10.8)$$

However, in the presence of polypyrrole, which was employed in the form of chloride ion–doped polymer, fluoride ions were removed from aqueous solutions through the replacement of doped ionizable chloride ions, which were present in it through ion-exchange mechanisms (Fig. 10.20). Thus, the aforementioned mechanisms were found to play a significant role in adsorption of fluoride ions from aqueous solutions.

Su et al. (2008) reported adsorption characteristics of Se(IV) and V(V) on activated alumina. The adsorption results suggested that adsorption of these species was strongly dependent on pH and surface loading. The pH study was performed in a pH range 1.5–12.0 with different initial analyte concentrations. The results demonstrated that at low initial analyte concentrations the maximum uptake amount for Se(IV) and V(V) was extended from pH 2.0–7.0. However at high initial analyte concentrations the pH range at which maximum adsorption occurred confined to pH 2.5–4.0. Effect of ionic strength on adsorption of Se(IV) and V(V) was investigated and no significant impact on the adsorption of these ions in the presence of interfering ions was detected, which may be due to the formation of inner sphere complexes with the active site on activated alumina. Further, it was also indicated that the electrostatic effect on the adsorption of these elements was relatively not important compared to surface chemical reactions.

The effect of surface loading on the adsorption process was also interpreted in the pH range of 1.5–12 under three surface loading conditions, 5, 20 and 50 mg/g sorbent. When the surface loading was increased to 40–50 mg/g sorbent, the pH range at which the adsorption maxima occurred decreased to a pH range of 2.5 to 4.

Saha et al. (2008b) explored the potential of trimesic acid (TMA)–coated alumina for adsorptive removal of Fe(III) and Fe (II) from aqueous solutions. It was reported that an organic acid–coated surface generates a negative charge on the surface, which can efficiently bind metal cations through surface complexation, thus on the basis of this hypothesis the adsorbent was synthesized. The adsorption study was optimized for various process variables, including initial ion concentration, time, pH, and temperature. The effect of solution pH on adsorption of Fe(III) and Fe(II) was investigated (Fig. 10.21). It was found that at pH ~ 1.0, Fe(III) uptake was very low at ~3.0 mg/g, which may be due to the competitive adsorption of H^+ with Fe(III). In the pH range of 1.5–3.0, the uptake amount of Fe(III) was increased from 12.6 mg/g to 19.5 mg/g, while no precipitation took place. The adsorption capacity of TMA-coated alumina for Fe(III) was high even at a pH of 1.5, when most of the carboxyl group of TMA was in protonated condition. At pH 3.0, some carboxyl groups of TMA exist in dissociated form and can strongly and preferentially bind Fe(III). This indicated that Fe(III) has high affinity for protonated as well as dissociated carboxyl groups of TMA that results in high adsorption capacity of Fe(III) in highly acidic pH condition. Precipitation of Fe(III) was observed beyond pH 3.0 and about 70–80% was precipitated at pH 4.0; as a result,

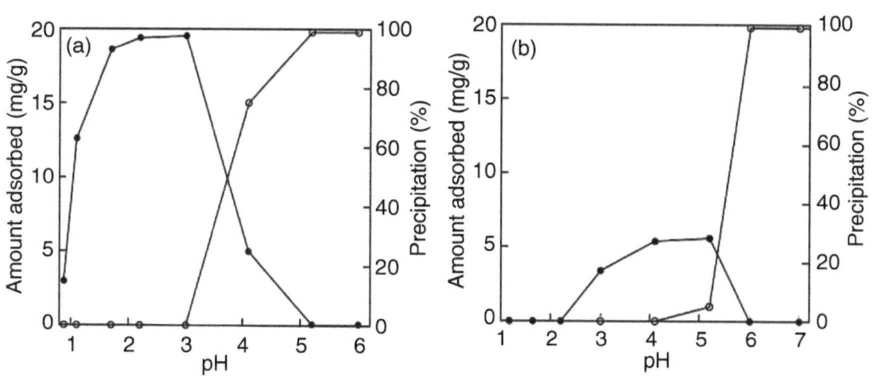

Figure 10.21 ***Effect of pH on adsorption and precipitation of 100 mg/L separate solution of (a) Fe(III) and (b) Fe(II) ions.*** *(Source: Saha et al., 2008b; © Elsevier).*

uptake amount declined remarkably. However, on the other hand, adsorption of Fe(II) was found to be much lower as compared to Fe(III). No adsorption was reported at pH 2.0, which indicated that highly acidic pH depressed the adsorption efficiency of the adsorbent. In equilibrium pH range of 3.0–5.0, the Fe(II) adsorbed only and was a mere 3.4 to 5.6 mg/g. Beyond pH 5.0, precipitation of Fe(II) started and almost 100% Fe(II) precipitated at pH 6.0.

Maximum metal uptake for Fe(III) and Fe(II) calculated from the Langmuir model was 26.6 and 8.4 mg/g, respectively. This indicates a higher adsorption capacity for Fe(III) compared to Fe(II). Mechanism of adsorption was also predicted on the basis of metal-ligand complex formation. Fe(III) was regarded as hard acid because of its small size and high charge, whereas Fe(II) was considered as borderline acid due to its relatively larger size and lower charge. Thus, as per Pearsons's rule, hard acid prefers to bind hard bases and soft acids prefer to bind soft bases. Thus TMA forms stable and strong complexation with hard acid Fe(III) as compared to Fe(II). Desorption of the adsorbent was tested using 0.2–1.0 M HCl and H_2SO_4. Desorption kinetics was very rapid as almost 100% desorption was completed within 10–15 min for both Fe(III) and Fe(II). Regenerated adsorbent was reused for three consecutive adsorption cycles. The adsorption capacity for Fe(III) was reduced from 20 to 15.5 mg/g after the third cycle, and for Fe(II), uptake value reduced from 5.4 to 4.0 mg/g after the third cycle.

Saha et al. (2008a) also reported adsorption of Cu(II) using similar adsorbent trimesic acid–coated alumina. pH-based removal study suggested that removal percentage of Cu(II) increased with the increase of pH. While increasing pH from 4 to 5, removal percent increased rapidly from 72 to 94%. At this pH, carboxyl groups of TMA exist in dissociative form and are complexed strongly with positively charged Cu(II), where the adsorbate binds with each carboxylate moiety of TMA. Maximum adsorption capacity for the Cu(II) ion was reported as 10.80 mg/g. The desorption study suggested that almost 100% recovery of the adsorbent was achieved using both HCl and H_2SO_4.

Adsorption of radionuclides on alumina nanoparticles was also reported by several workers. Qian et al. (2014) demonstrated adsorption-based removal of U(VI) ions from aqueous solutions using alumina nanoparticles and alpha alumina particles. An adsorption kinetics study revealed (Fig. 10.22) that the adsorption process attained equilibrium within 48 h of contact time and most of the U(VI) removal of about 100 and 70% was achieved within 6 h for nano-alumina and for alpha-alumina, respectively.

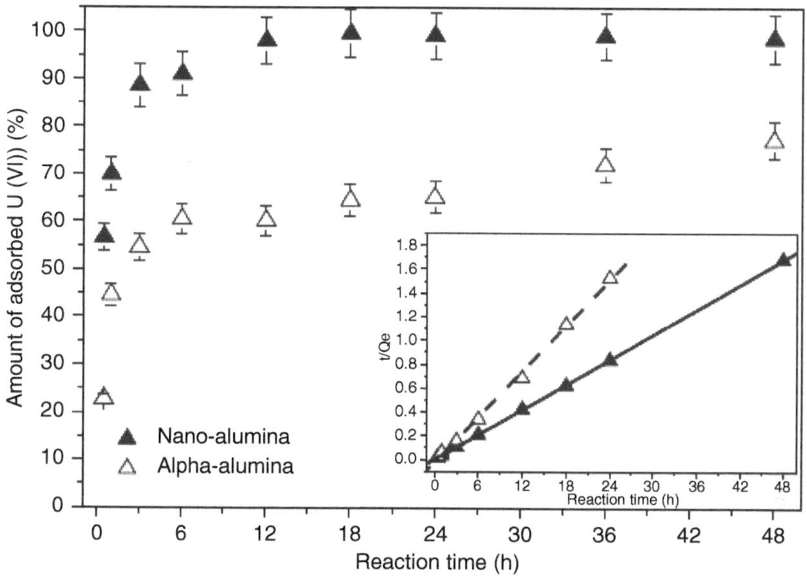

Figure 10.22 ***The adsorption kinetics of U(VI) on the nano-alumina and alpha-alumina (pH = 4.5, U(VI) = 0.1 mmol/L, T = 298 K.*** *(Source: Qian et al., 2014; © Elsevier).*

It was also presumed that the fast uptake of U(VI) on nano-alumina and alpha-alumina within 6 h could be due to fast sorption of U(VI) on the external surface of nano-alumina and alpha-alumina, whereas intraparticle diffusion of U(VI) ions can be distinguished from the plots corresponding to the contact time of 6–48 h, indicating slow sorption of U(VI). It was also observed that the reaction rate of nano-alumina for U(VI) was much higher than that of alpha-alumina, which could be attributed to the nanoscale space confinement effect. Effect of pH on adsorption of U(VI) on nano-alumina and alpha-alumina was also investigated. It was observed that removal of U(VI) was increased significantly from pH 3.0 to 6.5 and demonstrated maximum efficiency at pH 6.5–8.5 due to electrostatic interaction. However at pH > 8.5, decreases in U(VI) adsorption was observed; this may be attributed to predomination of negative U(VI) species at higher pH range. Further, the study on the effect on ionic strength on adsorption of U(VI) was also conducted. The study revealed that for nano-alumina adsorption was independent of ionic strength, while for alpha-alumina adsorption of U(VI) was decreased with the increase of ionic strength. This was due to formation of inner sphere complexes with nano-alumina and outer sphere complexes with alpha-alumina. The maximum adsorption capacity of the adsorbents, nano-alumina and alpha-alumina, was calculated from

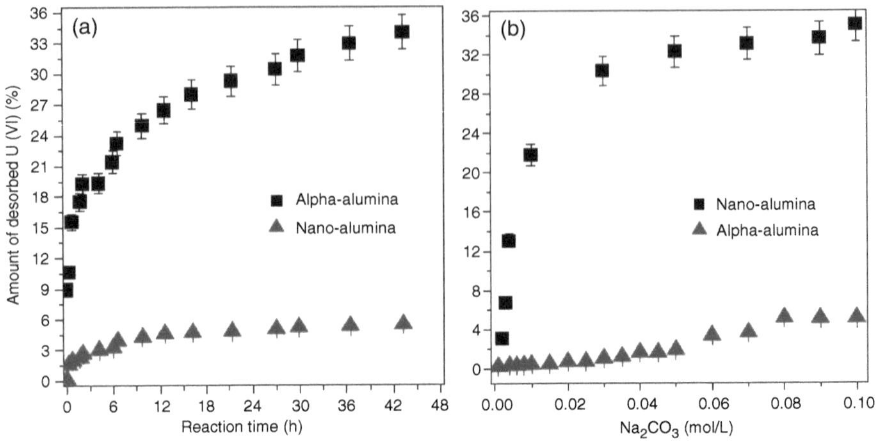

Figure 10.23 ***Desorption study of U(VI)*** (a) as a function of time, and (b) $NaHCO_3$ concentration on nano-alumina and alpha-alumina (adsorbent dose = 0.5 g/L, T = 298 K). *(Source: Qian et al., 2014; © Elsevier).*

the Langmuir model at pH 4.5 and T = 313 K as 129.87 and 78.13 mg/g, respectively.

Desorption study for nano-alumina and alpha-alumina was also reported as a function of extraction time and extractant concentration (Fig. 10.23). It was observed that with the increase of reaction time desorption of adsorbate increases, and within 48 h in the presence of 0.01 M $NaHCO_3$ solution approximately 35 and 5% of U(VI) were desorbed from alpha-alumina and nano-alumina, respectively, indicating the strong chemical interaction for nano-alumina due to the nanoscale space confinement effect. The effect of $NaHCO_3$ concentration on the desorption of U(VI) from alpha-alumina and nano-alumina was also investigated. It was observed that the desorption rates for alpha-alumina increased rapidly from 3 to 33%, while the desorption rates for nano-alumina were observed as 0.3–2.0% with the increasing $NaHCO_3$ concentration from 0.001 to 0.05 mol/L, respectively.

Another study reported isolation of radioactive iodine-131 (I-131) from fission products by adsorption onto silver-coated alumina (Mushtaq et al., 2014). It was reported that I-131 was produced as a by-product of molybdenum-99 production. However, most of the fission products such as uranium and actinides can be precipitated out after its treatment with basic solution, while radioisotopes of iodine and molybdenum remained in the filtrate. Thus, in order to isolate these isotopes the filtrate was further passed through the silver-coated alumina column. It was observed that sorption of Mo-99 on silver-coated alumina was negligible while I-131 demonstrated strong affinity for the adsorbent as more than 99% of I-131 was retained by the column. The I-131 saturated column was further desorbed using 1.0 M $Na_2S_2O_3$ (sodium thiosulfate) solution, and nearly 90% recovery of radioiodine was reported. The sorption–desorption mechanism is given as follows:

$$\text{Sorption} = Ag^+ + I^- \rightarrow AgI(ppt) \tag{10.9}$$

$$\text{Desorption} = AgI + 2\,Na_2S_2O_3 \rightarrow NaI + Na_3[Ag(S_2O_3)_2] \tag{10.10}$$

In alkaline medium, iodine reacts with silver and is precipitated as NaI, and with the addition of sodium thiosulfate, the sodium iodide is dissolved, whereas silver forms a soluble complex.

10.4 CONCLUSIONS

Alumina as an adsorbent material exhibited tremendous potential for removal of various organic and inorganic contaminants. Most importantly, nanostructured alumina revealed strong affinity for anionic pollutants due to the electron deficiency at its surface; therefore, the adsorbent can be effectively used for the scavenging of toxic metal oxyanions. However, alumina in its bare form displayed relatively poor adsorption ability for organic compounds due to hydrophilicity, which can be overcome through modification with some chemicals, which presumably introduced certain functional groups on alumina particles that promote adsorption capacity for organic compounds. However, most of the adsorption studies were performed in small scale and most often limited to laboratory domain only, thus the obtained result possibly deviates when a similar study was performed using large-scale waste effluents. Thus, it would be highly desirable for the researchers to scale up the potential of alumina-based adsorbents for real waste samples. Moreover, if the adsorption results for real water is appealing then the same could be successfully commercialized as a beneficiary to the society.

REFERENCES

Adak, A., Pal, A., 2006a. Removal kinetics and mechanism for crystal violet uptake by surfactant-modified alumina. J. Environ Sci. Health 41, 2283–2297.
Adak, A., Pal, A., 2006b. Removal of phenol from aquatic environment by SDS-modified alumina: batch and fixed bed studies. Sep. Purif. Technol. 50, 256–262.
Adak, A., Bandyopadhyay, M., Pal, A., 2005. Removal of crystal violet dye from wastewater by surfactant-modified alumina. Sep. Purif. Technol. 44, 139–144.

Afkhami, A., Tehrani, M.S., Bagheri, H., 2010. Simultaneous removal of heavy-metal ions in wastewater samples using nano-alumina modified with 2,4-dinitrophenylhydrazine. J. Hazard. Mater. 181, 836–844.
Afkhami, A., Bagheri, H., Madrakian, T., 2011. Alumina nanoparticles grafted with functional groups as a new adsorbent in efficient removal of formaldehyde from water samples. Desalination 281, 151–158.
Banerjee, S., Gautam, R.K., Jaiswal, A., Chattopadhyaya, M.C., Sharma, Y.C., 2015. Rapid scavenging of methylene blue dye from aqueous solutions by adsorption on nanoalumina. RSC Adv. 5, 14425–14440.
Bhargavi, R.J., Maheshwari, Utkarsh, Gupta, Suresh, 2015. Synthesis and use of alumina nanoparticles as an adsorbent for the removal of Zn(II) and CBG dye from wastewater. Int. J Ind. Chem. 6, 31–41.
Blomqvist, J., Salo, P., 2009. Adsorption of benzene, phenol, propane and carbonic acid molecules on oxidized Al(111) and α-Al2O3(0001) surfaces: a first-principles study. J. Phys. Condens. Matter 21, 1–18.
Castro, R.H.R., Ushakov, S.V., Gengembre, L., Gouvêa, D., Navrotsky, A., 2006. Surface energy and thermodynamic stability of γ-alumina: effect of dopants and water. Chem. Mater. 18, 1867–1872.
Danis, T.G., Albanis, T.A., Petrakis, D.E., Pomonis, P.J., 1998. Removal of chlorinated phenols from aqueous solutions by adsorption on alumina pillared clays and mesoporous alumina aluminum phosphates. Water Res. 32, 295–302.
Fan, A., Somasundaran, P., Turro, N.J., 1997. Adsorption of alkyltrimethylammonium bromides on negatively charged alumina. Langmuir 13, 506–510.
Gawade, A.S., Vanjara, A.K., Sawant, M.R., 2005. Removal of herbicide from water with sodium chloride using surfactant treated alumina for wastewater treatment. Sep. Purif. Technol. 41, 65–71.
Huang, L., Maltesh, C., Somasundaran, P., 1996. Adsorption behavior of cationic and nonionic surfactant mixtures at the alumina–water interface. J. Colloid Interface Sci. 177, 222–228.
Johnston, G.P., Muenchausen, R., Smith, D.M., Fahrenholtz, W., Foltyn, S., 1992. Reactive Laser Ablation Synthesis of Nanosize Alumina Powder. J. Am. Ceram. Soc. 12, 3293–3298.
Karthikeyan, M., Satheesh Kumar, K.K., Elango, K.P., 2009. Conducting polymer/alumina composites as viable adsorbents for the removal of fluoride ions from aqueous solution. J. Fluor. Chem. 130, 894–901.
Kato, E., Daimon, K., Nanbu, M., 1981. Decomposition of two aluminum sulfates and characterization of the resultant aluminas. J. Am. Ceram. Soc. 64, 436.
Khosla, E., Kaur, S., Dave, P.N., 2013. Mechanistic study of adsorption of acid orange-7 over aluminum oxide nanoparticles. J. Eng. 2013, 1–8.
Kim, Y., Kim, C., Choi, I., Rengaraj, S., Yi, J., 2004. Arsenic removal using mesoporous alumina prepared via a templating method. Environ. Sci. Technol. 38, 924–931.
Li, J., Shi, Y., Cai, Y., Moua, S., Jiang, G., 2008. Adsorption of di-ethyl-phthalate from aqueous solutions with surfactant-coated nano/microsized alumina. Chem. Eng. J. 140, 214–220.
Li, W., Cao, C.Y., Wu, L.Y., Ge, M.F., Song, W.G., 2011. Superb fluoride and arsenic removal performance of highly ordered mesoporous aluminas. J. Hazard Mater. 198, 143–150.
Mushtaq, A., Pervez, S., Hussain, S., Asif, M., Siddique, M.U., Mirza, J.A., Khan, M.M., Khalid1, U., Haq, M.A., Shahzad, K., 2014. Preparation and characterization of silver coated alumina for isolation of iodine-131 from fission products. J. Eng. Manuf. Technol. 2, 1–9.
Nadafi, K., Vosoughi, M., Asadi, A., Omidvar Borna, M., Shirmardi, M., 2014. Reactive red 120 dye removal from aqueous solution by adsorption on nano-alumina. J. Water Chem. Technol. 36 (3), 125–133.

Neupane, D., Park, J.W., 1999. Binding of dialkylated disulfonated diphenyl oxide surfactant onto alumina in the aqueous phase. Chemosphere 38, 1–12.
Neupane, D., Park, J.W., 2000. Partitioning of naphthalene to Gemini surfactant treated alumina. Chemosphere 41, 787–792.
Noda, H., Muramoto, K., Kim, H., 2003. Preparation of nano-structured ceramics using nanosized Al_2O_3 particles. J. Mater. Sci. 38, 2043.
Ogihara, T., Nakagawa, H., Yanagawa, T., Ogata, N., Yoshida, K., 1991. Preparation of Monodisperse, Spherical Alumina Powders from Alkoxides. J. Am. Ceram. Soc. 74, 2263–2269.
Parida, K.M., Pradhan, A.C., Das, J., Sahu, N., 2009. Synthesis and characterization of nano-sized porous gamma-alumina by control precipitation method. Mater. Chem. Phys. 113, 244–248.
Park, H.C., Kim, S.W., Lee, S.G., Kim, J.K., Hong, S.S., Lee, G.D., Park, S.S., 2003. Synthesis of α -alumina platelets using flux method in 2.45 GHz microwave field. Mater. Sci. Eng. 363, 330–334.
Patra, A.K., Dutta, A., Bhaumik, A., 2012. Self-assembled mesoporous Al_2O_3 spherical nanoparticles and their efficiency for the removal of arsenic from water. J. Hazard. Mater. 201–202, 170–177.
Qian, L., Ma, M., Cheng, D., 2014. The effect of water chemistry on adsorption and desorption of U(VI) on nano-alumina. J. Mol. Liq. 197, 295–300.
Rahmani, A., Mousavi, H.Z., Fazli, M., 2010. Effect of nanostructure alumina on adsorption of heavy metals. Desalination 253, 94–100.
Raut, P.A., Dutta, M., Sengupta, S., Basu, J.K., 2013. Alumina-carbon composite as an effective adsorbent for removal of Methylene Blue and Alizarin Red-s from aqueous solution. Indian J. Chem. Technol. 20, 15–20.
Renuka, N.K., Shijina, A.V., Praveen, A.K., 2012. Mesoporous γ-alumina nanoparticles: synthesis, characterization and dye removal efficiency. Mater. Lett. 82, 42–44.
Saha, S., Sarkar, P., 2012. Arsenic remediation from drinking water by synthesized nano-alumina dispersed in chitosan-grafted polyacrylamide. J. Hazard. Mater. 227-228, 68–78.
Saha, B., Chakraborty, S., Das, G., 2008a. Trimesic acid coated alumina: an efficient multi-cyclic adsorbent for toxic Cu(II). J. Colloid Interface Sci. 320, 30–39.
Saha, B., Chakraborty, S., Das, G., 2008b. A comparative metal ion adsorption study by trimesic acid coated alumina: a potent adsorbent. J. Colloid Interface Sci. 323, 26–32.
Shojaie-Bahaabad, M., Taheri-Nassaj, E., 2008. Economical synthesis of nano alumina powder using an aqueous sol–gel method. Mater. Lett. 62, 3364–3366.
Somasundaran, P., Fuerstenau, D.W., 1966. Mechanism of alkyl sulfonate adsorption at the alumina-water interface. J. Phys. Chem. 70 (1), 90–96.
Srivastava, V., Weng, C.H., Singh, V.K., Sharma, Y.C., 2011. Adsorption of nickel ions from aqueous solutions by nano alumina: kinetic, mass transfer, and equilibrium studies. J. Chem. Eng. Data 56, 1414–1422.
Su, T., Guan, X., Gu, G., Wang, J., 2008. Adsorption characteristics of As(V), Se(IV), and V(V) onto activated alumina: effects of pH, surface loading, and ionic strength. J. Colloid Interface Sci. 326, 347–353.
Sun, S., Jaffe, P.R., 1996. Sorption of phenanthrene from water onto alumina coated with dianionic surfactants. Environ. Sci. Technol. 30, 2906–2913.
Tang, Y., Guan, X., Su, T., Gao, N., Wang, J., 2009a. Fluoride adsorption onto activated alumina: modeling the effects of pH and some competing ions. Colloids Surf. A 337, 33–38.
Tarun, K.N., Ashim, K.B., Sudip, K.D., 2009. Adsorption of Cd(II) and Pb(II) from aqueous solutions on activated alumina. J. Colloid Interface Sci. 333, 14–26.
Tok, A.I.Y., Boey, F.Y.C., Zhao, X.L., 2006. Novel synthesis of Al_2O_3 nano-particles by flame spray pyrolysis. J. Mater. Process. Technol. 178, 270–273.
Varatharajan, K., Dash, S., Arunkumar, A., Nithya, R., Tyagi, A.K., Raj, B., 2003. Synthesis of Nanocrystalline α–Al2O3 by Ultrasonic Spray Pyrolysis. Mater. Res. Bull. 38, 577–583.

Wang, S., Li, X., Wang, S., Li, Y., Zhai, Y., 2008. Synthesis of γ-alumina via precipitation in ethanol. Mater. Lett. 62, 3552–3554.
Wu, J.M., 2001. Nano-sized amorphous alumina particles obtained by ball milling ZnO and Al powder mixture. Mater. Lett. 48, 324–330.
Zhang, L., Somasundaran, P., Mielczarski, J., Mielczarski, E., 2002. Adsorption mechanism of n-dodecyl-*β*-D-maltoside on alumina. J. Colloid Interface Sci. 256, 16–22.
Zhang, J., Zhou, Q., Ou, L., 2012. Kinetic, isotherm, and thermodynamic studies of the adsorption of methyl orange from aqueous solution by chitosan/alumina composite. J. Chem. Eng. Data 57, 412–419.
Zhou, J., Wang, L., Zhang, Yu, J., 2013. Facile synthesis of alumina hollow microspheres via trisodium citrate-mediated hydrothermal process and their adsorption performances for p-nitrophenol from aqueous solutions. J. Colloid Interface Sci. 394, 509–514.
Zhu, L.N., Ma, J., Yang, S.D., 2007. Removal of phenol by activated alumina bed in pulsed high-voltage electric field. J. Environ. Sci. 19, 409–415.
Zolgharnein, J., Asanjrani, N., Bagtash, M., Azimi, G., 2014. Multi-response optimization using Taguchi design and principle component analysis for removing binary mixture of alizarin red and alizarin yellow from aqueous solution by nano γ-alumina. Spectrochimica Acta Part A 126, 291–300.
Zolgharnein, J., Bagtash, M., Shariatmanesh, T., 2015. Simultaneous removal of binary mixture of Brilliant Green and Crystal Violet using derivative spectrophotometric determination, multivariate optimization and adsorption characterization of dyes on surfactant modified nano-γ-alumina. Spectrochimica Acta Part A 137, 1016–1028.

CHAPTER 11

Bimetallic Nanomaterials for Remediation of Water and Wastewater

11.1 INTRODUCTION

During recent years, zero valent metallic iron (Fe^0) and magnetic nanoparticles have been successfully utilized to remove dyestuffs, heavy metals, metalloids, phenols, and pathogenic bacteria from aqueous media due to their low cost, high efficiency, ease of operation, effectiveness, and biocompatibility in nature (Fan et al., 2009; Hu et al., 2007; Fang et al., 2011). Iron oxide–based nanomaterials can remediate a variety of contaminants when used in the presence of oxygen (Joo et al., 2005). Fan et al. (2009) applied nanoscale zero valent iron to degrade methyl orange in an aqueous solution with a decolorization efficiency of 100% within 60 min. Nurmi et al. (2005) stated that pH adjustment was unnecessary in decontamination of pollutants because of the high surface area, small particle size, and large density of intrinsic surface sites of the iron particles. Nanoscale iron particles have gained wide applications in the separation and purification of toxic metal ions, dyes, phenols, bacteria, and other organic compounds. However, a number of research papers published on the use of iron oxide materials revealed that the pollutant adsorption rate by iron oxide nanoparticles was very slow (Joo et al., 2005). In addition, the doping of a second catalytic metal such as Ni, Cu, Pd, and Pt might enhance the adsorption rate compared with nanoscale iron alone (Wang et al., 2004; Schrick et al., 2002) and also prevents oxidation in air. Therefore, a large number of combinations of nanoscale iron-based bimetallic compounds have been manufactured and utilized for the degradation of dyes, phenols, toxic metals, bisphenol A, and chlorinated organic compounds (Lee et al., 2014; He and Zhao, 2005; Lin et al., 2004; Wang et al., 2009). Ma et al. (2004) utilised an Fe–Cu bimetallic system for the decolorization of methylene blue in aqueous solution, and the degradation of orange G in aqueous solution was investigated using Fe–Ni bimetallic nanoparticles (Bokare et al., 2007).

Nanomaterials for Wastewater Remediation
http://dx.doi.org/10.1016/B978-0-12-804609-8.00011-X

11.2 APPLICATIONS OF BIMETALLIC NANOMATERIALS

11.2.1 Degradation of Chlorophenol

Wei et al. (2006) synthesized nanoscale Pd/Fe bimetallic particles and used them to hydrodechlorinate 2,4-dichlorophenol from aqueous solutions. In the catalytic degradation, 2,4-dichlorophenol was transformed to phenol and a small amount of 2-chlorophenol and 4-chlorophenol was produced in the reaction. The effects of various parameters such as Pd bulk loadings in the bimetals, temperatures, and pH conditions on the hydrodechlorination of 2,4-dichlorophenol were examined. The results showed that higher Pd bulk loadings, higher temperatures, and weak acid conditions are beneficial to the catalytic dechlorination of 2,4-dichlorophenol. The reaction pathway for the transformation of 2,4-dichlorophenol has been illustrated in Fig. 11.1. The degradation rate of 2,4-dichlorophenol and the yield of phenol increased with the increase of Pd bulk loading. The catalytic degradation rate of 2,4-dichlorophenol corresponding to the Pd bulk loading of 0.0233, 0.0541, 0.0672, and 0.0932% was 71.78, 98.57, 98.44, and 100%, respectively. The dechlorination of 2,4-dichlorophenol and produced intermediates were favorable in the weak acid conditions such as pH ~5.5.

Wang et al. (2013) studied the dispersity and physicochemical properties of palladium/iron nanoparticles (Pd/Fe), and modified Pd/Fe nanoparticles with poly(methylmethacrylate) (PMMA), polyacrylic acid (PAA), and cetyltrimethylammonium bromide (CTAB), respectively, and obtained three hybrid nanoparticles denoted M-Pd/Fe nanoparticles, A-Pd/Fe nanoparticles, and C-Pd/Fe nanoparticles. The three hybrid Pd/Fe nanoparticles were studied for the dechlorination of 2,4-dichlorophenol in aqueous

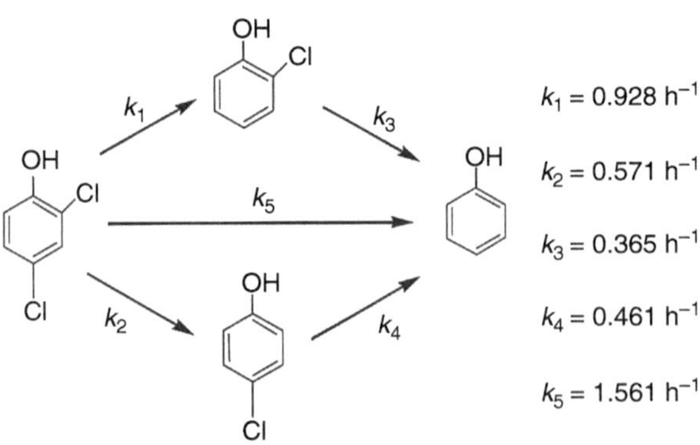

Figure 11.1 ***Reaction pathway of hydrodechlorination of 2,4-dichlorophenol by nanoscale Pd/Fe.*** *(Source: Wei et al., 2006; © Elsevier).*

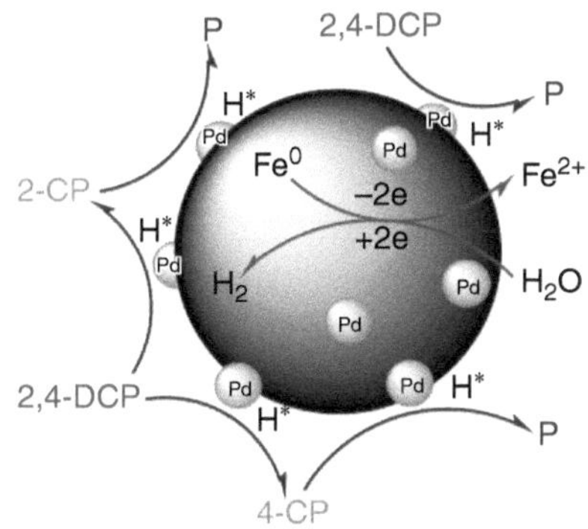

Figure 11.2 ***Schematic diagram of the mechanism of Pd/Fe nanoparticles reacting with 2,4-dichlorophenol.*** *(Source: Wang et al., 2013; © Elsevier).*

solution. The results showed that the catalytic dechlorination efficiency of 2,4-dichlorophenol with modified Pd/Fe nanoparticles increased by 23–58% within a given reaction period of 20 min. The kinetic studies of 2,4-dichlorophenol by Pd/Fe nanoparticles implied that their behaviors did not match the first- or pseudo-first-order kinetics: because the presence of oxidation phases on the surface of pristine Pd/Fe nanoparticles and the dispersants on the surface of nanoparticles could influence the diffusion of 2,4-dichlorophenol onto reactive sites, thus affecting the whole degradation process. A new revised kinetic was proposed for the degradation of 2,4-dichlorophenol by Pd/Fe nanoparticles. Basically, the dechlorination of 2,4-dichlorophenol by Pd/Fe or/and modified Pd/Fe involved the following multistep reactions: (1) the diffusion of 2,4-dichlorophenol onto the surface of Pd/Fe nanoparticles; (2) the adsorption of 2,4-dichlorophenol on the surface reactive sites of nanoparticles; (3) subsequent catalytic reduction reactions between 2,4-dichlorophenol adsorbed on the surface reactive sites and the Pd/Fe nanoparticles; and (4) the formation, desorption, and external diffusion of the final and intermediate products. Fig. 11.2 shows the schematic diagrams of the mechanism of Pd/Fe nanoparticles reacting with 2,4-dichlorophenol. The catalytic dechlorination of 2,4-dichlorophenol with modified Pd/Fe nanoparticles generated the intermediate by-products as 2-chlorophenol and 4-chlorophenol, which were then dechlorinated to phenol as end products.

With the aim to increase the dispersion of nanoparticles and enhance their reactivity, Su et al. (2011) synthesized a bentonite-supported Fe/Pd nanoparticle (B/Fe/Pd), and used it for the dechlorination of *p*-chlorophenol from aqueous solutions. The B/Fe/Pd was characterized with an SEM, EDS,

XRD, and BET surface area analyzer. It was confirmed from the characterization that the presence of bentonite led to a decrease in aggregation of Fe/Pd nanoparticles and enhanced their reactivity. More specifically, complete dechlorination was found by the B/Fe/Pd nanoparticles, while only 25.0 and 15.0% were obtained using B/Fe and bentonite, respectively. Kinetic studies revealed that the dechlorination rate of *p*-chlorophenol was followed by pseudo-first-order models, and the dechlorination rate increased with rising temperature and Pd loading. While the apparent activation energy for dechlorination *p*-chlorophenol using B/Fe/Pd was 40.1 kJ/mol, which shows the characteristic of a chemically surface-controlled reaction. The potential mechanism for degradation of *p*-chlorophenol was based on Fe acting as the reducing agent, while Pd acts as a catalyst with hydrogen generated from water and bentonite as a support.

Patel and Suresh (2006) obtained greater than 85% removal of pentachlorophenol (PCP) by magnesium/silver bimetallic systems in the presence of acetic acid. Dechlorination of PCP by magnesium/silver bimetallic systems was found to be sequential, and phenol was recorded as the ultimate hydrocarbon along with some accumulation of tetra-, tri-, and dichlorophenols. The catalytic degradation was fitted to pseudo-second-order kinetics. Low degradation rate was obtained for PCP by magnesium/silver bimetallic systems when the reaction was carried out in the absence of acid using the Mg^0/Ag system. Dechlorination studies on initial concentrations of 2,3,4,6-tetrachlorophenol and 2,4,5-trichlorophenol, under identical conditions as to PCP by the magnesium/silver bimetallic system, revealed that dechlorination efficiency and reaction rate constants decrease with decreasing number of chlorine atoms on the target compound. Fig. 11.3 shows the schematic representation of events occurring during dechlorination of PCP by magnesium/silver bimetallic systems.

Zhang et al. (2006) synthesized the Ni–Fe bimetallic particles and applied for the removal of PCP from aqueous solutions. The average particle size and surface area was found to be 30 nm and 20.9 m^2/g, respectively. The catalytic degradation of PCP by the Ni–Fe bimetallic nanoscale particles in aqueous solutions were carried out under the enhancement of ultrasound. Various degradation factors which influence the dechlorination efficiency, such as initial pH value, Ni content in the Ni–Fe particles, and output power of ultrasonic irradiation were investigated. The maximum removal percentage was 96% in the presence of ultrasound irradiance within 30 min of contact time. The degradation was greater in acidic conditions. It has been suggested that the low pH allows more of the iron surface to remain

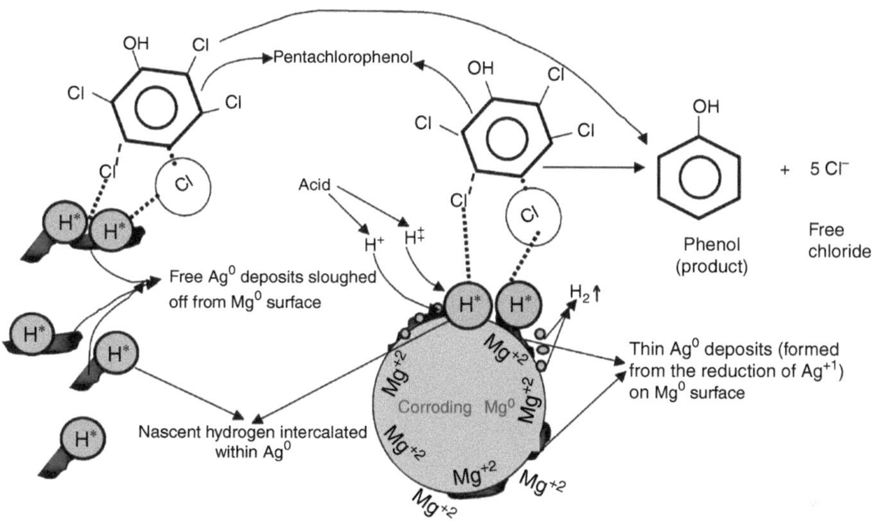

Figure 11.3 ***Schematic representation of events occurring during dechlorination of PCP by a magnesium/silver bimetallic system.*** *(Source: Patel and Suresh, 2006; © Elsevier).*

available for reaction with the chlorinated molecules or at least assists in the corrosion of the iron. While in alkaline medium, oxide and hydroxide coatings undoubtedly develop, which hinder access to the Fe^0 surface. In addition, the degradation of PCP was increased in ultrasound enhancement. The dechlorination of PCP in 20 min increased from 30 to 81% as ultrasonic power changed from 75 to 175 W, respectively. The degradation by-product contained less chlorinated phenols including tetrachlorophenol, trichlorophenol, dichlorophenol, and monochlorophenol during the initial reaction, and phenol was determined as the end product.

Sun et al. (2014) synthesized Pd–Ni bimetallic electrodes by the electrodeposition method assisted with surfactants of cetyl trimethyl ammonium bromide (CTAB), sodium lauryl sulfonate, or sodium dodecyl sulfate. Among the prepared electrodes the Pd–Ni(CTAB)/Ti has high electrochemical activity and was applied for the dechlorination of PCP in aqueous solution. Various parameters such as initial concentration, current, pH, and time were studied. It has been reported that complete removal was obtained at ambient temperatures under the conditions of constant current of 3 mA and initial pH value of 2.1 on the Pd–Ni(CTAB)/Ti electrode within 90 min. Approximately 100% removal of PCP was found at 90 min, when the current was 3 mA (Fig. 11.4). Solution pH plays a very important role in the degradation of PCP and maximum removal was obtained

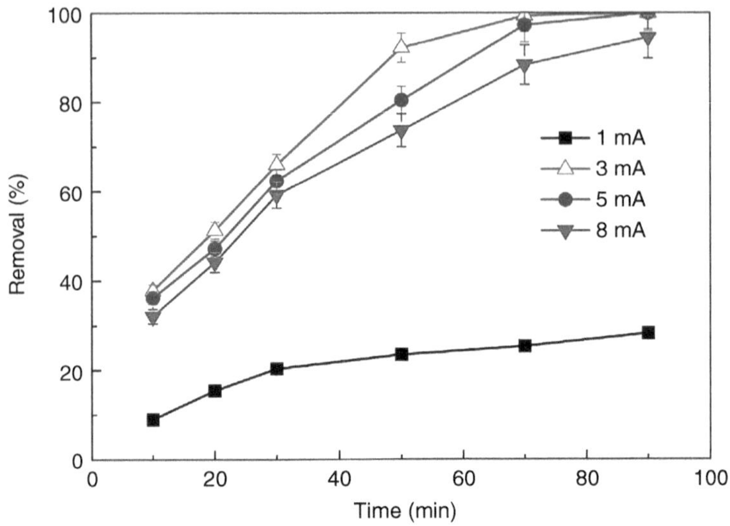

Figure 11.4 ***Effects of electrolysis current on PCP removal on Pd–Ni(CTAB)/Ti electrode.*** *(Sun et al., 2014; © Elsevier).*

in acidic medium. When the initial pH was in the range of 1.8 and 2.1, the removal of PCP reached 100% at 90 min (Fig. 11.5). It has been stated that in the acidic medium, hydrogen ions were preferentially reduced to form adsorbed H^+ easily. Moreover, in the neutral or alkaline range, the adsorbed H^+ was generated via the electrolysis of water. The intermediate products were determined as tetrachlorophenol, trichlorophenol, and cyclohexanone at 70 min. With the increase of time up to 180 min, only cyclohexanone and phenol as organics was detected in the solution.

Zhou et al. (2010) conducted the dechlorination of 4-chlorophenol (4CP); 2,4-dichlorophenol (24DCP); and 2,4,6-trichlorophenol (246TCP) with Pd/Fe nanoparticles. The reaction rates were significantly faster than those achieved with Pt/Fe, Ni/Fe, Cu/Fe, and Co/Fe nanoparticles. Pd/Fe nanoparticles were able to completely decompose the 4-chlorophenol to phenol as the end product. The kinetic modeling showed that the degradation of chlorophenols to phenol by Pd/Fe nanoparticles followed the pseudo-first order kinetics, and the rates followed the order of 246TCP < 24DCP < 4CP. The higher temperature increased the rate of decomposition of 4CP by Pd/Fe nanoparticles, and the activation energy was found to be 29.8 kJ/mol. Gradual increase of Pd loading in the Pd/Fe nanoparticles enhanced the degradation of chlorophenols, suggesting that a Pd catalyst was the only reactive site. On the behavior of chlorophenol

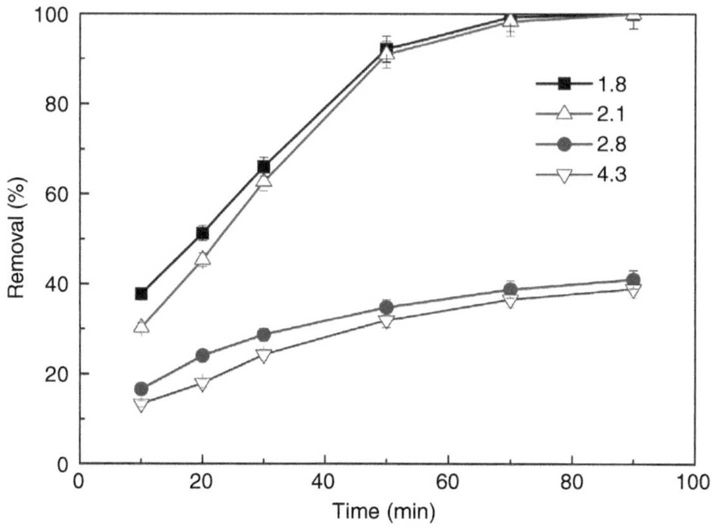

Figure 11.5 ***Effects of initial pH value on PCP removal on Pd–Ni(CTAB)/Ti electrode. Initial PCP concentration, 10 mg/L; current, 3 mA.*** *(Sun et al., 2014; © Elsevier).*

degradation by Pd/Fe nanoparticles, the transient formations, and the concentrations of various intermediates, a scheme of transformation pathways and the reaction mechanism were proposed to delineate 246TCP dechlorination to phenol (Fig. 11.6). The effect of solution pH directly influences the catalytic degradation of chlorophenols and maximum removal was obtained in weak acidic condition.

11.2.2 Degradation of Dyes

Bismuth tungstate (Bi_2WO_6) was synthesized with a facile hydrothermal route for the shape-controlled micro/nanostructured Bi_2WO_6, without adding surfactants or templates by Huang et al. (2012). The characterization results revealed the different morphologies of Bi_2WO_6, including coralloid spherical particles, packed nanosheets, fluffy microspheres, and plates, which were obtained by adjusting the pH values of the precursors. The as-prepared porous fluffy microspheres show the highest photocatalytic activity for the degradation of rhodamine under visible light irradiation in aqueous solutions. The photocatalytic degradation efficiency of rhodamine B was found to be as high as 99% after 20 min of light irradiation. Rhodamine B absorption peak rapidly decreases with a blue-shift from 554 to 498 nm, with enhancement of light irradiation time. The pink color of rhodamine B water solution totally disappeared after 20 min. It has been proposed that the

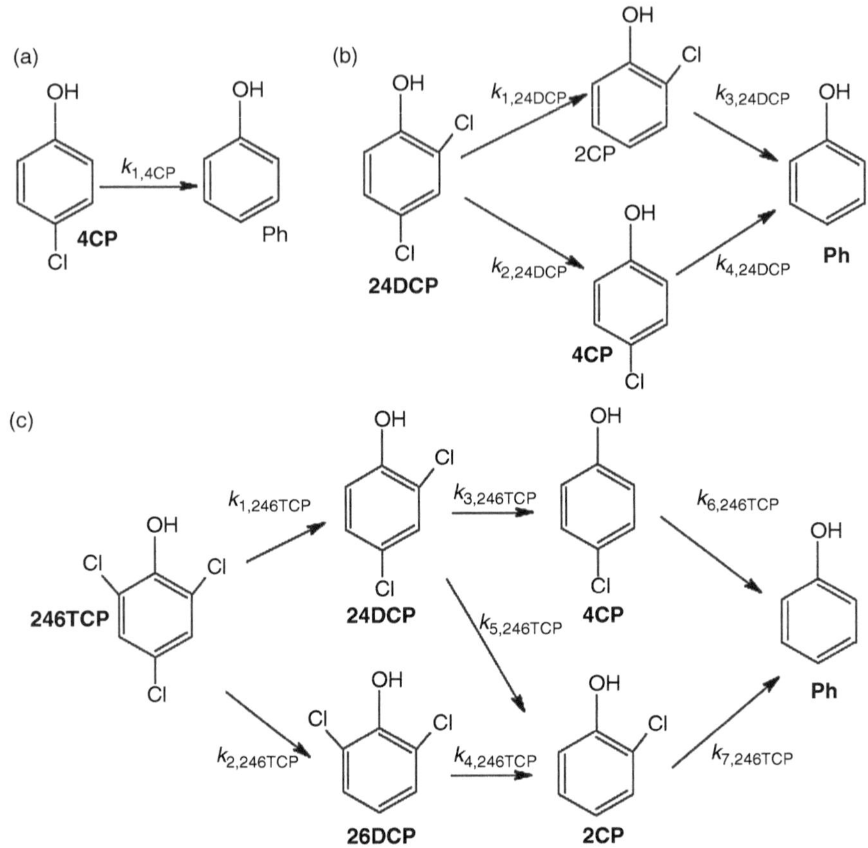

Figure 11.6 Possible reaction pathways for the dechlorination of (a) 4CP, (b) 24DCP, and (c) 246TCP with Pd/Fe nanoparticles. *(Source: Zhou et al., 2010; © Elsevier).*

degradation of rhodamine B in aqueous solution occurs via two competitive processes: de-ethylation and the destruction of the conjugated structure.

Lin et al. (2012) performed the degradation of scarlet 4BS in aqueous solution using Fe/Ni nanoparticles. Various optimization parameters such as the initial dye concentration, the solution pH, and the total Fe/Ni concentration and reaction temperature were studied. Highest degradation of scarlet 4BS in aqueous solution was found to be 84.5% when the catalyst dose was 8.0 g/L, pH 5.0, and 30°C for 3 h contact time. The degradation kinetics for the scarlet 4BS was fitted to the pseudo-first-order model. The degradation of scarlet 4BS proceeds through a reductive cleavage of the azo linkage, and that was the reason the degradation of scarlet 4BS was slow. Additionally, different substituent groups of azo dyes also affected the degradation rate. A degradation mechanism was proposed for scarlet 4BS removal using Fe/Ni nanoparticles involving the oxidation of iron and adsorption of hydrogen into nickel. As iron corrodes, protons of water were reduced to hydrogen atoms and then to molecular hydrogen. The formed hydrogen was adsorbed onto the catalytic nickel surface, and scarlet 4BS was then adsorbed onto the surface of Fe/Ni nanoparticles, hence leading to the breakdown of scarlet 4BS.

Xu and Sun (2012) prepared a series of Mo-Zn-Al-O catalysts with different Zn/Al molar ratios and applied for the removal of cationic red GTL dye wastewater under room condition. The catalytic degradation showed that the highest color removal efficiency and total organic carbon removal efficiency of cationic red GTL was achieved to 90.9% and 65.8% by the Mo-Zn-Al-O catalyst with 1:1 of Zn/Al molar ratio, respectively. Mo-Zn-Al-O catalyst with 1:1 of Zn/Al molar ratio with highest catalytic activity was attributed to its lowest zeta potential, largest specific surface area, special crystalline phases, special Mo valences, largest number of active adsorption sites, and the maximum active species of $^1O_2$ or $^{\bullet}OH$. Due to the lower leaching concentrations of Zn, Mo, and Al in the synthesized Mo-Zn-Al-O catalyst with Zn/Al, molar ratios from 1:1 to 1:3 have good stability, which is suitable for use in aqueous solutions. Fig. 11.7 shows the leaching behavior of Mo in cationic red GTL wastewater oxidized by Mo-Zn-Al-O catalysts prepared with different ratios of Zn and Al.

11.2.3 Remediation of Arsenic

Recently, Li et al. (2015) synthesized a series of iron and titanium copillared montmorillonites (Fe-Ti/MMT) using hydrolysis of inserted titanium and different iron content in montmorillonite, and the synthesized Fe-Ti/MMT was used to remove arsenite [As(III)] from aqueous solutions under different conditions. The Fe-Ti/MMT was characterized by XRF, BET surface area, XRD, SEM, and TEM and confirmed the effective insertion of Fe species and TiO_2 in the montmorillonite. The results show that the adsorption of As(III) was very fast, and the equilibrium reached within 60 min of contact time. The FTIR and XPS analysis revealed that the hydroxyl groups bonded to metal oxide played an important role in the adsorption of As(III). Fig. 11.8 shows the adsorption kinetics and isotherm analysis for iron and titanium copillared montmorillonite adsorption of As(III) with and without UV light irradiation. The kinetic studies show that the adsorption process was fitted to the pseudo-second-order model with higher correlation coefficients than the pseudo-first order model. It further revealed that the

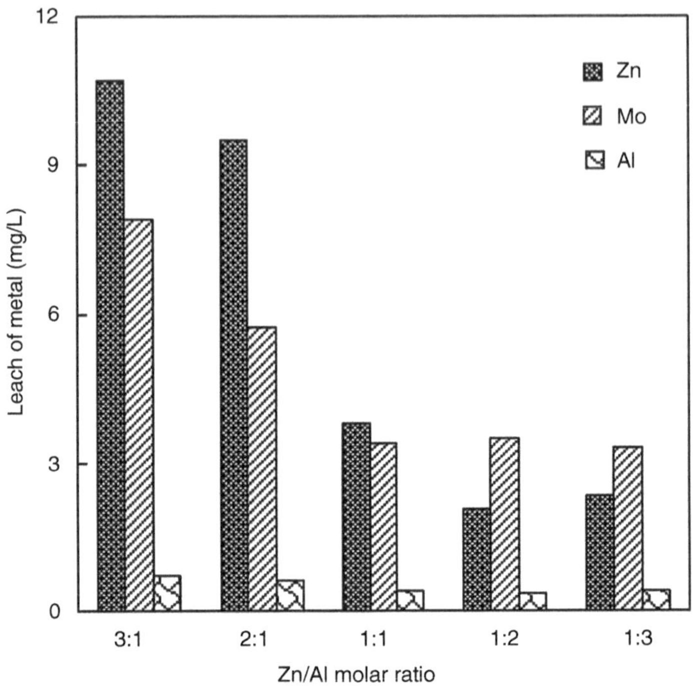

Figure 11.7 ***The leaching of Mo in cationic red GTL wastewater oxidized by Mo-Zn-Al-O catalysts prepared with different ratios of Zn and Al.*** *(Source: Xu and Sun, 2012; © Elsevier).*

adsorption mechanism was chemisorption controlled. Fig. 11.9 shows the schematic diagrams for the possible adsorption mechanism of As(III) on the iron and titanium copillared montmorillonites with or without UV irradiation.

11.2.4 Removal of Benzene and Volatile Organic Compounds

Xu et al. (2009) performed the catalytic dechlorination of *p*-nitrochlorobenzene by Ni/Fe bimetallic nanoparticles, and the result indicated that the catalytic reduction and dechlorination reaction was believed to take place on the surface site of the nanoscale catalysts. The catalytic performance of Ni/Fe bimetallic nanoparticles was very high, and 100% dechlorination efficiency was achieved for particles with 2.0% Ni after 300 min. The *para*-nitrochlorobenzene was first adsorbed on the surface of Ni/Fe bimetallic nanoparticles completely then quickly reduced to *p*-chloroaniline and aniline, and *p*-chloroaniline was detected as an intermediate. The degradation of *para*-nitrochlorobenzene by Ni/Fe bimetallic nanoparticles increased

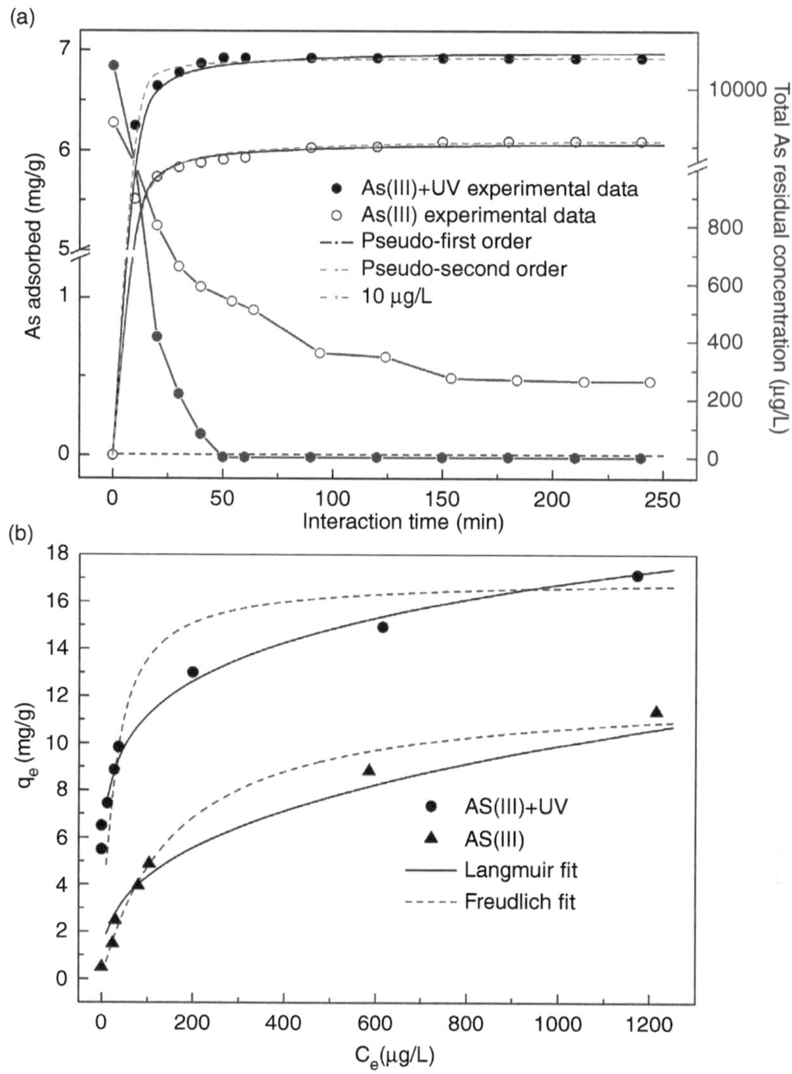

Figure 11.8 (a) Adsorption kinetics and (b) isotherm analysis for iron and titanium copillared montmorillonite adsorption of As(III) with and without UV light irradiation. *(Source: Li et al., 2015; © Elsevier).*

with the increasing load of Ni percentage in bimetallics at low pH and high temperature. The Ni/Fe bimetallic catalyst showed high efficiency for the reductive catalytic dechlorination of *p*-nitrochlorobenzene at ambient temperature and pressure as well as high removal efficiency of *p*-nitrochlorobenzene and *p*-chloroaniline in agricultural pesticide

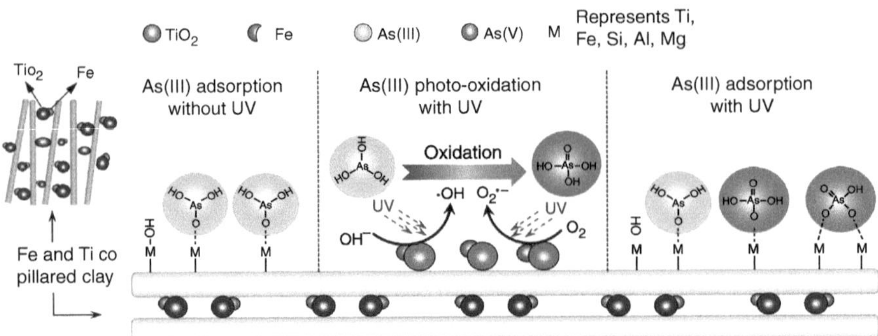

Figure 11.9 ***Schematic diagrams for the possible adsorption mechanism of As(III) on the iron and titanium copillared montmorillonites with or without UV irradiation.*** *(Source: Li et al., 2015; © Elsevier).*

wastewater. The catalytic reduction of *p*-nitrochlorobenzene by Ni/Fe bimetallic nanoparticles occurred as follows.

$$p-\text{nitrochlorobenzene}+6\text{H}^{+}+3\,\text{Fe}^{0}\rightarrow p-\text{chloroaniline}+3\,\text{Fe}^{2+}+2\,\text{H}_2\text{O}$$

$$2p-\text{chloroaniline}+\text{Fe}^{0}\rightarrow 2\,\text{aniline}+\text{Fe}^{2+}+2\,\text{Cl}^{-}$$

The reactivity of Fe–Pd bimetallic nanoparticles with sodium carboxymethyl cellulose as a stabilizer toward *p*-nitrochlorobenzene was investigated by Dong et al. (2011). The sodium carboxymethyl cellulose–stabilized Fe–Pd nanoparticles show greater reactivity than nonstabilized iron counterparts when used for the degradation of *p*-nitrochlorobenzene in aqueous solution. The kinetic study showed that the pseudo-first-order rate constant with the stabilized bimetallic particles was four times greater than that for nonstabilized iron. The experimental study showed that 100% degradation of *p*-nitrochlorobenzene was obtained by the Fe–Pd bimetallic nanoparticles at a dose of 0.2 g/L within 40 min of contact time; however, small amounts of *p*-chloroaniline as an intermediate were detected in the final stage of degradation, with no other products except aniline and inorganic chloride. Fig. 11.10 shows the degradation of *p*-nitrochlorobenzene and its by-products by the sodium carboxymethyl cellulose–stabilized Fe–Pd bimetallic nanoparticles. It has been noted that the catalytic degradation efficiency of *p*-nitrochlorobenzene increased with the increasing addition of Pd bulk loading, Fe–Pd mass concentration, and temperature and with the decrease of pH values and humic acid dosage.

Alumina-supported bimetallic catalysts $Cu_{1.25}Co_{3.75}/Al_2O_3$, $Cu_{2.5}Co_{2.5}/Al_2O_3$, and $Cu_{3.75}Co_{1.25}/Al_2O_3$ were studied in a pilot-scale incinerator

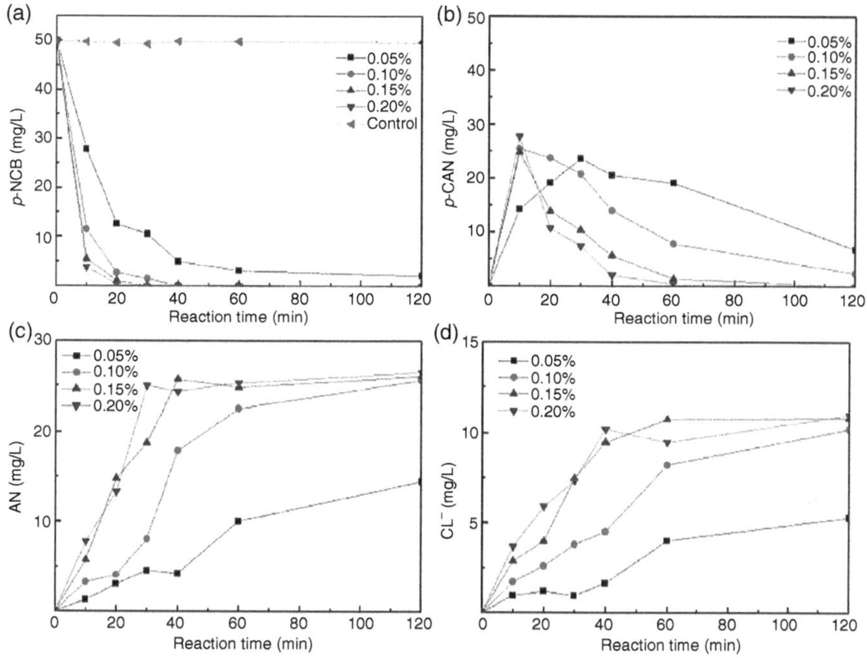

Figure 11.10 Effect of Pd bulk loading on (a) degradation of *p*-nitrochlorobenzene; (b) yield of *p*-chloroaniline; (c) yield of aniline; and (d) yield of chloride using sodium carboxymethyl cellulose stabilized Fe–Pd bimetallic nanoparticles. Initial concentration of *p*-nitrochlorobenzene = 50 mg/L, sodium carboxymethyl cellulose = 0.8% (w/w), bimetallic dose = 0.2 g/L, temperature = 25°C, and pH = 5.5. *(Source: Dong et al., 2011; © Elsevier).*

system to remove benzene, toluene, ethylbenzene, xylene, and polycyclic aromatic hydrocarbons (Lu et al., 2009).

Zhang et al. (2012) synthesized Ni/Fe bimetallic nanoparticles and applied for the treatment of Aroclor 1242 to evaluate their applicability for in situ remediation of groundwater and soil contaminated by polychlorinated biphenyls. The result shows that the total polychlorinated biphenyls concentration changed during the reduction of 3,5-dichlorobiphenyl, and biphenyl was produced as the final product. Originally, the concentration of 3-chlorobiphenyl was increased in the prophase reaction and then slowly decreased, which means that Aroclor 1242 was first adsorbed by Ni/Fe nanoparticles, and then the higher chlorinated congeners were converted gradually to the lower chlorinated congeners, and finally to biphenyl. The degradation rate of Aroclor 1242 by Ni/Fe nanoparticles reached approximately 80% at 25°C in just 5 h, then 95.6 and 95.8% in 10 h and 24 h, respectively. With the increasing initial Aroclor 1242 concentrations in aqueous solutions from 0.5, 1.0, 1.5, to 2.0 mg/L, the removal efficiencies of tetrachlorobiphenyl, trichlorobiphenyl, dichlorobiphenyl, and chlorobiphenyl were recorded as 98.2, 95.0, 96.4, and 90.9%; 97.7, 97.3, 96.9, and 94.3%; 98.4, 98.3, 98.1, and 96.2%; and 73.4, 86.8, 73.2, and 73.8%, respectively. The degradation results show that the high Ni/Fe nanoparticle dosage and high Ni content in Ni/Fe nanoparticles favor the catalytic dechlorination reaction.

Wang et al. (2011) performed a low-temperature catalytic oxidation of chlorobenzene (CB) at temperatures of 60–210°C using ozone over iron oxide and manganese oxide, respectively. It has been reported that conversion of chlorobenzene was below 10% in the absence of ozone at 200°C. However, the addition of 1200 ppm ozone resulted in a remarkable increase in chlorobenzene conversion, and the conversion reached 91.7% at 150°C for iron oxide, while 81.5% conversion was achieved with manganese oxide at 90°C. The activation energy of manganese oxide was slightly higher (48 kJ/mol) than that of iron oxide (43 kJ/mol) without ozone. Carbon monoxide and carbon dioxide were the only carbon-containing products detected in the effluent gas stream. It has been also reported that the long-term use of iron oxide with ozone does not reduce its performance. However, 3% reduction of chlorobenzene conversion was found in the case of manganese oxide with ozone. It was postulated that the slight deactivation might be attributed to a small amount of reaction by-products (carboxylic acid species) and residual chloride ($MnCl_2$) being deposited on the active sites of the catalysts.

Wang et al. (2008a) synthesized the poly-vinylidene fluoride microfiltration membrane successfully, hydrophilized by coating with a novel hydrophilic layer, which was mainly composed of polyvinyl alcohol, glutaraldehyde, and polyethylene glycol. Pd/Fe nanoparticles were prepared and immobilized in hydrophilized polyacrylic acid polyvinyl alcohol/polyvinylidene fluoride microfiltration membrane. It has been reported that the membrane-supported Pd/Fe nanoparticles exhibited high reactivity in the dechlorination of trichloroacetic acid. From the experimental results, it was found that the Pd/Fe nanoparticle membrane system showed high reactivity for catalytic dechlorination of trichloroacetic acid, and complete dechlorination of trichloroacetic acid was achieved by Pd/Fe nanoparticle membrane. More than 97.22% of trichloroacetic acid was dechlorinated within the first 120 min, with the rapid concomitant rise of dichloroacetic acid as the major product. The kinetic result showed that

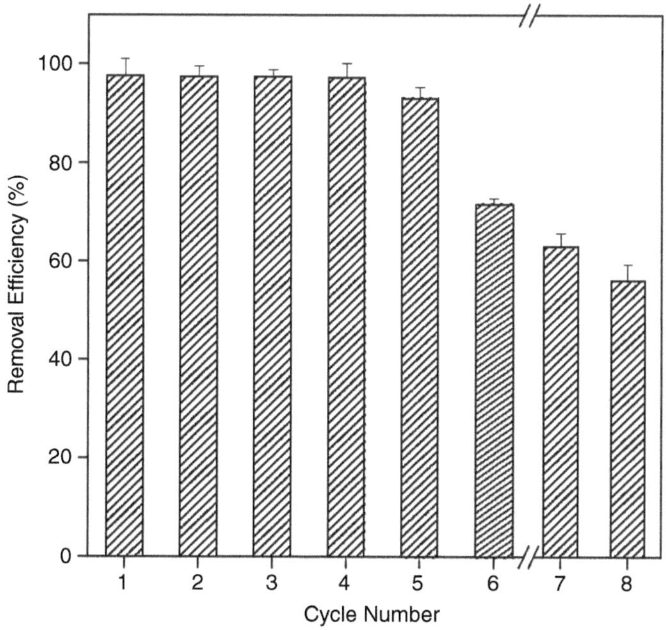

Figure 11.11 ***Eight-cycle dechlorination reaction of trichloroacetic acid.*** *(Source: Wang et al., 2008a; © Elsevier).*

the catalytic dechlorination of trichloroacetic acid by the Pd/Fe nanoparticle membrane system was found to be a pseudo-first order reaction. Comparing with nonimmobilized Pd/Fe nanoparticles, dechlorination rate of Pd/Fe nanoparticle membrane systems was not affected by diffusion resistance. Fig. 11.11 shows the reuse of the synthesized Pd/Fe nanoparticle membrane system up to an eight-cycle dechlorination reaction of trichloroacetic acid. The reuse of nanoparticle membrane was carried out for four consecutive cycles and after that the dechlorination percentage decreased.

Wang et al. (2014) synthesized the CeO_2–MnO_x composites possessing rod-like morphology through the hydrothermal method and chose them as supporters to load PdO nanoparticles (PdO/Ce_xMn_{1-x}). The catalytic behaviors of supported catalysts were studied in the degradation of benzene. The experimental outcome shows that the activities of supported catalysts were enhanced greatly as compared to unsupported, and the complete conversion of benzene was obtained. The obtained results also revealed that the interaction between PdO and supporter, and intrinsic properties of supporter resulted in the enhancement of catalytic abilities.

11.2.5 Dechlorination of Trichloroethylene

The highly reactive bimetallic Fe/Ni nanoparticles immobilized in nylon 66, and poly-vinylidene fluoride membranes were synthesized and applied for dechlorination of trichloroethylene (TCE) under anoxic conditions (Parshetti and Doong, 2009). The synthesized microfiltration membrane shows the particle sizes of bimetallic Fe/Ni in poly-vinylidene fluoride, and nylon 66 membranes were 81 ± 12 and 55 ± 14 nm with the Ni layers of 12 ± 3 and 15 ± 2 nm, respectively. The hydrodechlorination of TCE was very fast and ethane was obtained as the main end product by the immobilized Fe/Ni nanoparticles. The degradation kinetic for TCE was fitted to pseudo-first-order model. The obtained result shows that the efficiency and rate of TCE dechlorination increased upon increasing the mass loading of Ni, ranging between 2.5 and 20 wt.%, and then decreased when further increased the Ni loading to 25 wt.%. The stability and longevity of the immobilized Fe/Ni nanoparticles was evaluated by repeatedly injecting TCE into the solutions. Complete dechlorination of TCE by trace amounts of Fe/Ni nanoparticles was observed after 16 cycles of injection within 10 days.

Zhang et al. (2011) studied the reductive dechlorination of TCE sorbed in soils using carboxymethyl cellulose–stabilized Fe/Pd bimetallic nanoparticles. Various experimental parameters such as the effects of sorption, surfactants, and dissolved organic matter were determined through batch kinetic experiments. The degradation rate for TCE was strongly limited by desorption kinetics, especially for the soil, which has higher organic matter content.

The degradation rate for TCE was in the range of ~44–82%, with soil varying in different organic matter. However, it has been found that the dissolved organic matter inhibits the degradation of TCE. The degradation rate of TCE was also reduced by 34 and 67% in the presence of extracted soil organic matter at 40 ppm and 350 ppm, respectively. Four surfactants, namely, sodium dodecyl sulfate, sodium dodecyl benzene sulfonate, hexadecyltrimethylammonium, and Tween 80, were tested on the degradation behavior of TCE. All of them were observed to enhance TCE desorption, with the anionic surfactant SDS being most effective.

Recently, Xie et al. (2013) synthesized a cathode of Pd-loaded Ti/TiO_2 nanotubes ($Pd–Ti/TiO_2NTs$) for the electrocatalytic reduction of TCE in groundwater. Using $Pd-Ti/TiO_2NTs$ as the cathode in the electrolytic cell, the TCE was efficiently degraded to ethane as the end product. Maximum removal of 91% was found within 120 min under conditions of 100 mA and at neutral pH. The kinetic modeling showed that the degradation was fitted to pseudo-first-order model with the rate constant of 0.019 min^{-1}.

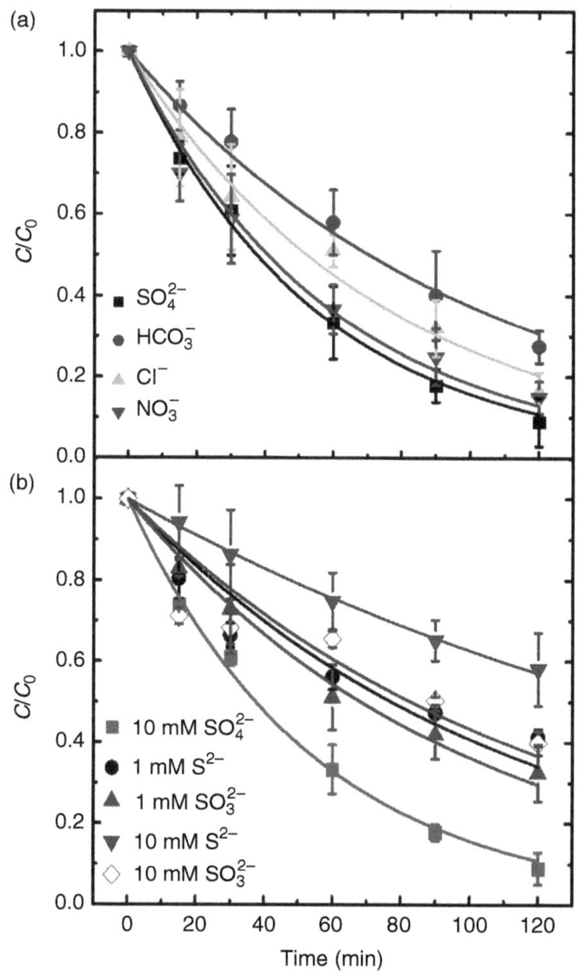

Figure 11.12 Effects of (a) anionic ions and (b) different concentrations of reduced sulfur compounds on TCE reduction on the Pd-Ti/TiO_2NTs cathode (reaction conditions: TCE = 163 mM, current = 100 mA, and pH = 7). Ion concentration was 10 mM unless otherwise specified. *(Source: Xie et al., 2013; © Elsevier).*

The effects of chloride, bicarbonate, sulfur, and nitrogen compounds were studied on the degradation of TCE in wastewater (Fig. 11.12). The presence of chloride or bicarbonate decreased the catalytic degradation of TCE in aqueous solutions. The degradation of TCE in low currents was mainly due to the reduction by direct electron transfer on the Pd-Ti/TiO_2NTs cathode. Moreover, the increase in current shows a significant hydrodechlorination of the TCE in aqueous solutions due to the indirect reduction mechanism. The comparison with other cathodes such as Ti-based mixed metal oxides, graphite and Pd/Ti, and Pd-Ti/TiO_2NTs cathodes showed superior performance for TCE reduction.

Wu and Ritchie (2006) studied cellulose acetate–supported Ni/Fe bimetallic nanoparticles for the removal of TCE from water. The effects of metal particle composition in cellulose acetate–supported, post-coated Ni/Fe nanoparticles were used to investigate the degradation of TCE in aqueous solutions. The results showed that the metal mass normalized observed reduction rate constant was proportional to the Ni content in the post-coated Ni/Fe nanoparticles in the range of 0–14.3 wt.%. The degradation of TCE was maximum in the range 14.3 and 21.4 wt.%, and further addition of Ni decreased the degradation of pollutants. The effects of TCE reduction by the nanoparticles followed the pseudo-first-order kinetic model. The catalytic degradation showed that the removal of TCE was very fast in initial contact time and reached the equilibrium within 15 min. The first phase represented the sorptive and chemical reduction of TCE, and the second phase was attributed to the effect of reduction by the cellulose acetate–supported Ni/Fe bimetallic nanoparticles.

The degradation of TCE by Fe(0) in the presence of Co(II), Cu(II), and Ni(II) was studied by Doong and Lai (2006). It has been found that the addition of metal ions enhanced the dechlorination efficiency of TCE. The maximum removal of 86 and 93% of TCE were observed within 120 h in the presence of 0.1 mM Co(II) and Cu(II), respectively. More over the addition of Ni(II) has a great effect on the dechlorination of TCE, and a nearly complete dechlorination of TCE was observed within 15 h when 0.1 mM Ni(II) was added into the solutions containing untreated Fe(0).

11.2.6 Dechlorination of Endocrine-Disrupting Chemicals

Recently, Begum and Gautam (2011) studied the performance of the $Mg^0/ZnCl_2$ bimetallic system for its efficiency to dechlorinate endosulfan and lindane in the aqueous phase. It was found that the presence of acetone in the reaction mixture played an important role by increasing the solubilities of both pesticides and thereby accelerating its mass transfer. Results show that water acetone ratio of 2:1 and 1:1 (v/v) was found optimum for the dechlorination of endosulfan and lindane, respectively. The degradation for endosulfan and lindane was found to be greater than 96 and 98%, respectively, at $Mg^0/ZnCl_2$ bimetal dose of 5/1 mg/mL within 30 min of reaction. Dechlorination kinetics for endosulfan and lindane was fitted to the first-order kinetic model for each initial concentration of pesticides.

The results show that the endosulfan and lindane were completely dechlorinated into their hydrocarbon skeletons, namely, bicyclo (2,2,1) hepta 2-5 diene and benzene, respectively.

Polychlorinated dibenzo-*p*-dioxins are toxic and widespread persistent organic pollutants. Wang et al. (2008b) studied the dechlorination rates of 1,2,3,4-tetrachlorodibenzo-*p*-dioxin (1,2,3,4-TCDD) in the presence of zero-valent zinc under aqueous conditions. It has been found that the reductive dechlorination of 1,2,3,4-tetrachlorodibenzo-*p*-dioxin was stepwise and complete to dibenzo-*p*-dioxin mainly via 1,2,4-trichlorodibenzo-*p*-dioxin (1,2,4-TrCDD); 1,3-dichlorodibenzo-*p*-dioxin (1,3-DCDD); and 1-chlorodibenzo-*p*-dioxin (1-MCDD) to dibenzo-*p*-dioxin, and via 1,2,4-TrCDD; 2,3-dichlorodibenzo-*p*-dioxin (2,3-DCDD); and 2-chlorodibenzo-*p*-dioxin (2-MCDD) to dibenzo-*p*-dioxin. Experimental results showed that the observed half-lives of 1,2,3,4-TCDD; 1,2,3-TrCDD; 1,2,4-TrCDD; 1,2-DCDD; 1,3-DCDD; 1,4-DCDD; and 2,3-DCDD were 0.56, 2.62, 5.71, 24.93, 41.53, 93.67 and 169.06 h, respectively.

Wang et al. (2010) investigated the reductive dechlorination of 1,2,3,4-TCDD in the presence of Pd/Fe, Cu/Fe, and Ni/Fe bimetallic catalysts or zero-valent iron under aqueous and ambient temperature conditions. The results showed that the Pd/Fe catalyst was highly effective in the transformation of 1,2,3,4-TCDD to chlorine-free dibenzo-*p*-dioxin with >95% conversion and half lives on the order of 5 h, whereas Cu/Fe, Ni/Fe, and zero-valent iron were not reactive over 10 d. The degradation of 1,2,3,4-TCDD in the Pd/Fe systems was found to be stepwise, with 1,2,3-tri (TrCDD); 1,2-di (DCDD); and mono-(MCDD) chlorodibenzo-*p*-dioxins as the three intermediate by-products at low concentrations, and dibenzo-*p*-dioxin as the final and dominant product. The increase of palladium on iron particle surface from 0.0060 to 0.0228 mol% rapidly decreased the dechlorination rate of 1,2,3,4-TCDD. Their study suggested that Pd/Fe bimetallic catalyst-based reactions may be suitable for rapid decontamination of polychlorinated dibenzo-*p*-dioxin/dibenzofuran polluted water, soils, and sediments.

Chen et al. (2015) applied a novel Cu@Pd bimetallic catalyst loaded on Ti plate for electrocatalytic hydrodehalogenation of atrazine. The synthesized bimetallic catalyst with uniformly dispersed Pd nanoparticles possessed a large electrochemically active surface area of 572 cm^2. The electrocatalytic efficiency toward atrazine reduction by Cu@Pd/Ti cathode was higher than the individual Pd/Ti or Cu/Ti cathodes and achieved up to 91.5% within 120 min under a current density of 1 mA/cm^{-2}. Fig. 11.13 shows

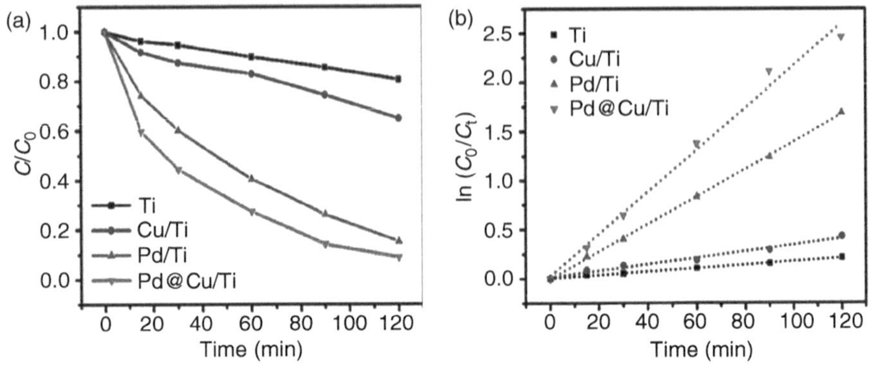

Figure 11.13 ***Dechlorination performance of atrazine by different bimetallic nanoparticles.*** (a) Dechlorination of atrazine on the Ti, Cu/Ti, Pd/Ti and Cu@Pd/Ti electrodes and (b) pseudo-first-order model for atrazine dechlorination. Current density: 1 mA/cm, catholyte: 5 mg/L atrazine and 10 mM Na_2SO_4 background electrolyte, and anolyte: 100 mM H_2SO_4. *(Source: Chen et al., 2015; © Elsevier).*

the dechlorination performance of atrazine by different bimetallic nanoparticles. The atrazine reduction followed pseudo-first-order kinetics with a rate constant of 0.0214 min^{-1}. Current density was found to have a critical influence on the atrazine reduction due to the competitive hydrogen evolution reaction at a higher current density. Fig. 11.14 depicts the schematic illustration of the electrocatalytic dechlorination process of atrazine on a Cu@Pd/Ti cathode.

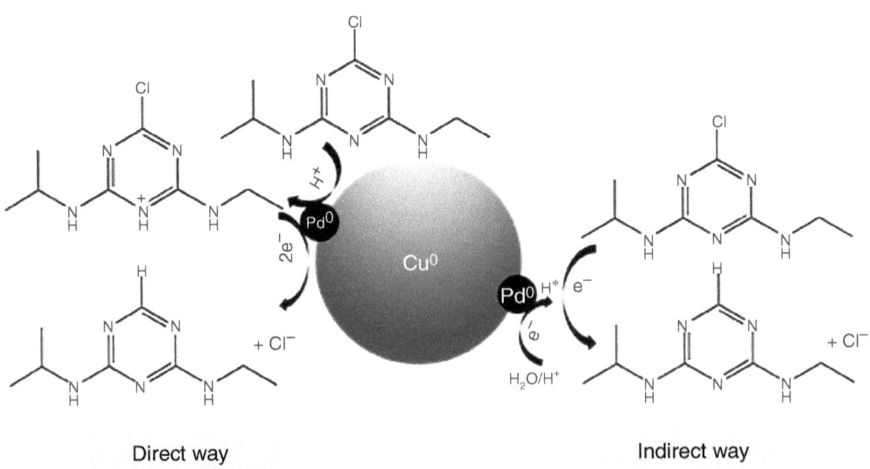

Figure 11.14 ***Schematic illustration of the electrocatalytic dechlorination process on the Cu@Pd/Ti cathode.*** *(Source: Chen et al., 2015; © Elsevier).*

Kustov et al. (2011) synthesized a series of chitosan-stabilized Pd-NZVI catalysts for dechlorination with variation in their composition and in the nature of the polymer. The presence of O and N atoms in the first coordination shell of the central Fe atom in Fe–Pd/chitosan proved the binding of the Fe metal particles with the chitosan surface via OH and NH_2 groups. The synthesized chitosan-stabilized Pd-NZVI catalysts were used for the catalytic degradation of perchloroethene from aqueous solutions.

11.3 CONCLUSIONS

Bimetallic nanomaterials have been effectively applied in the field of remediation of dyes, volatile organic pollutants, benzenes, phenols, chlorophenols, insecticides, and pesticides. These materials are highly promising due to their high surface area, excellent catalytic properties, stability in aqueous medium, ease in synthesis, and low cost of maintenance. Recently, various methods such as coprecipitation, hydrothermal, sol-gel, and electrochemical formation have been utilized for the synthesis of bimetallic nanoparticles. All of them have their own drawbacks and benefits. However, coprecipitation and hydrothermal methods have been widely used. Many sophisticated instruments such as SEM, TEM, XRD, FTIR, Raman spectroscopy, XPS, and VSM have been used to characterize and decipher their chemistry. Results on the use of bimetallic nanoparticles have shown that these materials can be applied to the real wastewater effluents. Due to high surface area and excellent catalytic properties the bimetallic nanoparticles have effectively degraded the phenol and chlorophenols from high concentration to ppb level. The degradation of environmental contaminants by bimetallic nanoparticles is very fast, and in most cases pseudo-first-order kinetics was suitable to explain the degradation mechanism. In extremely harsh environmental conditions such as highly acidic or basic aqueous medium the bimetallic nanoparticles are easily oxidized, and their catalytic efficiency is reduced. To overcome these issues the bimetallic nanoparticles have functionalized with organic stabilizers to increase their stability and lifetime. However, stabilization of bimetallic nanoparticles needs to be further researched, and there is also a need to conduct more studies in the area of recycling and reuse of low-cost catalytic bimetallic nanoparticles. Toxicity is an important issue, and it should be well researched before the application of bimetallic nanomaterials in real wastewater treatment.

REFERENCES

Begum, A., Gautam, S.K., 2011. Dechlorination of endocrine disrupting chemicals using $Mg^0/ZnCl_2$ bimetallic system. Water Res. 45, 2383–2391.

Bokare, A.D., Chikate, R.C., Rode, C.V., Paknikar, K.M., 2007. Effect of surface chemical of Fe–Ni nanoparticle on mechanistic pathway of azo dye degradation. Environ. Sci. Technol. 41, 7437–7443.

Chen, Y.L., Xiong, L., Song, X.N., Wang, W.K., Huang, Y.X., Yu, H.Q., 2015. Electrocatalytic hydrodehalogenation of atrazine in aqueous solution by Cu@Pd/Ti catalyst. Chemosphere 125, 57–63.

Dong, T., Luo, H., Wang, Y., Hu, B., Chen, H., 2011. Stabilization of Fe–Pd bimetallic nanoparticles with sodium carboxymethyl cellulose for catalytic reduction of *para*-nitrochlorobenzene in water. Desalination 271, 11–19.

Doong, R., Lai, Y., 2006. Effect of metal ions and humic acid on the dechlorination of tetrachloroethylene by zerovalent iron. Chemosphere 64, 371–378.

Fan, J., Guo, Y., Wang, J., Fan, M., 2009. Rapid decolorization of azo dye methyl orange in aqueous solution by nanoscale zerovalent iron particles. J. Hazard. Mater. 166, 904–910.

Fang, Z., Qiu, X., Chen, J., Qiu, X., 2011. Debromination of polybrominated diphenyl ethers by Ni/Fe bimetallic nanoparticles: influencing factors, kinetics, and mechanism. J. Hazard. Mater. 185, 958–969.

He, F., Zhao, D., 2005. Preparation and characterization of a new class of starch-stabilized bimetallic nanoparticles for degradation of chlorinated hydrocarbons in water. Environ. Sci. Technol. 39, 3314–3320.

Hu, J., Lo, I.M.C., Chen, G., 2007. Comparative study of various magnetic nanoparticles for Cr(VI) removal. Sep. Purif. Technol. 56, 249–256.

Huang, H., Chen, H., Xia, Y., Tao, X., Gan, Y., Weng, X., Zhang, W., 2012. Controllable synthesis and visible-light-responsive photocatalytic activity of Bi2WO6 fluffy microsphere with hierarchical architecture. J. Colloid Interf. Sci. 370, 132–138.

Joo, S.H., Faitz, A.J., Sedleck, D.L., Waite, T.D., 2005. Quantification of the oxidizing capacity of nanoparticulate zero valent iron. Ind. Environ. Sci. Technol. 39, 1263–1268.

Kustov, L.M., Finashina, E.D., Shuvalova, E.V., Tkachenko, O.P., Kirichenko, O.A., 2011. Pd–Fe nanoparticles stabilized by chitosan derivatives for perchloroethene dechlorination. Environ. Int. 37, 1044–1052.

Lee, H., Yoo, H.Y., Choi, J., Nam, I.H., Lee, S., Lee, S., Kim, J.H., Lee, C., Lee, J., 2014. Oxidizing capacity of periodate activated with iron-based bimetallic nanoparticles. Environ. Sci. Technol. 48, 8086–8093.

Li, Y., Cai, X., Guo, J., Zhou, S., Na, P., 2015. Fe/Ti co-pillared clay for enhanced arsenite removal and photooxidation under UV irradiation. Appl. Surf. Sci. 324, 179–187.

Lin, C.J., Lo, S.L., Liou, Y.H., 2004. Dechlorination of trichloroethylene in aqueous solution by noble metal-modified iron. J. Hazard. Mater. B116, 219–228.

Lin, Y., Chen, Z., Megharaj, M., Naidu, R., 2012. Degradation of scarlet 4BS in aqueous solution using bimetallic Fe/Ni nanoparticles. J. Colloid. Interf. Sci. 381, 30–35.

Lu, C.Y., Tseng, H.H., Wey, M.Y., Liu, L.Y., Kuo, J.H., Chuang, K.H., 2009. Al_2O_3-supported Cu–Co bimetallic catalysts prepared with polyol process for removal of BTEX and PAH in the incineration flue gas. Fuel 88, 340–347.

Ma, L.M., Ding, Z.G., Gao, T.Y., Zhou, R.F., Xu, W.Y., Liu, J., 2004. Discoloration of methylene blue and wastewater from a plant by a Fe/Cu bimetallic system. Chemosphere 55, 1207–1212.

Nurmi, J.T., Tratnyek, P.G., Sarathy, V., Baer, D.R., Amonette, J.E., Pecher, K., Wang, C., Linehan, J.C., Matson, D.W., Penn, R.L., Driessen, M.D., 2005. Characterization and properties of metallic iron nanoparticles: spectroscopy, electrochemistry, and kinetics. Environ. Sci. Technol. 39, 1221–1230.

Parshetti, G.K., Doong, R., 2009. Dechlorination of trichloroethylene by Ni/Fe nanoparticles immobilized in PEG/PVDF and PEG/nylon 66 membranes. Water Res. 43, 3086–3094.

Patel, U., Suresh, S., 2006. Dechlorination of chlorophenols by magnesium–silver bimetallic system. J. Colloid Interf. Sci. 299, 249–259.

Schrick, B., Blough, J.L., Jones, A.D., Mallouk, T.E., 2002. Hydrodechlorination of trichloroethylene to hydrocarbons using bimetallic nickel–iron nanoparticles. Chem. Mater. 14, 5140–5147.

Su, J., Lin, S., Chen, Z., Megharaj, M., Naidu, R., 2011. Dechlorination of *p*-chlorophenol from aqueous solution using bentonite supported Fe/Pd nanoparticles: synthesis, characterization and kinetics. Desalination 280, 167–173.

Sun, Z., Wei, X., Shen, H., Hu, X., 2014. Preparation of palladium–nickel loaded titanium electrode with surfactant assistance and its application in pentachlorophenol reductive dechlorination. Sep. Purif. Technol. 124, 224–230.

Wang, J., Blowers, P., Farell, J., 2004. Understanding reduction of carbon tetrachloride at nickel surfaces. Environ. Sci. Technol. 38, 1576–1581.

Wang, X., Chen, C., Liu, H., Ma, J., 2008a. Preparation and characterization of PAA/PVDF membrane-immobilized Pd/Fe nanoparticles for dechlorination of trichloroacetic acid. Water Res. 42, 4656–4664.

Wang, Z., Huang, W., Fennell, D.E., Peng, P., 2008b. Kinetics of reductive dechlorination of 1,2,3,4-TCDD in the presence of zero-valent zinc. Chemosphere 71, 360–368.

Wang, X., Chen, C., Chang, Y., Liu, H., 2009. Dechlorination of chlorinated methanes by Pd/Fe bimetallic nanoparticles. J. Hazard. Mater. 161, 815–823.

Wang, Z., Huang, W., Peng, P., Fennell, D.E., 2010. Rapid transformation of 1,2,3,4-TCDD by Pd/Fe catalysts. Chemosphere 78, 147–151.

Wang, H.C., Liang, H.S., Chang, M.B., 2011. Chlorobenzene oxidation using ozone over iron oxide and manganese oxide catalysts. J. Hazard. Mater. 186, 1781–1787.

Wang, X., Zhu, M., Liu, H., Ma, J., Li, F., 2013. Modification of Pd–Fe nanoparticles for catalytic dechlorination of 2,4-dichlorophenol. Sci. Total Environ. 449, 157–167.

Wang, Z., Yang, M., Shen, G., Liu, H., Chen, Y., Wang, Q., 2014. Catalytic removal of benzene over CeO_2–MnOx composite oxides with rod-like morphology supporting PdO. J. Nanopart. Res. 16, 2367.

Wei, J., Xu, X., Liu, Y., Wang, D., 2006. Catalytic hydrodechlorination of 2,4-dichlorophenol over nanoscale Pd/Fe: reaction pathway and some experimental parameters. Water Res. 40, 348–354.

Wu, L., Ritchie, S.M.C., 2006. Removal of trichloroethylene from water by cellulose acetate supported bimetallic Ni/Fe nanoparticles. Chemosphere 63, 285–292.

Xie, W., Yuan, S., Mao, X., Hu, W., Liao, P., Tong, M., Alshawabkeh, A.N., 2013. Electrocatalytic activity of Pd-loaded Ti/TiO_2 nanotubes cathode for TCE reduction in groundwater. Water Res. 47, 3573–3582.

Xu, Y., Sun, D., 2012. Structure and catalytic activity of Mo-Zn-Al-O catalyst for degradation of cationic red GTL under room conditions. Chem. Eng. J. 183, 332–338.

Xu, X., Wo, J., Zhang, J., Wu, Y., Liu, Y., 2009. Catalytic dechlorination of *p*-NCB in water by nanoscale Ni/Fe. Desalination 242, 346–354.

Zhang, W., Quan, X., Wang, J., Zhang, Z., Chen, S., 2006. Rapid and complete dechlorination of PCP in aqueous solution using Ni–Fe nanoparticles under assistance of ultrasound. Chemosphere 65, 58–64.

Zhang, M., He, F., Zhao, D., Hao, X., 2011. Degradation of soil-sorbed trichloroethylene by stabilized zero valent iron nanoparticles: effects of sorption, surfactants, and natural organic matter. Water Res. 45, 2401–2414.

Zhang, Z., Hu, S., Baig, S.A., Tang, J., Xu, X., 2012. Catalytic dechlorination of Aroclor 1242 by Ni/Fe bimetallic nanoparticles. J. Colloid Interf. Sci. 385, 160–165.

Zhou, T., Li, Y., Lim, T.T., 2010. Catalytic hydrodechlorination of chlorophenols by Pd/Fe nanoparticles: comparisons with other bimetallic systems, kinetics and mechanism. Sep. Purif. Technol. 76, 206–214.

CHAPTER 12

Desorption, Regeneration, and Reuse of Nanomaterials

12.1 INTRODUCTION

Desorption is very necessary when the material synthesis, processing, and fabrication are costly. Regenerability of loaded nanomaterials is a key factor for improving the economy of adsorption processes. The economic feasibility of using adsorbents based on nanomaterials to remove dye, phenols, heavy metals, metalloids, and organic pollutants from aqueous solution relies on their regeneration ability during multiple adsorption/desorption cycles. It is possible to decrease the process cost and also the dependency of the process on a continuous supply of nanomaterials through desorption. A successful desorption process requires the proper selection of elutants, which strongly depends on the type of material and the mechanism of adsorption (Das, 2010). Elutant must be (1) nondamaging to the nanomaterials, (2) less costly, (3) ecofriendly, and (4) effective.

12.2 REGENERATION OF PHOTOCATALYSTS

The recycling and stability of photocatalysts are very important for high photocatalytic degradation and practical applications in real wastewater. After each photocatalytic process, the graphene/titanium dioxide nanotube photocatalyst was irradiated with UV light for 2 h and washed with a large amount of water, then filtered several times and dried at 100°C (Tao et al., 2015). Thus the regenerated photocatalysts were again used for the degradation of acetaminophen. The regenerated photocatalyst efficiency for the degradation was still maintained above 90% after five cycles. However, an insignificant decrease of degradation efficiency was due to the absorbed intermediates on the surface of the catalyst and to a few graphene/titanium dioxide nanotubes that were leached into the solution during the photodegradation or cleaning process.

Nanomaterials for Wastewater Remediation
http://dx.doi.org/10.1016/B978-0-12-804609-8.00012-1

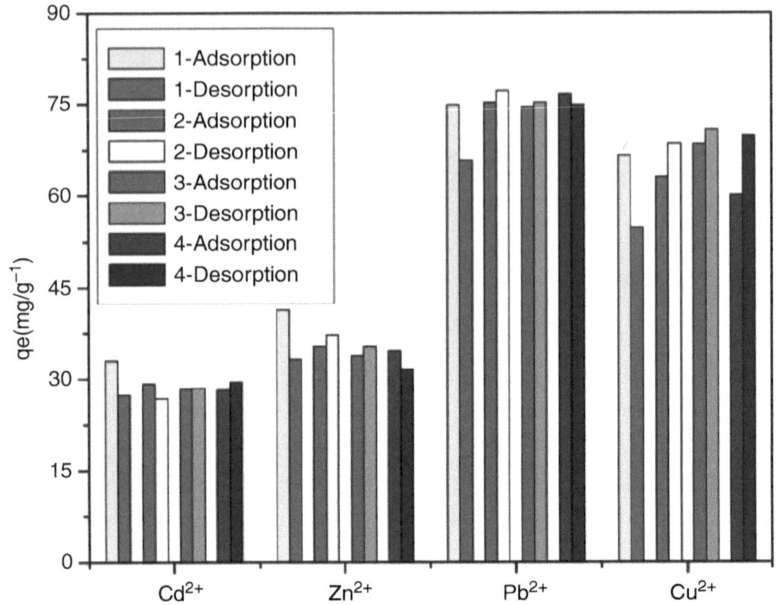

Figure 12.1 ***The performance of Fe_3O_4@APS@AA-co-CA by multiple regeneration cycles.*** *(Source: Ge et al., 2012; © Elsevier).*

12.3 RECOVERY OF METALS AND REGENERATION OF MAGNETIC NANOPARTICLES

Recently, Ge et al. (2012) synthesized the novel Fe_3O_4 magnetic nanoparticles (MNPs) modified with 3-aminopropyltriethoxysilane (APS) and copolymers of acrylic acid (AA) and crotonic acid (CA). The authors explored the ability of the MNPs for removing heavy metal ions (Cd^{2+}, Zn^{2+}, Pb^{2+}, and Cu^{2+}) from aqueous solution. Furthermore, they explored desorption and reuse of MNPs. Fe_3O_4@APS@AA-*co*-CA MNPs were excellent for the removal of heavy metal ions such as Cd^{2+}, Zn^{2+}, Pb^{2+}, and Cu^{2+} from aqueous solution (Fig. 12.1). The Fe_3O_4@APS@AA-*co*-CA MNPs did not significantly adsorb metal ions at pH < 2, which suggested that the adsorbed metal ions could possibly be desorbed in a solution with such pH values. The authors investigated different concentrations of H^+ to determine the best conditions for reusing MNPs (Fig. 12.2). The concentration of metal ions increased with increasing H^+ in water but changed little with 0.1 mol/L H^+. Meanwhile, the adsorbent was corroded by the H^+, which could be seen from the content of the Fe^{3+}. The adsorbent completely disappeared at 2 mol/L H^+ after 3 h. Considering the desorption efficiency, and the reusability cycles of the adsorbents, the initial concentration of

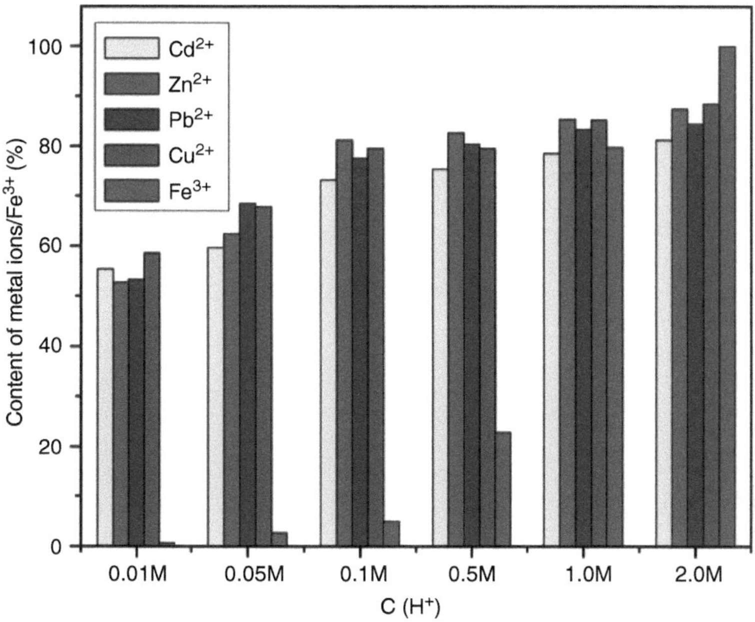

Figure 12.2 ***The contents of the metal ions and Fe^{3+} at different concentrations of H^+.*** *(Source: Ge et al., 2012; © Elsevier).*

0.1 M H^+ was better than others. The metal ion adsorption capacity of Fe_3O_4@APS@AA-*co*-CA MNPs remained almost constant for the four cycles, which indicates no irreversible sites on the surface of Fe_3O_4@APS@AA-*co*-CA MNPs for desorption with 0.1 mol/L H^+, and the reusability of the adsorbents was satisfactory.

Chitosan-coated magnetic nanoparticles (CCMNPs) modified with a biodegradable and ecofriendly biologic reagent, α-ketoglutaric acid, was used as a magnetic nanoadsorbent to remove toxic Cu^{2+} ions from aqueous solution (Zhou et al., 2009). Various operating parameters such as initial metal concentration, contact time, and pH have been studied to attain highest removal efficiency. The Langmuir maximum adsorption capacity for Cu^{2+} ions on α-ketoglutaric acid–modified CCMNPs was estimated to be 96.15 mg/g, which was higher than that of pure CCMNPs. Several eluents have been applied to regenerate the adsorbents and Cu^{2+} recovery. After the optimization of the eluent solutions, the regeneration cycles were repeated for 7 cycles using 50 mg of α-KA-CCMNPs and 10 mL of 200 ppm $CuSO_4$ solution for the adsorption process, and the desorption process was carried out with 10 mL of 100 mmol/L Na_2EDTA. The Cu^{2+} adsorption and desorption capacity of α-KA-CCMNPs was almost stable

Table 12.1 Eluents for desorption, and the desorption efficiency of various pollutants

Eluents	Pollutants	Desorption efficiency (%)	References
100 mmol/L Na_2EDTA	Cu^{2+}	91.5	Zhou et al. (2009)
25 mmol/L Na_2EDTA	Cu^{2+}	61.5	Zhou et al. (2009)
100 mmol/L HCl	Cu^{2+}	26.5	Zhou et al. (2009)
25 mmol/L HCl	Cu^{2+}	24	Zhou et al. (2009)
100 mmol/L citric acid	Cu^{2+}	24	Zhou et al. (2009)
25 mmol/L citric acid	Cu^{2+}	21.5	Zhou et al. (2009)
100 mmol/L acetic acid	Cu^{2+}	49	Zhou et al. (2009)
25 mmol/L acetic acid	Cu^{2+}	26.5	Zhou et al. (2009)
1 M HCl	Cr(VI)	65	Lasheen et al. (2014)
1 M HNO_3	Cr(VI)	90	Lasheen et al. (2014)
5 M HCl	Cr(VI)	75	Lasheen et al. (2014)
5 M HNO_3	Cr(VI)	94	Lasheen et al. (2014)
10 mM HNO_3	Pb(II)	90	Nassar (2010)
2 M NaOH	As(III,V)	90	Tang et al. (2013)

at seven cycles without loss of adsorption efficiency and after three cycles desorption efficiency was slightly decreased. Table 12.1 shows the used eluents for desorption and the desorption efficiency of various pollutants from aqueous solutions.

The 10 mL of 1 and 5 M of HNO_3 and HCl separate eluent were mixed with Cr-adsorbed magnetite for 2 h to desorb the adsorbed Cr(VI) and to regenerate the magnetite (Lasheen et al., 2014). The desorption efficiencies of HNO_3 and HCl were found to be 90 and 65% at 1 M concentration, respectively. As the solution's acidic strength was increased at 5 M concentration of HNO_3 and HCl solution the desorption efficiency reached to 94 and 75%, respectively. The reusability of magnetite and Cr(VI) readsorption were carried out up to five consecutive cycles. After each cycle of adsorption–desorption, the adsorbent was washed thoroughly with distilled water to neutrality, then dried and reconditioned for adsorption in the succeeding cycle. The results showed that the efficiencies of the recycled magnetite for removing Cr(VI) are nearly the same as those for the fresh one even after three times recycling.

Nassar (2010) used magnetic nanoparticles for the removal of Pb(II) ions from water as an effective adsorbent that provides high adsorption capacity, rapid adsorption rate, simple separation, and regeneration. The author studied the desorption and regeneration extensively to recover the Pb(II) ions and to reuse the magnetic nanoparticles. It has been established that

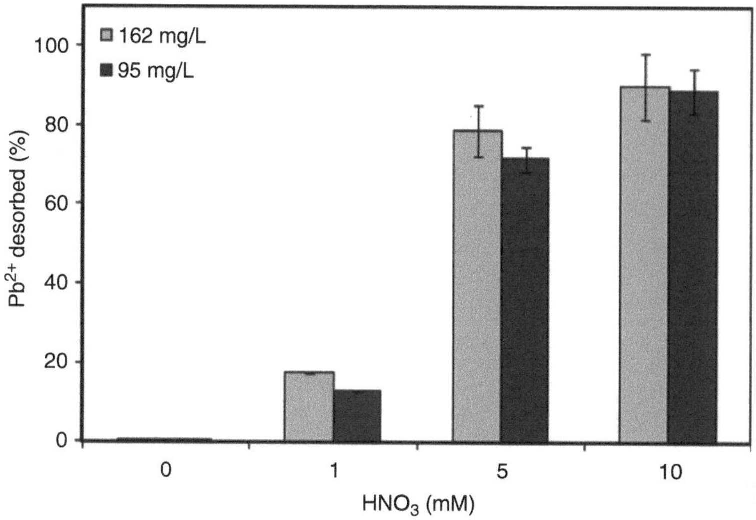

Figure 12.3 ***Desorption of Pb(II) from Pb(II)-loaded Fe_3O_4 nanoadsorbents using different concentrations of HNO_3 solution at fixed initial concentration.*** *(Source: Nassar, 2010; © Elsevier).*

adsorption of Pb(II) ions was highly dependent on increasing the pH of the solution. Therefore, the decrease in pH of the solution was highly helpful in the desorption of Pb(II). Fig. 12.3 shows that the desorption efficiency increased with the increase in HNO_3 concentration, because the increase in acid concentration significantly decreases the pH of the solution. However, a very insignificant desorption was achieved without addition of HNO_3; while almost 90% recovery was obtained at 10 mM HNO_3, and it was independent of the adsorbed Pb(II) concentration. It is important to note that no significant decrease in the adsorption capacity of Fe_3O_4 nanoadsorbents during the five cycles was observed.

Tang et al. (2013) synthesized superparamagnetic magnesium ferrite nanoadsorbent and applied it to the removal of arsenic from water, and the adsorption–desorption behavior was extensively studied. After the separation from water, arsenic species adsorbed on the ultrafine $Mg_{0.27}Fe_{2.50}O_4$ nanoadsorbent was desorbed by washing with 2 M NaOH solution. Hence more than 90% arsenic species were desorbed, and the regenerated ultrafine $Mg_{0.27}Fe_{2.50}O_4$ nanoadsorbent was reused for arsenic removal. Fig. 12.4 demonstrates the arsenic removal percentage of the regenerated ultrafine $Mg_{0.27}Fe_{2.50}O_4$ nanoadsorbent on natural water samples from Lake Yangzonghai for five times. After being regenerated four times, the regenerated ultrafine $Mg_{0.27}Fe_{2.50}O_4$ nanoadsorbent could still keep up to 70–80% of its arsenic removal capability.

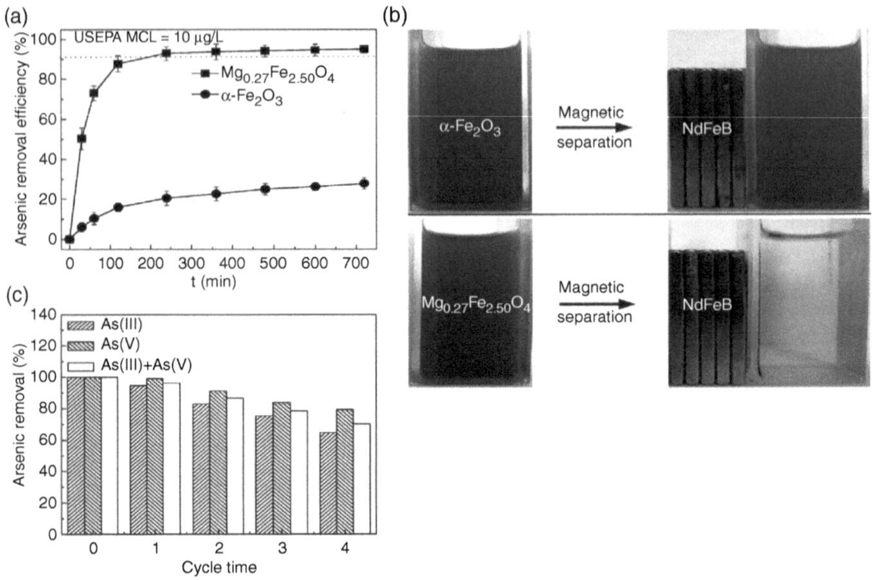

Figure 12.4 (a) Arsenic removal percentages from natural water samples of Lake Yangzonghai by the ultrafine $Mg_{0.27}Fe_{2.50}O_4$ nanoadsorbent and its comparison with the α-Fe_2O_3 nanoadsorbent. (b) Images of ultrafine $Mg_{0.27}Fe_{2.50}O_4$ and α-Fe_2O_3 nanoadsorbents dispersed in water and their separation by external magnetic field, and (c) the arsenic removal percentage of the regenerated ultrafine $Mg_{0.27}Fe_{2.50}O_4$ nanoadsorbent on natural water samples from Lake Yangzonghai for five times. *(Source: Tang et al., 2013; © Elsevier).*

Singh et al. (2011) used surface-engineered magnetic nanoparticles for removal of toxic metal ions and bacterial pathogens. Magnetic nanoparticles show high adsorption capacity of metals and bacterial pathogens from water, and due to its magnetic nature the used magnetic nanoparticles were easily separable from water through an external magnetic field, which would be helpful in lowering the capital and operational costs for industrial applications. Fig. 12.5 shows (a) the desorption ratio of adsorbed toxic metal ions using 10 mL of 0.1 M HCl and (b) the removal efficiency of toxic metal ions during three different adsorption cycles by amine MNP at pH 8. All the metal ions showed over 85% of desorption ratio in their first adsorption–desorption cycle. With increasing the number of cycles the removal efficiency of metal ions is reduced gradually; however, the removal efficiency of all metal ions was above 80% after three cycles. The gradual decrease in the removal efficiency could be attributed to incomplete release of adsorbed metal ions. In order to avoid the leaching of magnetic nanoadsorbents, 0.1 M HCl concentration was preferred.

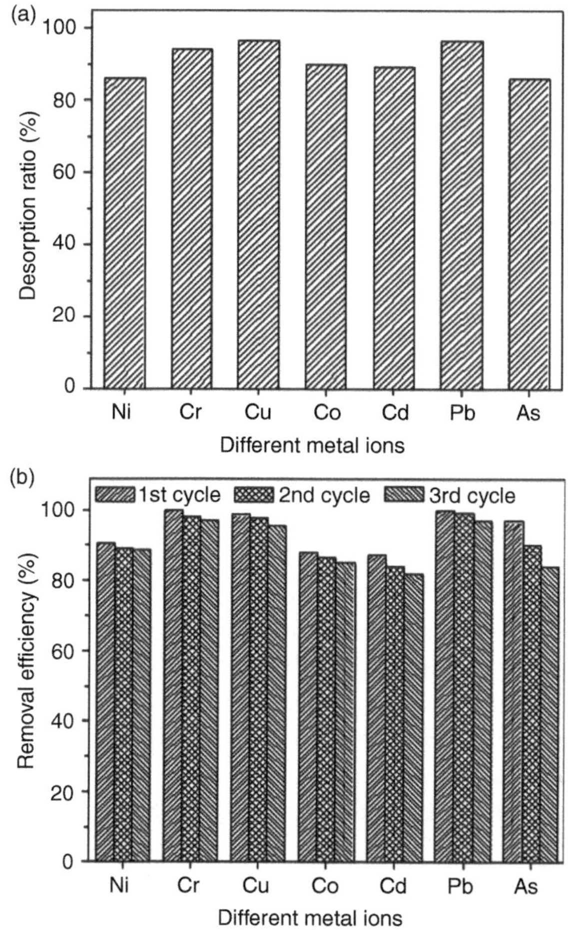

Figure 12.5 (a) Desorption ratio of adsorbed toxic metal ions using 10 mL of 0.1 M HCl and (b) removal efficiency of toxic metal ions during three different adsorption cycles by amine magnetic nanoparticles at pH 8. *(Source: Singh et al., 2011; © Elsevier).*

12.4 REGENERATION OF GRAPHENE-BASED NANOCOMPOSITES

Recently, graphene and graphene oxide has been widely used in the separation and purification of environmental pollutants from aquatic media, and many research papers have been published in this field in the removal of phenols, chlorophenols, benzenes, pharmaceuticals, dyes, metal ions, and metalloids from water and wastewater. Although, graphene or graphene oxide shows low capacity of removal percentage for the contaminants, some major efforts have been made to synthesize graphene-based nanocomposites such as graphene-based iron oxide, graphene-based titanium oxide, and graphene-based zinc oxide. The doping or functionalization of graphene and graphene oxide with organic or inorganic moieties tremendously enhances the adsorption capacity and reduces the contact time between adsorbate and adsorbent. Graphene magnetic nanocomposites have been found most suitable in the removal of heavy metals and dyes from various sources to reduce the hazardous impact.

12.5 REGENERATION OF NANOSORBENTS USED IN DYE REMOVAL

Recently, Fan et al. (2013) synthesized the magnetic β-cyclodextrin–chitosan/graphene oxide materials (MCCG) through a facile chemical route, and their application as excellent adsorbents for methylene blue dye removal were investigated. Due to the valuable surface property of graphene oxide nanocomposites with hydrophobicity of β-cyclodextrin, the abundant amino and hydroxyl functional groups of chitosan, and the magnetic property of Fe_3O_4; the adsorbent possesses quite a good and versatile adsorption capacity to the dye and was easily and rapidly extractable from water by an external magnetic field. To make the adsorbent feasible for real wastewater application, the recycling experiments were conducted by washing MCCG under magnetic stirring with ethanol for 12 h. The adsorption–desorption were conducted by regenerated adsorbents up to the five cycles, and the result shows that the adsorption capacity decreased for each new cycle after desorption with five cycles. Meanwhile, the adsorption capacity remained at 30 mg/g after five cycles of the desorption–adsorption.

Superparamagnetic mesoporous silica microspheres embedded with a silica-coated superparamagnetic iron oxide nanoparticle core and mesoporous silica shell were prepared using stearyl trimethyl ammonium bromide as surfactant template and modified with anhydride functionalized silane to graft carboxylic groups, and the adsorbent was used for the removal of methylene blue and acridine orange from wastewaters (Fu et al., 2011). The used adsorbent was regenerated by simple washing with acid solution to recover both the adsorbents and the adsorbed dyes. More than 88% of adsorption efficiency was obtained by regenerated adsorbent even after use for seven times. Desorption efficiency were 90 and 97% at pH 3 and 2, respectively. However, highly acidic pH value may lead to the leaching of Fe_3O_4 from magnetic microspheres. Hence, desorption was carried out at pH 3. The desorption percentage has been shown in Fig. 12.6 for methylene blue

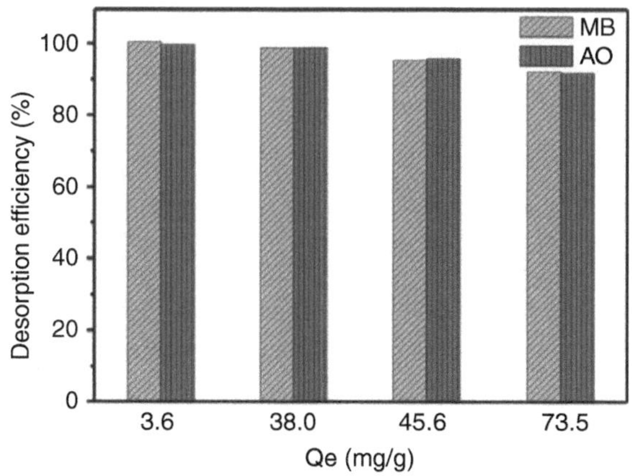

Figure 12.6 ***Desorption percentage for methylene blue (MB) and acridine orange (AO) from magnetic microspheres.*** *(Source: Fu et al., 2011; © Elsevier).*

and acridine orange. The adsorption on regenerated adsorbents was repeated up to seven cycles, and the removal efficiency in each cycle is shown in Fig. 12.7. The adsorption efficiency was greater than 88% in the final cycle.

Catalytic performance of the magnetically separable Fe-MCM-41 catalyst was explored through the oxidation of phenol in aqueous solution (Xia et al., 2011). The catalytic activity of Fe-MCM-41 in the presence of H_2O_2 showed a significant total organic carbon reduction, achieving a conversion value of c. 78% after 2 h of reaction. The regeneration experiments were carried out in three consecutive runs at the same reaction conditions. When the equilibrium was achieved, the catalysts were separated by an external

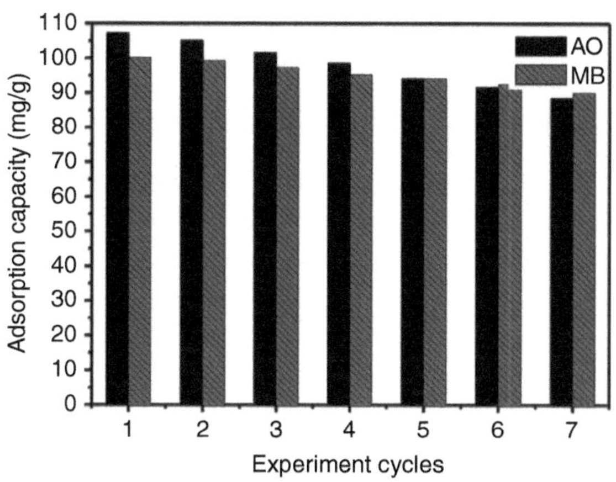

Figure 12.7 ***Adsorption capacity of magnetic microspheres in different cycles for methylene blue (MB) and acridine orange (AO).*** *(Source: Fu et al., 2011; © Elsevier).*

magnet and washed to remove the intermediates. It was found that the total organic carbon removal efficiency decreased from 78 to 65% after three times recycle reaction.

12.6 DESORPTION AND REGENERATION OF INORGANIC SOLID WASTES

The adsorption capacity of seven inorganic solid wastes, namely, air-cooled blast furnace slag, water-quenched BF slag, steel furnace slag, coal fly ash, coal bottom ash, water treatment (alum) sludge, and seawater-neutralized red mud for Cd^{2+}, Cu^{2+}, Pb^{2+}, Zn^{2+}, and Cr^{3+} was determined and their desorption behavior was investigated (Zhou and Haynes, 2011). Relative desorption of previously adsorbed Pb and Cd in 0.01 M $NaNO_3$ was very low for all three materials tested, and the largest values were recorded for Cd sorbed to water treatment sludge. Removal of Pb and Cd was greater using 0.10 M than 0.01 M HNO_3, and for higher concentration, the bulk of the metal was removed in the first desorption cycle.

The adsorption of As(III), As(V), Se(IV), and Se(VI) by seawater-neutralized red mud and alum water treatment sludge was investigated, and the regeneration of adsorbent was carried out in batch adsorption technique (Zhou and Haynes, 2012). The adsorption of metalloids was highly dependent on the pH of the solution. It has been found that the NaOH was more effective at removing adsorbed metals from both adsorbents than HNO_3. Water treatment sludge maintained its As(III) and Se(IV) adsorption capability at greater than 70% of that added over eight successive cycles of adsorption–regeneration using 0.5 M NaOH as a regenerating agent.

12.7 MANAGEMENT OF SPENT ELUENTS

Desorption and regeneration is a very crucial process for any adsorbent used in the separation, purification, and recovery of precious metals and environmental contaminants from aquatic media. This process is directly dependent on the type of materials used in the recovery of metals and concentration of environmental pollution in wastewater. Nanomaterials such as TiO_2, iron oxide nanoparticles, zinc oxide nanoparticles, and alumina nanoparticles have been extensively used in the separation of contaminants and have been regenerated by mild acids such as HCl, HNO_3, acetic acid, acetone, citric acid, Na_2EDTA or mild base of NaOH, and KOH. However, the regeneration of adsorbents is cost effective in the real application of nanosorbents, and the recovery of metals is also profitable to the industries, but the regeneration of nanomaterials consumes a lot of eluents, which further may generate a secondary pollution to the environment. By-products of the degradation of organic pollutants also possess a threat to the flora and fauna if discharged into the nearby water bodies. Thus there is an urgent need to collect the used eluents in the regeneration process that can be used for other purposes. Mild acids can be used in cleaning of glasswares in small-scale laboratories and factories. Further, incineration, distillation, or thermal degradation can be used for the full disposal of waste eluents. However, this practice can be avoided if the eluent is used in the separation of pollutants such as choline compounds or volatile organic pollutants as they may generate dioxins or furans to the environment (Gómez-Pastora et al., 2014).

12.8 MANAGEMENT OF SPENT NANOMATERIALS

Nanomaterials have been extensively explored in the removal of pollutants from aquatic media, and a variety of contaminants were removed in a number of studies. The used nanomaterials have been widely regenerated and reused for the economic point of view in industrial applications. The nanosorbents such as alumina, layered double hydroxide, magnetic nanoparticles, titanium dioxide, graphene, metal organic frameworks, and graphene based nanocomposites have been applied in the remediation of dyes, heavy metals, and organic pollutants, and regenerated and reused several times. Sometimes it is very difficult to regenerate the used adsorbent due to high cost, low adsorption capacity, and economic point of view. Hence, the activated carbons have been disposed by incinerating or composting below the soil. However, in this process the adsorbed pollutants may be released or leached into the ground water and may pose secondary pollution.

Recently, a major focus has been on synthesizing the magnetic nanoparticles and nanocomposites of high adsorption capacity, infinite surface sites, increased functional groups, high physicochemical stability in harsh environmental conditions, and easy separation from water/wastewater by external magnetic field. The application of magnetic nanoparticles to remediation of ground water is helpful to decontaminate the pollution; however, at the same time the magnetic nanoparticles may be corroded in the acidic environment and the iron leached to the ground water. Magnetic nanoparticles may also be influenced by the gravitational force in the ground water and may move farther distances in the surface water table of ground water.

The nanomaterial adsorbents should be collected properly after their use in wastewater treatment and treated appropriately to dispose the exhausted nanosorbents. Incineration or thermal decomposition should be applied to the disposal of exhausted adsorbents, or they should be compost in the soil by making proper chambers so that they cannot be released into nearby water bodies.

Solidification can also be helpful to minimize the hazardous impact of exhausted adsorbents and also would be cost effective if they were converted to bricks or concrete for road making (Gómez-Pastora et al., 2014).

12.9 CONCLUSIONS

Desorption and regeneration is a crucial step in the separation of environmental contaminants from aquatic media. Various nanoadsorbents such as alumina, layered double hydroxide, magnetic nanoparticles, titanium dioxide, graphene, metal organic frameworks, and graphene-based nanocomposites have been extensively used in water treatment. These adsorbents show high adsorption capacity within low contact time and are highly stable. Their regeneration has been performed by mild acids or bases and recovery of precious metals and compounds have been carried out. In most of the experiments the acids such as HCl, HNO_3, acetic acids, and citric acids have been applied to regenerate the nanosorbents and recover the adsorbed materials. However, acid treatment may corrode the nanosorbents and leaching may occur if the nanosorbents are prone to acidic environments. Iron oxide and magnetic nanoparticles are highly sensitive to acid treatment as they are easily leached into acidic conditions. So there is a need to develop some novel adsorbents which could sustain in the acidic or basic medium, as the regeneration and recovery is a crucial part of treatment technologies to reduce operating costs. Graphene-based nanocomposites are helpful to overcome this gap, but extensive experiments need to be performed in this area.

REFERENCES

Das, N., 2010. Recovery of precious metals through biosorption—a review. Hydrometallurgy 103, 180–189.

Fan, L., Luo, C., Sun, M., Qiu, H., Li, X., 2013. Synthesis of magnetic β-cyclodextrin–chitosan/graphene oxide as nanoadsorbent and its application in dye adsorption and removal. Colloids Surf. B 103, 601–607.

Fu, X., Chen, X., Wang, J., Liu, J., 2011. Fabrication of carboxylic functionalized superparamagnetic mesoporous silica microspheres and their application for removal basic dye pollutants from water. Microporous Mesoporous Mater. 139, 8–15.

Ge, F., Li, M.-M., Ye, H., Zhao, B.-X., 2012. Effective removal of heavy metal ions Cd^{2+}, Zn^{2+}, Pb^{2+} and Cu^{2+} from aqueous solution by polymer-modified magnetic nanoparticles. J. Hazard. Mater. 211–212, 366–372.

Gómez-Pastora, J., Bringas, E., Ortiz, I., 2014. Recent progress and future challenges on the use of high performance magnetic nano-adsorbents in environmental applications. Chem. Eng. J. 256, 187–204.

Lasheen, M.R., El-Sherif, I.Y., Sabry, D.Y., El-Wakeel, S.T., El-Shahat, M.F., 2014. Removal and recovery of Cr (VI) by magnetite nanoparticles. Desalin. Water Treat. 52, 1–13.

Nassar, N.N., 2010. Rapid removal and recovery of Pb(II) from wastewater by magnetic nanoadsorbents. J. Hazard. Mater. 184, 538–546.

Singh, S., Barick, K.C., Bahadur, D., 2011. Surface engineered magnetic nanoparticles for removal of toxic metal ions and bacterial pathogens. J. Hazard. Mater. 192, 1539–1547.

Tang, W., Su, Y., Li, Q., Gao, S., Shang, J.K., 2013. Superparamagnetic magnesium ferrite nanoadsorbent for effective arsenic (III,V) removal and easy magnetic separation. Water Res. 47, 3624–3634.

Tao, H., Liang, X., Zhang, Q., Chang, C.T., 2015. Enhanced photoactivity of graphene/titanium dioxide nanotubes for removal of Acetaminophen. Appl. Surf. Sci. 324, 258–264.

Xia, M., Chen, C., Long, M., Chen, C., Cai, W., Zhou, B., 2011. Magnetically separable mesoporous silica nanocomposite and its application in Fenton catalysis. Microporous Mesoporous Mater. 145, 217–223.

Zhou, Y.F., Haynes, R.J., 2011. A comparison of inorganic solid wastes as adsorbents of heavy metal cations in aqueous solution and their capacity for desorption and regeneration. Water Air Soil Pollut. 218, 457–470.

Zhou, Y.F., Haynes, R.J., 2012. A comparison of water treatment sludge and red mud as adsorbents of as and se in aqueous solution and their capacity for desorption and regeneration. Water Air Soil Pollut. 223, 5563–5573.

Zhou, Y.T., Nie, H.L., Branford-White, C., He, Z.Y., Zhu, L.M., 2009. Removal of Cu^{2+} from aqueous solution by chitosan-coated magnetic nanoparticles modified with α-ketoglutaric acid. J. Colloid Interf. Sci. 330, 29–37.

CHAPTER 13

Nanomaterials in the Environment: Sources, Fate, Transport, and Ecotoxicology

13.1 INTRODUCTION

The fundamental understanding of nanomaterial toxicology has become crucial from the nanomaterials standpoint as well as from the applications point of view. There has been a gradual enhanced production of nanostructured commercial products from medical tablets to talcum powders, hence toxicological evaluation of nanostructured materials should receive greater attention than ever before, whether it is from industrial workers, general public, governmental organizations, or private sectors involved in the expansion of nanomaterials. In the area of environmental science, the outcome learned from these nanotoxicological studies should help researchers and users to better select the type of nanostructure materials for their application in environmental remediation at realistic levels, so the risk and toxicity associated with the introduction of nanomaterials should be minimized. Seeking the paucity of data associated with nanotoxicology, we will review the literature on how this new area of nanotoxicology has developed, the various measures that can be taken to solve the emerging problems will be highlighted, and an action plan will be further discussed.

13.2 RELEASE OF NANOMATERIALS INTO THE ENVIRONMENT

The recent advances in magnetic nanoparticles and their application in removal of toxic metal ions from wastewater have resulted in uncertainties regarding environmental impacts. The objective of this section is to introduce the key aspects of existing nanomaterials in the environment and to discuss what is known concerning their fate, behavior, and toxicity. The increasing production and application of nanomaterials of all types, the potential for their release in the environment, and subsequent effects on ecosystem health is becoming an increasing concern that needs to be addressed.

Nanomaterials for Wastewater Remediation
http://dx.doi.org/10.1016/B978-0-12-804609-8.00013-3

The biological and ecological toxicity of nanomaterials is actively being addressed across the world. Studies show that nanomaterials can be toxic to plants, animals, and microbes. Toxicity to individual organisms is only one piece of the puzzle, and the ultimate ecological impact of chronic nanomaterial contamination will be governed by the origin and magnitude of nanomaterial contamination, how they move through the environment, and their exposure to plants and animals. According to Posner (2009) this system of processes is called the *source-pathway-receptor-impact* paradigm. A source is a point where nanomaterials originate and then they travel, transport, and accumulate in pathways (eg, atmosphere, ground, or aqueous ecosystem) and finally encounter a receptor and ultimately have an impact on the ecosystem. Fig. 13.1 shows the important processes and pathways of nanoparticles in the environment.

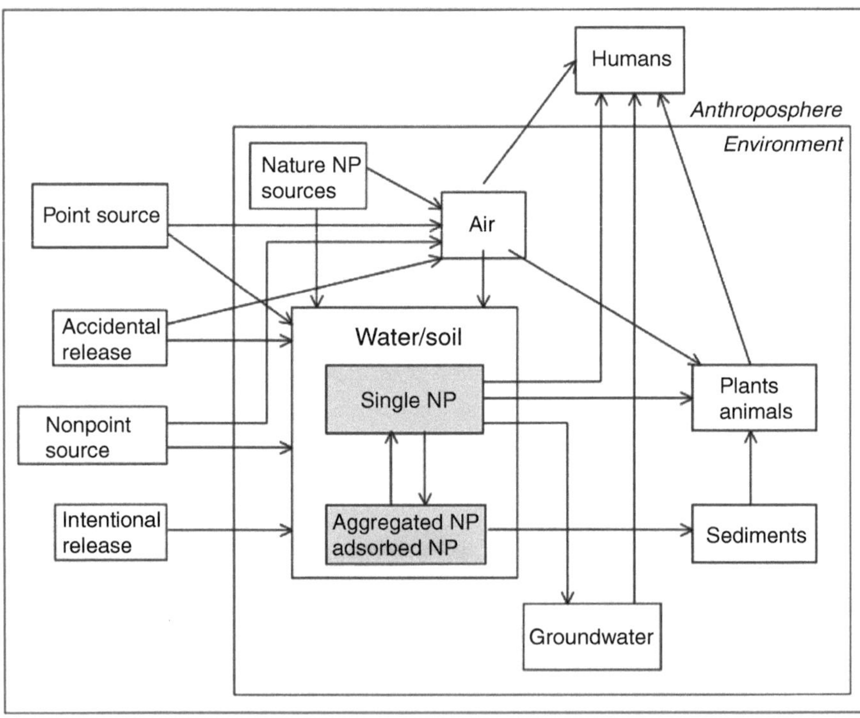

Figure 13.1 ***Nanoparticle pathways from the anthroposphere into the environment, reactions in the environment, and exposure to human beings.*** Release of nanoparticles may come from point sources such as production and manufacturing facilities, landfills or wastewater treatment plants, or from nonpoint sources such as wear from materials containing nanoparticles. Accidental release of nanoparticles during production or transport is also possible. In addition to the unintentional release, there are also nanoparticles released intentionally into the environment. For example, release of nano zero valent iron (nZVI) directly into the groundwater containing or polluted with chlorinated solvents. *(Source: Nowack and Bucheli, 2007; © Elsevier).*

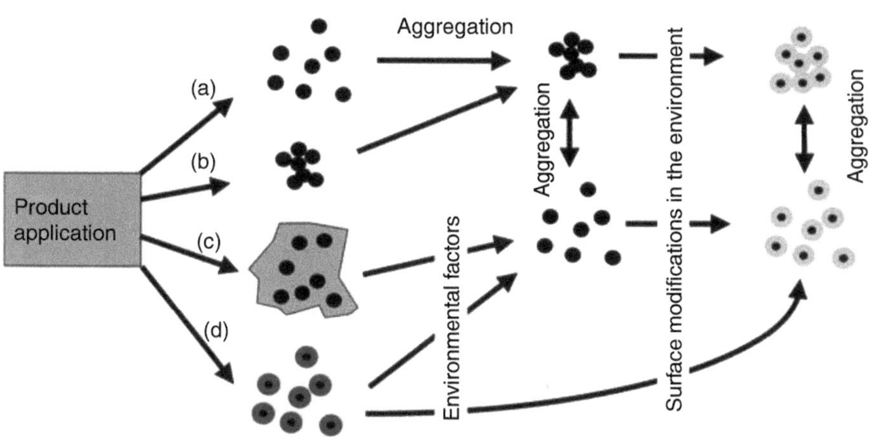

Figure 13.2 ***Release of nanoparticles from products and (intended or unintended) applications.*** (a) Release of free nanoparticles, (b) release of aggregates of nanoparticles, (c) release of nanoparticles embedded in a matrix, and (d) release of functionalized nanoparticles. Environmental factors (eg, light, microorganisms) result in formation of free nanoparticles that can undergo aggregation reactions. Moreover, surface modifications (eg, coating with natural compounds) can affect the aggregation behavior of the nanoparticles. *(Source: Nowack and Bucheli, 2007; © Elsevier).*

Whether the nanoparticles are released directly into water/soil or the atmosphere, they all end up in soil or water, either directly or indirectly, for instance, via sewage treatment plants, waste handling, or aerial deposition. In the environment the formation of aggregates and therefore of larger particles that are trapped or eliminated through sedimentation affects the concentrations of free nanoparticles (Nowack and Bucheli, 2007). Fig. 13.2 shows the release of nanoparticles from products and (intended or unintended) applications.

It is somehow complicated to predict the nature of nanoparticles when they will lose, change, or gain new coatings, including natural organic matter, during their transport through aquatic environments. Certainly, microbial action may play an important role in modifying nanoparticle surface coatings. Functionalization of nanoparticle surface with varying concentrations of natural organic matter in river water has been found to result in dramatic shifts between aggregation and stabilization of nanoparticles. In rivers, aggregation and sedimentation of nanoparticles results in their storage in river-bed sediments, which slows their downstream transport, thus putting greater risk of nanoparticles for exposed organisms that live on, or in, the river bed (Jarvie and King, 2010). Modification of nanoparticles with functional materials might influence their environmental pathways and ecological exposure routes. For example, in the case of primary wastewater treatment, changes in the surface coatings of nanoparticles might be exploited to bring about a shift from an effluent (aquatic) pathway, to a terrestrial pathway, via sludge disposal to land. However, simply changing the pathway of the nanomaterials into the environment does not mitigate ecotoxicological risk; it simply changes the target (Jarvie and King, 2010). The widespread application of sewage sludge as an agricultural fertilizer raises important issues of interactions with soil biota and the potential for uptake by plants and crops. Moreover, nanoparticles applied to the land in sewage sludge may ultimately be transported into surface water bodies via "diffuse" or "nonpoint" pathways, mobilized by runoff during rainfall events, or leached into groundwater. However, within each environmental compartment that a nanoparticle moves through, it will be subject to a whole host of interactions with the surrounding water, soil, and biotic diversity. These processes may promote or retard nanoparticle transport or translocation through each environmental compartment and are primarily mediated by surface chemical interactions.

13.3 TITANIUM DIOXIDE

Titanium dioxide (TiO_2) has been noted to be a safe physical sunscreen, because it reflects and scatters UVB and UVA in sunlight (Dunford et al., 1997). However, TiO_2 absorbs about 70% of incident UV, and in aqueous environments this leads to the generation of hydroxyl radicals which can initiate oxidations. Dunford et al. (1997) showed that the sunlight-illuminated TiO_2 catalyzes DNA damage both in vitro and in human cells by generating the superoxide or hydroxyl radicals in the biological cells.

The wide use of metal oxide nanoparticles will inevitably increase their environmental release into soil, which consequently raises concerns about their environmental impacts and ecological risks. Xu et al. (2015) studied the effects of two typical metal oxide nanoparticles (TiO_2 and CuO) in different doses (0, 100, 500, and 1000 mg/kg soil) and evaluated their effects on microbes in flooded paddy soil. The negative effects of CuO nanoparticles were stronger than that of TiO_2 nanoparticles on soil microbes, as reflected by the significant decline in soil microbial biomass and enzyme activities, including urease, phosphatase, and dehydrogenase.

Recently, Yu et al. (2015) studied the effects of nanoscale TiO_2 (n-TiO_2), CeO_2 (n-CeO_2), and ZnO (n-ZnO) and bulk ZnO on a model ammonia oxidizing bacterium, *Nitrosomonas europaea*. They observed that the n-ZnO displayed the most serious cytotoxicity while n-TiO2 was the least toxic. The stress of nanoparticles on the change of cell morphologies, the retardance of specific oxygen uptake rates and ammonia oxidation rates, and the depression of amoA gene expressions were studied. The nanoparticle-impacted cells revealed the increase of the corresponding intracellular Ti, Ce, or Zn contents and suggested the intracellular nanoparticle accumulation in the bacterium. It was found that the elevation of intracellular S contents accompanied with higher K contents implied the possible activation of thiol-containing glutathione and thioredoxin production for nanoparticle stress alleviation. However, it was reported that the nanoparticle cytotoxicity was not always a function of nanoparticle concentration. When the concentration of n-TiO_2 or n-CeO_2 increases from 20 to 200 mg/L, the impacted cells display the higher amoA gene expression.

Despite the fact that the aquatic environment is often the ultimate recipient of all contaminants, there is a paucity of data pertaining to the potential detrimental effects of nanoparticles on aquatic organisms. TiO_2 nanoparticles have been reported to cause oxidative stress–related effects, including inflammation, cytotoxicity, and genomic instability, either alone or in the presence of UVA irradiation in mammalian cells (Reeves et al., 2008). Therefore, the effect of TiO_2 nanoparticles on goldfish skin cells (GFSk-S1) was studied to evaluate the potential cytotoxic and genotoxic effects, either alone or in combination with UVA radiation. It has been observed that for cell viability, TiO_2 alone (0.1–1000 μg/mL) had little effect whereas coexposure with UVA (0.5–2.0 kJ/m^2) caused a significant dose-dependent decrease, which was dependent on both the concentration of TiO_2 and the dose of UVA administered. For the comet assay, doses of 1, 10, and 100 μg/mL in the absence of UVA caused elevated levels of Fpg-sensitive sites, which further indicated the oxidation of purine DNA bases by TiO_2 nanoparticles. In addition, the UVA irradiation of TiO_2-treated cells caused further increases in DNA damage. It was concluded that the observed toxic effects of TiO_2 nanoparticles were most likely due to hydroxyl radical (•OH) formation.

13.4 SILICON DIOXIDE

Silica (SiO_2) is the most abundant compound in the earth's crust, and it can be divided into crystalline or amorphous (Park and Park, 2009). Inhalation of the crystalline form of silica has been a well-known exposure route and historically associated with the development of a severe respiratory disease, silicosis, which is lung-pneumoconiosis characterized by alveolar proteinosis and diffused fibrosis in biological cells (Hamilton et al., 2008). Recently, Park and Park (2009) evaluated the oxidative stress and inflammatory responses induced by silica nanoparticles both in mice and in RAW264.7 cell line. Single treatment of silica nanoparticles (50 mg/kg) led to the activation of peritoneal macrophages, the increased blood level of IL-1β and TNF-α, and the increased level of nitric oxide released from the peritoneal macrophages. mRNA expressions of inflammation-related genes such as IL-1, IL-6, TNF-α, iNOS, and COX-2 were also elevated in the cultured peritoneal macrophages harvested from the treated mice. Cell proliferation without cytotoxicity was shown in the group treated with a relatively low dose of 50 mg/kg. In vivo and in vitro study using RAW 264.7 cell line was carried out to elucidate the proinflammatory mechanism of silica nanoparticles, and exposure of silica nanoparticles to the cultured RAW264.7 cells led to the reactive oxygen species (ROS) generation with a decreased intracellular glutathione. It has been found that in accordance with ROS generation, silica nanoparticles increased the level of nitric oxide released from the cultured macrophage cell line. Thus silica nanoparticles generate ROS, and the generated ROS may trigger the proinflammatory responses both in vivo and in vitro in biological cells. In order to elucidate the nanoparticle-induced cytotoxicity and its mechanism, the effects of 20 and 50 nm silica nanoparticles on cultured human embryonic kidney (HEK293) cells were investigated (Wang et al., 2009). Several parameters such as cell viability, mitochondrial function, cell morphology, ROS, glutathione, thiobarbituric acid reactive substance, cell cycle, and apoptosis were monitored under control and silica-exposed conditions. It has been reported that the exposure to 20 or 50 nm SiO_2 nanoparticles at dosage levels between 20 and 100 μg/mL decreased cell viability in a dose-dependent manner, and the lethal dose of 24 h exposure was 80.2 $\pm$ 6.4 and 140.3 $\pm$ 8.6 μg/mL for 20 and 50 nm SiO_2 nanoparticles, respectively. The effects of SiO_2 nanoparticle exposure revealed cell shrinkage, nuclear condensation, increase in intracellular ROS level, and reduction in glutathione content. However, continuous increase in the amount of thiobarbituric acid reactive substance suggested an elevated level of lipid peroxidation. Similarly, flow cytometric analysis showed that SiO_2 nanoparticles can cause G2/M phase arrest and apoptotic sub-G1 population increase in a dose-dependent manner. Hence, exposure to SiO_2 nanoparticles resulted in a dose-dependent cytotoxicity in cultured HEK293 cells due to the increased oxidative stress.

13.5 IRON OXIDE NANOPARTICLES

Superparamagnetic iron oxide nanoparticles have been successfully applied for in vivo applications such as in magnetic resonance imaging, tumor therapy, and cardiovascular disease diagnosis (Yang et al., 2013). Iron oxide nanoparticles are very promising nanomaterials as they can be controlled by an external magnet. However, their role in the toxicity in cellular damage has been an immense area of research, although very few studies are reported on their toxicity. It has been estimated that the iron oxide damages the DNA, and oxidative stress mediated by reactive oxygen species (Mbeh et al., 2012). The toxicity of magnetic nanoparticles is strongly dependent on the cell type and plays a significant role in the toxicity (Mahmoudi et al., 2011; Kumari et al., 2011). It has also been established that the cytotoxicity and genotoxicity was influenced by particle size and shape (Hong et al., 2011). The exposure to magnetic nanoparticles causes inflammation, the formation of apoptotic bodies, impaired mitochondrial function, membrane leakage of lactate dehydrogenase generation of reactive oxygen species, increase in micronuclei, and chromosome condensation (Singh et al., 2010).

Recently, differential cellular responses to the surface-functionalized magnetic nanoparticles were investigated in normal fibroblasts versus fibrosarcoma cells (Yang et al., 2013). It has been found that all magnetic nanoparticles induced cytotoxicity and genotoxicity in fibrosarcoma cells at lower than 500 μg/mL; APTMS-coated magnetic nanoparticles resulted in greater than 10% toxicity against normal cells. Although, the genotoxicity of magnetic nanoparticles was dependent on their dose, size, and surface charge, showing that positively charged (APTMS- or TEOS/APTMS-coated) magnetic nanoparticles induced appreciable DNA aberrations irrespective of cell type. However, magnetic nanoparticles with smaller size and positive charges (APTMS-coated) led to more severe toxicity in normal cells than their cancer counterparts. Hence, it can be concluded that the functional groups and sizes of magnetic nanoparticles are critical determinants of degrees of cytotoxicity and genotoxicity, and potential mechanisms of toxicity.

13.6 GRAPHENE-BASED MATERIALS AND THEIR TOXICITY

Zanni et al. (2012) evaluated the toxicity of graphite nanoplatelets in the nematode *Caenorhabditis elegans*. They observed the life expectancy as well as reproductive capability in terms of acute or chronic toxicity. It has been shown that no effect on *C. elegans* reproductive potential was found.

However, good spatial distribution of the graphite nanoplatelets inside the nematode was reported.

The cytocompatibility tests of graphene oxide and reduced graphene oxide under the stability of the suspensions in a wide range of concentrations (3.125–100 μg/mL) of three different dispersants ware studied (Wojtoniszak et al., 2012). Polyethylene glycol (PEG), polyethylene glycol–polypropylene glycol–polyethylene glycol (Pluronic P123), and sodium deoxycholate (DOC) were investigated as the dispersants for the cytocompatibility tests of graphene oxide and reduced graphene oxide. It has been found that the toxicity depends on the type of dispersant and concentration of the nanomaterials in the suspensions. However, detailed analysis suggested that graphene oxide functionalized with PEG in the concentration range between 3125 and 25 μg/mL exhibited the best biocompatibility with mice fibroblast cells (line L929).

Zhang et al. (2011) determined the distribution and biocompatibility of graphene oxide (GO) in mice by using radiotracer technique and a series of biological assays. Results showed that graphene oxide was predominantly deposited in the lungs, where it was retained for a long time. There were no pathological changes observed in examined organs exposed to 1 mg/kg body weight of graphene oxide for 14 days. Furthermore, graphene oxide showed good biocompatibility with red blood cells in the mice. However, due to its high accumulation and long-time retention, significant pathological changes, including inflammation cell infiltration, pulmonary edema, and granuloma formation were found at the dosage of 10 mg/kg body weight.

13.7 METAL AND SEMICONDUCTOR NANOPARTICLES

The toxicity of cobalt-containing nanoparticles may be attributed either to the direct uptake of the nanoparticles by cells or to the dissolution of the nanoparticles leading to the increased level of cationic ions in the growth media, with consequent effects on cells (Horev-Azaria et al., 2011). The toxicological effects of cobalt nanoparticle (Co-NPs) aggregates were examined and compared with those of cobalt ions (Co-ions) using six different cell lines representing lung, liver, kidney, intestine, and the immune system (Horev-Azaria et al., 2011). Dose-response curves were studied in the concentration range of 0.05–1.0 mM, employing 3-(4,5-dimethylthiazol-2-Yl)-2,5-diphenyltetrazolium bromide test, neutral red, and Alamar blue as end-point assays following exposures for 48 and 72 h. It was established that concentration is the highest rank parameter, and the second-rank parameter emerged to be either the compound type (Co-ions or Co-NPs) or the cell model, depending on the concentration range for the toxicity. However, the exposure duration was the third and lowest rank in the toxicity of the studied model. The hierarchy of cell sensitivity toward cobalt ions was found to obey the following sequence of cell lines: A549 > MDCK > NCIH441 > Caco-2 > HepG2 > dendritic cells, with A549 being the most sensitive cell line, and primary dendritic cells were the least sensitive ones. On the other hand, a different hierarchy pattern emerged for Co-NPs: A549 = MDCK = NCIH441 = Caco-2 > dendritic cells > HepG2. The overall findings were in line with the hypothesis that the toxic effects of aggregated cobalt nanoparticles (CdS, CdSe, ZnO, MnO, and Co) were mainly due to cobalt ion dissolution from the aggregated nanoparticles.

13.8 COPPER NANOPARTICLES

The bioavailability of CuO nanoparticles is thought to induce the major toxicity to microbes in the flooded paddy soil, as determined by the increased Cu contents in the soil extractions and the microbial cells. The elevated stress ratio values demonstrated that CuO nanoparticles may also indirectly affect soil microbes by changing nutrient bioavailability.

13.9 NICKEL NANOPARTICLES

Nickel nanoparticles have been widely used in catalysis, sensing, electrical, and environmental applications. While nanomaterials may produce some negative effects on the biological system, it becomes important to determine the fate and toxicity of any particular nanomaterials due to their different size and shape. Three different size nickel nanoparticles of 30, 60, and 100 nm and larger particle clusters of aggregated 60 nm entities with a dendritic structure were synthesized and exposed to zebrafish embryos assessing mortality and developmental defects to examine the toxicity behavior (Ispas et al., 2009), and the nickel nanoparticle exposure was compared to soluble nickel salts. It was found that all three 30, 60, and 100 nm nickel nanoparticles were equal to or less toxic than soluble nickel while dendritic clusters were more toxic in the biological cells. With an increase in the lethal dose from LD10 to LD50, the thinning of the intestinal epithelial cells first occurred, and skeletal muscle fiber separation occurred at exposure to LD50. However, the soluble nickel does not cause intestinal defects while skeletal muscle separation occurred at concentrations well over LD50. Zebrafish embryos exposed to nickel nanoparticles first developed defects in the intestine with minimal folding and thin cuboidal epithelial cells beginning at LD10 concentrations, and on the fifth day of embryogenesis, the intestine was highly folded and epithelial cells were columnar. In the initial days intestinal defects were not immediately lethal, because embryos survive on yolk throughout embryogenesis and do not require a functional digestive system. The results of this study suggested that configuration of nanoparticles may affect toxicity more than size, and defects from nickel nanoparticle exposure occurred by different biological mechanisms than soluble nickel.

13.10 SILVER NANOPARTICLES

Silver nanoparticles easily get transformed in the environment, which modifies their properties and alters their transport, fate, and toxicity (Levard et al., 2012). Hence, it becomes essential to consider such transformations when assessing the potential toxicological effects of silver nanoparticles. Recently, Levard et al. (2012) reviewed the major transformation processes of silver nanoparticles in various aqueous environments, particularly transformations of the metallic silver cores caused by reactions with (in)organic ligands, and the effects of such transformations on physicochemical stability and toxicity. It has been proposed that silver binds with sulfur in aquatic environments as well as in wastewater treatment plants. Sulfidation of silver nanoparticles may decrease in the toxicity due to the lower solubility of silver sulfide, potentially limiting their short-term environmental impact.

13.11 MAGNETIC NANOPARTICLES IN THE ENVIRONMENT

This decade has seen revolutionary developments in the field of nanotechnology, with newer and diverse applications of nanoparticles appearing every day. Novel nanomaterials are emerging with different characteristics and compositions for specific applications. However, little attention is being paid to understand, access, and manage the environmental impact and adverse effects of nanoparticles. Currently the information about the toxicity of nanoparticles and their environmental fate in air, water, soil, and tissues is severely lacking. Inhalation, ingestion, and dermal penetration are the potential exposure routes for nanoparticles (Hoet et al., 2004), whereas particle size, shape, surface area, and surface chemistry collectively define the toxicity of nanoparticles. Nanoparticles can potentially cause adverse effects on organ, tissue, cellular, subcellular, and protein levels due to their unusual physicochemical properties (Alt et al., 2004; Poon and Burd, 2004; Drake and Hazelwood, 2005; Kim et al., 2006; Ryman-Rasmussen et al., 2007). Several studies have shown excessive generation of reactive oxygen species as well as transient or persistent inflammation following exposure to various classes of nanoparticles. Increased production and intentional or unintentional exposure to nanoparticles increases the possibilities of adverse health effects. Thus, the novel nanomaterials need to be biologically characterized for their health hazards to ensure risk free and sustainable implementation of nanotechnology. Currently there is little literature available in this specific area to cover toxicological aspects of nanoparticles.

13.12 ENVIRONMENTAL AND SAFETY CONCERNS TOWARD NANOMATERIALS

The wide application of engineered nanoparticles and their entry into the environment, the study of their impact on the ecosystem, and a growing concern in society regarding the possible adverse effects of manufactured nanoparticles has been raised in recent years. Therefore, it is necessary to study their release, uptake, and mode of toxicity in organisms. Furthermore, to understand the long-term effects of nanoparticles on the ecosystem, substantial information is required regarding their persistence and bioaccumulation. The use of engineered nanoparticles in the environment as a consequence of the development of nanotechnology is a serious case of concern for environmental biologists worldwide. However, a few studies have already demonstrated the toxic effects of nanoparticles on various organisms, including mammals. Nanotechnology is still in the discovery phase in which novel materials are first synthesized in small scale in order to identify new properties and further applications (Toksha et al., 2008; Amara et al., 2009; Christian et al., 2008; Arora et al., 2012; Blaser et al., 2007; Cushen et al., 2011).

Perception and knowledge are important parts of public understanding of nanotechnology. They can be influential for achievable benefit obtained and the possible risks and hazards. Therefore, detailed understanding of their sources, release, interaction with environment, and possible risk assessment would provide a basis for safer use of engineered nanoparticles with minimal or no hazardous impact on the environment. Thus, ecotoxicology of nanoparticles will be closely related to their intrinsic properties as shown in Table 13.1 and Fig. 13.3. Increase surface activity, mobility, and diffusion and adsorption ability are some other effects.

Table 13.1 Biological effects due to physicochemical properties related to ecotoxicology of nanomaterials

Physicochemical properties	Toxicological findings	Biological effects
Size	Affects reactivity and permeability of cells and organs	Increases biodistribution of nanoparticles in environmental system
Surface/volume ratio	Higher reactivity	Inflammatory effects
Chemical composition	Increase in UVA absorption, higher activation of reactive oxygen species in cell media	Cancerogenic, cell proliferation reduced
Aggregation state	More pronounced cytotoxic effects	Cytotoxicity
Surface charge	Charged nanoparticles present higher deposition degree in tissues	Bioaccumulation in brain, lungs, and others

A further comprehension of the structure–function relationship in nanomaterial matter could lead to new protocols for nanomaterial manufacturing wherein high precision and low waste methods are included. Some criteria should be taken into account when referring to nanoparticle release and the effects on nature, including human beings. Nanoparticle effects should be

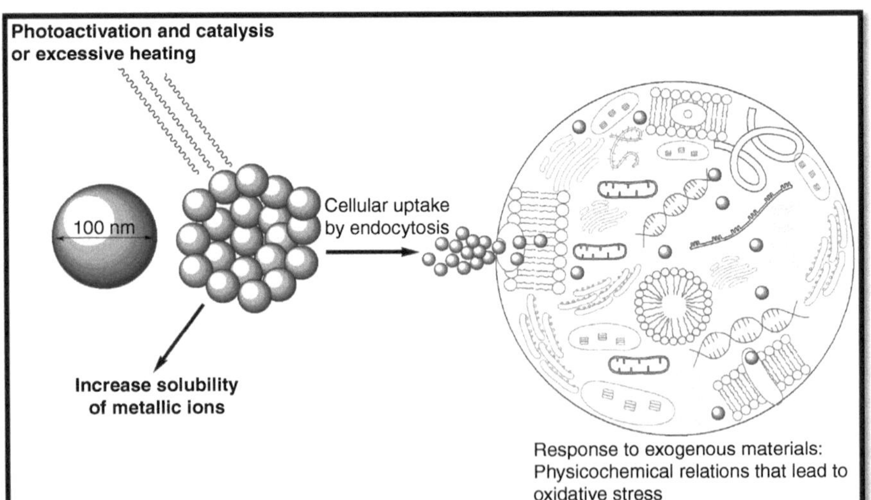

Figure 13.3 *Toxic effects of nanoparticles on biological systems due to their size and physicochemical properties.*

scale dependent and not the same in larger scale or agglomerates. This means that effects may be quite different to adopt specific and more appropriate regulations. The differences in nanomaterials are generally based on size, surface chemistry, and other specific interactions depending on the scale. Thus, the same material may have different regulations through the different sizes presented. Nanomaterial effects must be conclusive to those products in which commercialization is imminent. So, the nanoparticles presented in the final product may be the ones on which the studies should focus. For the novelty of some materials, data to extrapolate environmental effects are difficult to obtain, so enhanced simulation systems are needed. This may be obtained by the various computer-based simulation systems.

13.13 CHALLENGES IN CERTAIN AREAS

Nanomaterials have been widely used in the field of environmental remediation such as the adsorption of heavy metals, dyes, and organic pollutants from wastewater effluents. As the application increases, the risk of nanomaterial contamination and its toxicity in the flora and fauna also enhances day by day. Extremely small amounts of nanoadsorbents may release into the environment via various sources and enter into the food cycles of animals and humans. The nanotoxicity is a new challenge to the coming generations as the major cause will be the nanomaterial-based compounds and products. The extent of nanotoxicity directly depends on the mode of entrance, dose, size of materials, aggregation behavior, chemical composition, solubility, type of nanomaterials, and their distribution in the biological systems. Some nanomaterials may pose a risk of mutagenicity to carcinogenicity in animals and human beings. The major challenges in this field are the collection of released nanomaterials from various sources, identification of nanomaterials whether they are natural or anthropogenic in origin, extraction and sample preparation for analysis, and the analysis of their toxicity behavior on plants and animals.

13.14 PROPOSED ACTIONS TO ADDRESS THESE CHALLENGES

As the market of nanotechnology is increasing with tremendous growth in the consumable products, water treatment, and cosmetic industries, their mode of risk is also increasing. There is urgent need to identify the mode of actions and exposures which may pose toxicity and risk to humans and animals. The nanomaterials may also cause stress in living plants, may alter their photosynthetic rate, and may chock the transport and translocation of water and mineral in the xylem and phloem through the aggregation of nanomaterials. Thus there is urgent need to address these challenges in the application of nanotechnology in the real world. Therefore, advanced analytical methods should be developed to identify and characterize the nanomaterials and their behavior in the environment. The analytical methods should also develop to recognize the toxicity of engineered nanomaterials and must be able to differentiate the toxicity properties from natural nanomaterials. Advanced instrumentation techniques should be developed to characterize the physicochemical properties of the engineered nanomaterials without the destruction of nanomaterials. Hence, there should be standard criteria for regulation of nanoparticles, risk identification, and use in consumer products. In the water treatment process the use of bare nanomaterials should be avoided, as they are very prone to corrosion and may leach into the groundwater. The leached nanomaterials in groundwater may produce secondary pollutants, which will be very difficult to remediate as there is still no option for the remediation of secondary pollutants generated into the groundwater by nanoobjects. During the in situ treatment of groundwater, more attention should be paid to the release of nanomaterials into the groundwater. Further, the nanomaterial movement should be checked properly to understand the gravitational force and flow of water in the case of groundwater.

13.15 CONCLUSIONS

Cost-effective treatment poses a challenge for environmental engineers, scientists, and others in the development of effective risk management strategies. Pollutants that are highly toxic, persistent, and difficult to treat present particular challenges. Environmental scientists support research that addresses new treatment approaches that are more effective in reducing contaminant levels and more cost effective than currently available techniques. For example, nanotechnology research that results in improved treatment options might include removal of the finest contaminants from water and "smart" materials or reactive surface coatings that destroy or immobilize toxic compounds.

REFERENCES

Alt, V., Bechert, T., Steinrucke, P., Wagener, M., Seidel, P., Dingeldein, E., Domann, E., Schnettler, R., 2004. An in vitro assessment of the antibacterial properties and cytotoxicity of nanoparticulate silver bone cement. Biomaterials 18, 4383–4391.

Amara, D., Felner, I., Nowik, I., Margel, S., 2009. Synthesis and characterization of Fe and Fe_3O_4 nanoparticles by thermal decomposition of triiron dodecacarbonyl. Colloids Surf. A Physicochem. Eng. Asp. 339, 106–110.

Arora, S., Rajwade, J.M., Paknikar, K.M., 2012. Nanotoxicology and in vitro studies: the need of the hour. Toxicol. Appl. Pharmacol. 258, 151–165.

Blaser, S.A., Scheringer, M., Macleod, M., Hungerbühler, K., 2007. Estimation of cumulative aquatic exposure and risk due to silver: contribution of nano-functionalized plastics and textiles, Europe. Sci Total Environ. 390, 396–409.

Christian, P., Von der Kammer, F., Baalousha, M., Hofmann, T., 2008. Nanoparticles: structure, properties, preparation and behaviour in environmental media. Ecotoxicology 17, 326–343.

Cushen, M., Kerry, J., Morris, M., Cruz-romero, M., Cummins, E., 2011. Nanotechnologies in the food industry—recent developments, risks and regulation. Trends Food Sci. Technol. 24, 30–46.

Drake, P.L., Hazelwood, K.J., 2005. Exposure-related health effects of silver and silver compounds: a review. Ann. Occup. Hyg. 7, 575–585.

Dunford, R., Salinaro, A., Cai, L., Serpone, N., Horikoshi, S., Hidaka, H., Knowland, J., 1997. Chemical oxidation and DNA damage catalysed by inorganic sunscreen ingredients. FEBS Lett. 418, 87–90.

Hamilton, Jr., R.F., Thakur, S.A., Holian, A., 2008. Silica binding and toxicity in alveolar macrophages. Free Radical Biol. Med. 44, 1246–1258.

Hoet, P.H.M., Brueske-Hohlfeld, I., Salata, O., 2004. Nanoparticles-known and unknown health risks. J. Nanobiotoxicol. 2, 1–2.

Hong, S.C., Lee, J.H., Lee, J., Kim, H.Y., Park, J.Y., Cho, J., Lee, J., Han, D.-W., 2011. Subtle cytotoxicity and genotoxicity differences in superparamagnetic iron oxide nanoparticles coated with various functional groups. Int. J. Nanomedicine 6, 3219–3231.

Horev-Azaria, L., Kirkpatrick, C.J., Korenstein, R., Marche, P.N., Maimon, O., Ponti, J., Romano, R., Rossi, F., Golla-Schindler, U., Sommer, D., Uboldi, C., Unger, R.E., Villiers, C., 2011. Predictive toxicology of cobalt nanoparticles and ions: comparative in vitro study of different cellular models using methods of knowledge discovery from data. Toxicol. Sci. 122, 489–501.

Ispas, C., Andreescu, D., Patel, A., Goia, D.V., Andreescu, S., Wallace, K.N., 2009. Toxicity and developmental defects of different sizes and shape nickel nanoparticles in zebrafish. Environ. Sci. Technol. 43, 6349–6356.

Jarvie, H.P., King, S.M., 2010. Just scratching the surface? New techniques show how surface functionality of nanoparticles influences their environmental fate. Nano Today 5, 248–250.

Kim, J.S., Yoon, T.J., Yu, K.N., Kim, B.G., Park, S.J., Kim, H.W., Lee, K.H., Park, S.B., Lee, J.K., Cho, M.H., 2006. Toxicity and tissue distribution of magnetic nanoparticles in mice. Toxicol. Sci. 89, 338–347.

Kumari, A., Kumar, V., Yadav, S.K., 2011. Therapeutic nanoparticles and associated toxicity. Curr. Nanosci. 7, 389–395.

Levard, C., Hotze, E.M., Lowry, G.V., Brown, Jr., G.E., 2012. Environmental tTransformations of silver nanoparticles: impact on stability and toxicity. Environ. Sci. Technol. 46, 6900–6914.

Mahmoudi, M., Laurent, S., Shokrgozar, M.A., Hosseinkhani, M., 2011. Toxicity evaluations of superparamagnetic iron oxide nanoparticles: cell "vision" *versus* physicochemical properties of nanoparticles. ACS Nano 5, 7263–7276.

Mbeh, D.A., França, R., Merhi, Y., Zhang, X.F., Veres, T., Sacher, E., Yahia, L., 2012. In vitro biocompatibility assessment of functionalized magnetite nanoparticles: biological and cytotoxicological effects. J. Biomed. Mater. Res. A 100, 1637–1646.

Nowack, B., Bucheli, T.D., 2007. Occurrence, behavior and effects of nanoparticles in the environment. Environ. Pollut. 150, 5–22.

Park, E.J., Park, K., 2009. Oxidative stress and pro-inflammatory responses induced by silica nanoparticles in vivo and in vitro. Toxicol. Lett. 184, 18–25.

Poon, V.K.M., Burd, A., 2004. In vitro cytotoxity of silver: implication for clinical wound care. Burns 30, 140–147.

Posner, J.D., 2009. Engineered nanomaterials: where they go, nobody knows. Nano Today 4, 114–115.

Reeves, J.F., Davies, S.J., Dodd, N.J.F., Jha, A.N., 2008. Hydroxyl radicals (•OH) are associated with titanium dioxide (TiO_2) nanoparticle-induced cytotoxicity and oxidative DNA damage in fish cells. Mutat. Res. 640, 113–122.

Ryman-Rasmussen, J.P., Riviere, J.E., Monteiro-Riviere, N.A., 2007. Surface coatings determine cytotoxicity and irritation potential of quantum dot nanoparticles in epidermal keratinocytes. J. Invest. Dermatol. 127, 143–153.

Singh, N., Jenkins, G.J.S., Asadi, R., Doak, S.H., 2010. Potential toxicity of superparamagnetic iron oxide nanoparticles (SPION). Nano Rev. 1, 5358.

Toksha, B.G., Shirsath, S.E., Patange, S.M., Jadhav, K.M., 2008. Structural investigations and magnetic properties of cobalt ferrite nanoparticles prepared by sol–gel auto combustion method. Solid State Commun. 147, 479–483.

Wang, F., Gao, F., Lan, M., Yuan, H., Huang, Y., Liu, J., 2009. Oxidative stress contributes to silica nanoparticle-induced cytotoxicity in human embryonic kidney cells. Toxicol. in Vitro 23, 808–815.

Wojtoniszak, M., Chen, X., Kalenczuk, R.J., Wajda, A., Łapczuk, J., Kurzewski, M., Drozdzik, M., Chu, P.K., Borowiak-Palen, E., 2012. Synthesis, dispersion, and cytocompatibility of graphene oxide and reduced graphene oxide. Colloids Surf. B 89, 79–85.

Xu, C., Peng, C., Sun, L., Zhang, S., Huang, H., Chen, Y., Shi, J., 2015. Distinctive effects of TiO_2 and CuO nanoparticles on soil microbes and their community structures in flooded paddy soil. Soil Biol. Biochem. 86, 24–33.

Yang, W.J., Lee, Jong Ho, Hong, S.C., Lee, J., Lee, J., Han, D.W., 2013. Difference between toxicities of iron oxide magnetic nanoparticles with various surface-functional groups against human normal fibroblasts and fibrosarcoma cells. Materials 6, 4689–4706.

Yu, R., Fang, X., Somasundaran, P., Chandran, K., 2015. Short-term effects of TiO_2, CeO_2, and ZnO nanoparticles on metabolic activities and gene expression of Nitrosomonas europaea. Chemosphere 128, 207–215.

Zanni, E., De Bellis, G., Bracciale, M.P., Broggi, A., Santarelli, M.L., Sarto, M.S., Palleschi, C., Uccelletti, D., 2012. Graphite nanoplatelets and *Caenorhabditis elegans*: insights from an in vivo model. Nano Lett. 12, 2740–2744.

Zhang, X., Yin, J., Peng, C., Hu, W., Zhu, Z., Li, W., Fan, C., Huang, Q., 2011. Distribution and biocompatibility studies of graphene oxide in mice after intravenous administration. Carbon 49, 986–995.

SUBJECT INDEX

B

D

G

H

O

R

S

FSC
www.fsc.org
MIX
Papier | Fördert gute Waldnutzung
FSC® C083411

Kolibri 360 GmbH
Ferdinand-Jühlke-Straße 7
99095 Erfurt, Deutschland
produktsicherheit@kolibri360.de